PHILOSOPHY OF BIOLOGY

PHILOSOPHY OF BIOLOGY

Edited by
Michael Ruse

Prometheus Books
59 John Glenn Drive
Amherst, NewYork 14228-2197

Published 1998 by Prometheus Books

02 01   5 4 3

Library of Congress Cataloging-in-Publication Data

Philosophy of biology / edited by Michael Ruse.
        p.   cm.
        Previously published: New York : Macmillan ; London : Collier Macmillan, c1989.
        Includes bibliographical references.
        ISBN 1-57392-185-8 (pbk. : alk. paper)
        1. Biology–Philosophy. I. Ruse, Michael.
QH331.P468   1998
570'.1–dc21
                                                      97–36460
                                                         CIP

Printed in the United States of America on acid-free paper

PREFACE

Many of us pay lip service to the dream of being truly interdisciplinary, but few of us are prepared to put the effort into making that dream a reality. I cannot pretend that I have succeeded entirely in this collection; however, the guiding principle throughout has been to make something equally accessible and useful to the student of philosophy and to the student of biology. I ask for no special indulgence, but experts in the respective fields should keep this fact in mind. In making this selection I am happy to acknowledge the great help I have had from distinguished scholars in biology and philosophy who share my belief that a meeting of the two disciplines can only benefit both sides: Francisco J. Ayala, John Beatty, Arthur Caplan, David Hull, and Ernst Mayr. Most importantly, I want to thank the general editor of the series, Paul Edwards. I have not had so demanding a teacher since I was in primary school. If you learn half as much from reading this collection as I have in preparing it, your gain will be great indeed.

M.R.

CONTENTS

THE CHALLENGE OF PUNCTUATED EQUILIBRIUM

PROBLEMS OF CLASSIFICATION

TELEOLOGY: HELP OR HINDRANCE?

MOLECULAR BIOLOGY

THE RECOMBINANT DNA DEBATE

HUMAN SOCIOBIOLOGY

Philosophy of Biology

INTRODUCTION

WHAT IS PHILOSOPHY? What is biology? What is the philosophy of biology? Let me start with the second question, which might seem the easiest. Biology, surely, is the science of life, of living things. But, what is a living thing? Examples come readily to mind. Humans are living things, living humans, that is! And so are snakes and oak trees and the bacteria that make food go bad. On the other hand, planets are not living things, and neither are shovels or spades or rocks or mountains or the sea. "Or the sea?" Is it irrational to talk about "the living ocean" and how it ebbs and flows, attacks and protects, helps and hinders? Or, what about a thunderstorm, if you do not believe the sea has the kind of self-maintenance often associated with life? A thunderstorm picks up energy as it goes along, throwing out waste and destruction behind it, just like an elephant, only more so.

Now, at this moment, I do not want to press this particular train of thought. Apart from anything else, we shall be coming back to the notion of "life." Rather, I want to use the discussion to make a point. What have we been doing in the last paragraph? We have certainly not been practicing biology—no dead frogs were laid out on a table. Rather, we have been thinking *about* biology. And this provides a clue to our other questions: to philosophy in general and to the philosophy of biology in particular. Philosophy is a second-order inquiry that looks at other subjects—politics, art, religion, or, in our case, biology—and asks questions about them. The biologist *qua* biologist dissects and studies frogs and humans and oak trees and bacteria. But when it comes

1

to questions about what the biologist is doing, then we are in the realm of philosophy.

Is this not all somewhat presumptuous? A nonexpert in the field calmly enters it and tells its practitioners what they are doing and how they should behave. Sometimes, I confess it is presumptuous; unless the philosopher makes some effort to understand the field being studied, grief will not be far behind. However, just as the experienced auto mechanic knows that there are certain principles common to all cars and that a detached general knowledge is not necessarily bad, so the experienced philosopher learns that certain principles apply through all human endeavors and various problems and concerns keep arising. A detached general knowledge is not necessarily bad.

Moreover, apart from the intrinsic interest of philosophical problems, some outside help can often be of much use to the success of a first-order subject. An example illustrates this point perfectly for biology. In recent years various extreme evangelical Christians, those who insist on taking every word of the Bible absolutely literally (so-called fundamentalists), have been trying hard to have their (Creationist) views inserted into biology curricula. Whatever the ultimate resolution of these efforts, what has become apparent is that there is more here than straight science. Can claims that the earth is only 6000 years old and that human beings were created miraculously by God ever, even in principle, be part of geology and biology, or are they necessarily religious? These, clearly are questions about science rather than questions within science. In other words, they are philosophical questions.

Since even the least reflective of us are bound to encounter philosophical problems sooner or later, we had better get them out on the table and look at them explicitly. But, how should we set about doing this, particularly in biology? My experience is that approaching philosophy is similar to driving a car: There is only so much talking you can do about it; it is far better to get behind the wheel and try it yourself. This is the approach I have taken in this volume, a collection of readings by biologists and philosophers, past and present. I have divided my selection of readings into sections, trying as much as possible to let people of different opinions each have their say; I myself simply try to offer a few guidelines to disentangle the various threads.

There is a reasonable continuity to the sections, but feel free to sample those subjects of particular personal interest. I suspect, however, that the topics with the greatest intellectual appeal, such as human nature, morality, and God, which I have put toward the end, will take on a fuller and richer meaning when some of the earlier sections have been covered. I should add that I have not chosen readings because they are simple. My guide has been to select topics and contributions that are interesting and important. But, I have tried as far as possible to keep matters jargon-free. In addition, at the end I have provided some suggestions for further reading.

The eighteenth-century English man of letters, Dr. Samuel Johnson, is reputed to have said that although he had tried to practice philosophy on several occasions, he could never stay miserable long enough! This image of philosophy persists to this day. I cannot pretend that this collection is a barrel of laughs, but I shall be very disappointed if the reader at no time is gripped by the material. The ideas are important and there is a real thrill in the cut and thrust of intellectual debate. Philosophy may not be funny, but it can be fun.

1. What Is Life?

This topic is a good one with which to begin. Let us pick up where we left off. For all the similarities, we probably do not want to say that the ocean or a thunderstorm is a living thing. But, wherein does lie the difference between life and nonlife? It is tempting to suppose that living things must be made of different substances from nonliving things — and this is often true. The body of Abraham Lincoln was not the same substance as the statue of Lincoln in Washington. Yet, we know now that living beings are made, ultimately, from the same minerals as nonliving things. Moreover, since the early nineteenth century it has been possible to synthesize the compounds of living beings from inert substances.

This all leads one to think — it has certainly led many philosophers to think — that living things must have something special, *in addition*, to make them living, a sort of life force or fluid, akin perhaps to the mind but living in or through all life and separating the quick from the dead. Such a force has been called, naturally, a "vital" force (and its enthusiasts "vitalists"). This view goes back to Aristotle, who was as distinguished a biologist as he was a philosopher.

I can think of no better person than Aristotle to begin this **collection**, and so our first reading is from one of his best known works, *On the Generation of Animals*, in which he wrestles with the way in which the sperm or the seed affects the growing organism. Aristotle thought that the male parent provides the form of the offspring whereas the female parent provides the substance, and he argues that somehow this form gets transmitted through "soul." Precisely what he means by this notion has engaged scholars for over two thousand years. It is certainly not the Christian notion, but represents some sort of animating force that gives vitality and feeling.

Everyone who reads Aristotle comes away with tremendous respect for his biological acumen. Nevertheless, I suspect that in this doctrine of soul he would find few modern followers. Let me therefore say that there is nothing at all foolish about such a position. After all, there is something different about the living and the nonliving. Nor is it significant that we can never see the soul or vital force. If I open your skull, I doubt I will see your mind; but, I do not doubt you have one. However, this said, the trouble with vital forces is they do not seem to do much. Or, perhaps, the problem is that they do too much. The more biologists delve into living things, the more they find that not only are the substances the same as for the nonliving, but the workings are not so very different either. A bird stays aloft because it exploits the powers of nature — gravity, winds, heat, and the like — no less (and no more) than does a cloud or a handful of sand in a storm.

This is an exaggeration, of course. The bird puts things together much better than does a sandstorm: a point that brings me to our second reading, by J. B. S. Haldane. He was (as we shall learn later) a distinguished and creative biologist who thought and worked in the second quarter of this century. He was also an ardent Marxist, and the piece I have chosen appeared during the Second World War in the *Daily Worker*, the English communist newspaper. (This explains the reference to Engels and perhaps also the enthusiasm for *Alexander Nevsky*, a film by the great Russian director, Serge Eisenstein.) Haldane rejects vital forces, as I have just rejected them. But he suggests that there is more to the organization of life and, in particular, what this organization does, than a trivial and slighting glance would indicate. In essence, he invites us to start shifting our gaze from what life *is* to what life *does*. We must stop thinking in terms of the static and start thinking in terms of the dynamic.

Whether this approach will work is another matter. Certainly, we are going to need much more detail than Haldane's sketch provides. But, surely, Haldane is onto something. After all, what is life if it is not activity, be this eating, drinking, defecating, copulating, or practicing biology or philosophy? For this reason, we should not dismiss Haldane's suggestion out of hand, particularly when, nearly a half century later, we see that much of his science is no less out of date than Aristotle's. Haldane thought that proteins are key substances in life. Now we know that that is only part of the story. Yet, the thrust of Haldane's suggestion that "life is a pattern of chemical processes" makes just as much sense now as it did then — indeed, with our deeper understanding of the chemistry of life, it might perhaps make more sense.

2. Explaining Design

Even if we agree with Haldane that the key question does not concern the composition of organisms but rather how they work and how they are organized, we have to go further to distinguish the quick from the dead or rather the living from the inert. What is it about biological organization that makes it special? Continuing a tradition that goes back to the Ancient Greeks, many believe today that the distinguishing property of life is that it shows "irrefragable evidence of creative forethought," as the nineteenth-century British anatomist Richard Owen once put it. Unlike shovels or spades or rocks or mountains—or even the sea or thunderstorms—snakes and oak trees and bacteria uniquely show the marks of design. They look as if they were planned and created by a thinking intelligence.

In the past, for most people the appearance of design meant only one thing: Organisms look designed because they *are* designed—by God! This section, therefore, opens with the classic exposition of the Argument for Design (or "Teleological Argument"), first published in 1802 by the English clergyman, Archdeacon William Paley. Yet, for all the authority and support of the very greatest theologians, and for all the brilliance of his own arguments, Paley's position leaves the modern mind dissatisfied. His solution to the mysteries of organic nature takes almost all the important and interesting questions out of the domain of science. Basic questions about organic origins and organic nature can be answered only by religion. And this, in the opinion of many— including sincere believers—is altogether too severe a constriction on biological understanding.

Things have changed greatly since the time of Paley. Biological opinion today is that life neither sprang out of nowhere nor was the result of some instantaneous creation out of nonliving substances. Rather, it is believed that life "evolved," that is, it developed continuously by natural processes, from simple forms (probably, ultimately, from basic minerals), and that such development has produced the complex diversity we see today. The Englishman Charles Robert Darwin (1809– 1882) played a key role in advancing the concept of evolution— primarily in his major work, published in 1859, *On the Origin of Species.*

However, even today there is controversy. All sorts of different and conflicting claims are made: about evolution being "only a theory not a fact," about gaps in the fossil record, about Darwin's theory concerning biology being unduly teleological (whatever that might mean), and so forth. In this and the next few sections, I hope we can progress in analyzing some of the problems and answering some of the questions just posed, as well as others.

As a start toward understanding, let me begin by making a three-part division. When people talk about "evolution" they mean several things, and although there are certainly connections, evidence for one is not necessarily evidence for another. First, there is evolution as *fact*, that is, that species are not fixed but developed out of other species. This assertion is of course in conflict with the teachings of the Bible (taken absolutely literally) and of most philosophers and biologists before the nineteenth century. Second, there is evolution as *path*. This involves the actual routes that evolution took. (The technical term for such a route is "phylogeny.") For example, some people think the birds evolved from the dinosaurs. Others are not so sure. These are questions about evolution as path. Third, there is evolution as *mechanism* or *theory*. What was the force or cause or power behind evolutionary change?

Now, relating these three things to Darwin, in the *Origin* he made a strong case for evolution as fact. So strong that this is the case we accept today. How Darwin did it is not often realized by nonbiologists. It is usually thought that evolution stands or falls by the fossils, but although fossils do play an important role in evolutionary theorizing, to establish the fact of evolution Darwin took another route. Realizing that he had to persuade people of the truth of something they will never see, he did what we always do when we are dealing with the unknown, he used circumstantial evidence. Just as in a court of law you establish a suspect's guilt through clues, so Darwin looked through the living world for clues of evolution.

And he found them. Consider the remarkable similarities between the forelimbs of animals, even though those limbs are used for such different purposes. (See Figure 1, p. 7. Such similarities are known technically as "homologies.") How could this be? As a result of evolution from a shared ancestor, answered Darwin. He was greatly impressed, for example, by the similarities among tortoises on different islands of the Galapagos archipelago. (See Figure 2, p. 8.) How could these similarities have come into existence except through evolution from a common ancestor? And so the questions were asked and the answers given. The clues of life point to evolution, and, conversely, belief in the fact of evolution explains the clues of life. (This appeal to all of the evidence is known technically as a "consilience of inductions." In Darwin's day, it was a method of proof much favored by the philosopher William Whewell, and there is good historical evidence that Darwin learned it from him. See Figure 3, p. 9.)

Is evolution just a theory not a fact? It depends what you mean by "theory" and what you mean by "fact." If by "theory" you mean that a mechanism is postulated, then (as you will see shortly) evolution is a theory. If you mean by "theory," as we sometimes do, "a bit of an iffy hypothesis" — "I've got a theory about Kennedy's assassination" — then evolution is not a theory. Darwin showed it to be a fact. Do I mean that logically, eternally, absolutely without possibility, biologists could never be wrong on this point? No, one never means this in science. But, as in a court of law, the fact of evolution has been established beyond a reasonable doubt. That is good enough for rational people.

Darwin said comparatively little about the path of evolution. Obviously, our knowledge of it does depend, very crucially, on the fossil record, and since this record is very incomplete (not to mention the difficulties in finding what there is of it), we shall always be ignorant about much of life's history. Within limits, however, these difficulties can be circumvented. You can compare similarities (homologies) between organisms, living and fossilized, and work out what would be the most reasonable links. Recently, our much-improved understanding of

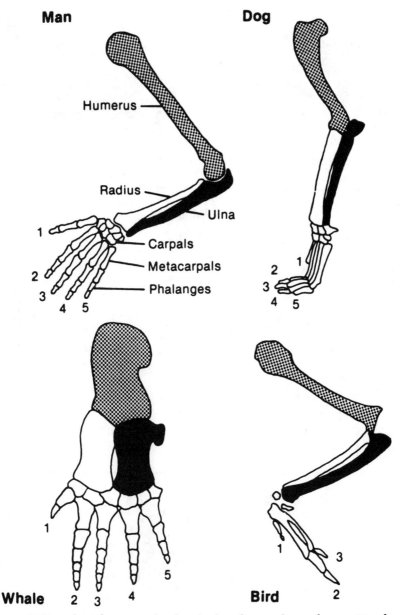

Man

Humerus

Radius

Ulna

Carpals

Metacarpals

Phalanges

1

2

3

4 5

Dog

2 1

3

4 5

Whale 2 3 4

5

1

Bird

1 3

2

Figure 1. Homology between the fore-limbs of several vertebrates. Numbers refer to digits. (Adapted with permission from Dobzhansky *et al.*, 1977, p. 41.)

the molecules of the body has made these techniques of comparison very powerful indeed. Ten years ago, on the basis of a spotty fossil record, it was thought that the human line had broken from the ape line at least 15 million years ago. Now, because of evidence based on the molecules, it seems that we broke from the chimpanzees a mere 6

Testudo microphyes, **Isabela I.**   *Testudo abingdonii,* **Pinta I.**

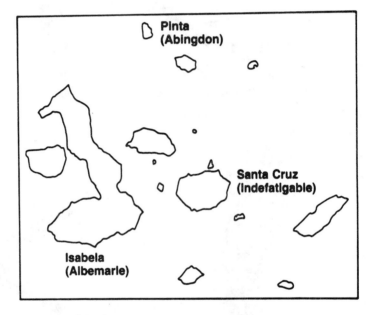

Testudo ephippium, **Santa Cruz I.**

Figure 2. Three different tortoises from three different islands of the Galapagos. (Adapted with permission from Dobzhansky *et al.*, 1977.)

million years ago, and the gorillas broke from both of us about 9 million years ago. (This remarkable finding means that we are more closely related to chimpanzees than they are to gorillas.)

Third, we come to evolution as mechanism. Darwin thought that he found this mechanism in what he called "natural selection." More

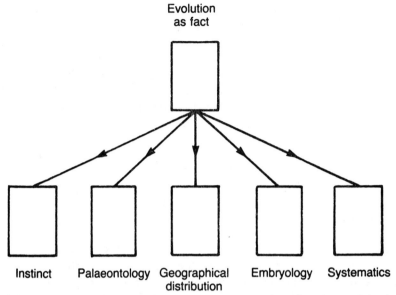

Figure 3. The structure of Darwin's argument for the fact of evolution. The fact explains and unifies claims made in the subdisciplines (only some of which are shown), which later in turn yield the 'circumstantial evidence' for the fact itself.

organisms are born than can survive and reproduce, survival is the result of special features (not possessed by the losers), and given enough time this all adds up to evolution. But, evolution of what kind? Answers make proper sense only if you know what questions they are addressing. This is particularly true for Darwin and natural selection. As will be seen from the beginning of our selection from the *Origin of Species*, Darwin stresses that he is trying to explain how organisms are *adapted*. For him, and for today's biologists, living beings are not just thrown together randomly. They work, they function, and they are adapted. They have "adaptations," such as the hand and the eye. Natural selection was supposed to address this.

Why did Darwin take adaptation so seriously? Many people think that he was writing against a background of Creationism, with everybody believing in the absolutely unchanged word of the Bible, particularly Genesis. This is not true. By the nineteenth century, educated people realized that the early chapters of Genesis have to be understood metaphorically. Nor did they consider this a strain on their faith. (More on this in the final section.) It was, rather, the Argument from Design that Darwin was addressing. He was nurtured on Paley and accepted the theologian's key premise, namely that the organic world is *as if* designed. What Darwin wanted to provide was an alternative explanation in terms of natural causes. This he thought he could do through natural selection.

Did Darwin succeed? Many people have questioned whether natural selection really explains adaptation, and the final readings of the section address the most important question of all. Darwin was always adamant that the variations on which selection works, the "raw stuff" of evolution, have no inherent direction or purpose. They just occur without any concern for the needs of their possessors. And the followers of Darwin today agree that the changes in the heritable links of life ("mutations") occur randomly with respect to needs. But, can a process of blind law, working on random differences between organisms, really lead to so much intricacy? Many people think not. Of course, in theory, a monkey randomly striking the keys of a typewriter could produce *Hamlet.* In practice not. In his discussion, the Australian molecular biologist Michael Denton shows just why he thinks randomness of the kind presupposed by selection could never lead to full functioning organisms.

The English biologist Richard Dawkins, one of today's most imaginative thinkers, believes that we read the monkey analogy incorrectly. He argues in the concluding reading of this section that, properly understood, random striking can indeed lead to *Hamlet* — or to organisms. Dawkins may or may not make his case. What impresses me is how he subtly uses one of the familiar symbols of our time, the home computer, to try to make common sense of what many people find to be very uncommon sense. In science, as in literature, the right analogy or metaphor is often the key to a successful argument.

3. Darwinism and the Tautology Problem

Having a mechanism is only the first step in building a complete theory. In the *Origin*, Darwin tried to use natural selection as the foundation for a comprehensive explanation of the phenomenon of evolution. But, as critics noted from the start, Darwin was hampered by his lack of understanding of the true nature of heredity. Why do pigs have piglets and not calves, and why do pink pigs tend to have pink piglets and not black ones, but not always? Without such understanding, selection falters. No matter how successful an organism may be in life's struggles, without an adequate method of transmitting its features its efforts are for naught. Moreover, we need some way of introducing new variations in each generation. Without a constant supply of building blocks, selection will exhaust the potential and evolution will grind to a halt.

Today, because of studies in heredity we know much about the origination and transmission of organic features. Most importantly, we know that nature precisely follows principles that make selection effective. These principles are collectively known as "Mendelian genetics" after the Moravian monk Gregor Mendel (a contemporary of Darwin) who first described them. In recent years genetics has increasingly been placed on a molecular basis, relating heredity to information carried on the macromolecules of nucleic acid. (See the later reading by John Beatty for an exact expression of one of Mendel's laws and discussion of its molecular foundation.)

However, at the beginning of this century, it was thought that genetics and Darwinian selection were rival causes of evolutionary change, and it was not until around 1930 that a number of thinkers (including J. B. S. Haldane) saw that they complemented each other, providing different parts of the overall picture. Rapidly then, because of the work of Theodosius Dobzhansky (in the United States) and Julian Huxley (in England), "neo-Darwinism" or the "synthetic theory of evolution" was born. It is in the light of these developments in Darwin's theory that one should approach the analysis of Darwinism given by the philosopher A. G. N. Flew in the first reading of this section.

Naturally, the mathematics of evolutionary theory is much more sophisticated than Flew demonstrates. One important finding of the way that the units of heredity ("genes") function in groups is that selection can promote diversity as well as eliminate it! A simple example follows from the supposition that there is a biological advantage to being rare (say, thus escaping a predator's eye). Selection will make the rare less rare and the common less common, creating a balance of forms. Obviously this conclusion is crucially important for the Darwinian. Even if new variations do not appear in order, that such variation is always being held in *groups* of organisms (balanced by selection) means that there will usually be an adequate supply of usable material, as times change and new evolutionary demands need to be met. Selection does not wait for the appropriate new variation, but generally has a veritable library on which to draw.

But, Darwinism is not yet out of the woods. As indicated, some critics fear that any claim about natural selection cannot be a genuine empirical proposition. It is rather a truism or a tautology, with this latter term being used in the logician's sense of a truth by logical necessity. Just as the claim that "all bachelors are male" holds universally because if something is a bachelor it is necessarily male, Darwin's theory of natural selection is really a tautology, true by definition.

How can this be? Critics seize on natural selection's *alter ego*, the "survival of the fittest." Initially, this was the term of Darwin's contemporary, Herbert Spencer, although Darwin introduced it into later editions of the *Origin*. Both by tradition and in fact it is, literally, natural selection by another name. But now one asks — who are the fittest? And the critics answer that the only way of determining "fitness" is in terms of survival. In other words, the fittest are those who survive. Hence, natural selection collapses into "those who survive are those who survive."

It is precisely this charge that Tom Bethell makes in "Darwin's Mistake." He maintains that Darwinism — that is, any evolutionary theory that depends crucially on the causal mechanism of natural selection—has to be fatally flawed. However, Bethell is called to account by today's most brilliant biological essayist, Stephen Jay Gould. As we shall learn, Gould is no supporter of ubiquitous adaptationism. He deplores seeking of design-like effects in each and every corner of the living world. For this reason, we shall see Gould draw back from a blanket Darwinism. He nevertheless regards Darwin's theory as genuinely empirical and basically sound. Gould thinks that Bethell, and all similar critics, miss the point and force of natural selection. It is true that the only way we can identify fitness is through survival and reproduction, but this in no way indicates a straight tautological identification. (Note incidentally how Bethell surely misreads Flew. The latter asserts explicitly that the propositions of Darwinism are empirical and nontautological. He sees logically necessary connections *between* the statements of the theory.)

4. The Challenge of Punctuated Equilibrium

I think it is fair to say that neo-Darwinism is still the theory that lies behind evolutionary thought today, although in certain respects it is being challenged, extended, revised, and rejected. This, of course, is the mark of living science. One of the orthodoxy's most persistent critics is Gould, whom we just met in the last section. There he was defending natural selection, so he is certainly no outright opponent of neo-Darwinism. However, as he explains in the first reading of this section, he thinks that the Darwinian sees altogether too much adaptation in the living world and he suggests that we must free ourselves from this. This opens the way for Gould to introduce his own major contribution to evolutionary thought, the theory of "punctuated equilibrium," which he formulated with his fellow paleontologist Niles Eldredge. Unlike the gradual change seen by the Darwinian, Gould and Eldredge see change on the large scale ("macroevolution") as going in fits and starts. The staccato fossil record is less a function of inadequacy and more one of the way evolution actually occurs.

Of course, if the pattern of evolution is one long period of nonchange ("stasis"), followed by rapid switches from one form to another, the question of what fuels these changes arises. Gould and Eldredge deny vehemently that they want to go beyond the Mendelian pale, arguing rather that in small isolated ("peripatric") populations of organisms change can be much more sudden than is usually the case. (This is known as the "founder principle.") The important consequence of this is that one cannot hope to understand the overall processes of evolution, "macroevolution," simply from the small-scale causes, "microevolution." Other phenomena such as external events must be taken into account, thus preventing a simplistic "reduction."

In response, we have an article by Francisco J. Ayala, one of today's most prominent neo-Darwinians. In his analysis of Gould's thesis, he tries to differentiate between those parts he thinks no one would want to deny and those parts he thinks everyone should want to deny. Ayala's objection to Gould is primarily concerned with the notion of "reduction," which he sees as conflating three separate issues. First we have the question of the underlying processes and of whether they operate at all times and levels. Second, there is the question of the adequacy of these processes to explain all of the phenomena of evolution, particularly phylogenies. Third, and most stringently, we have the question of whether everything that happens at the macro level can be shown to be a straight deductive consequence of happenings at the micro level. With this trichotomy made explicit, Ayala is ready to answer Gould.

Details apart, it is clear that Gould and Ayala have a fundamentally different attitude toward the significance and extent of organic adaptation. In some respects, the clash between Darwinians and critics such as Gould resembles conflicting approaches at the beginning of the nineteenth century, when there were two major schools of natural theology: The Paleyites, with their stress on adaptation, and continental thinkers such as the German poet Goethe, forerunner of the school known as *Naturphilosophie*, which always stressed the structure of organisms as much as it stressed their workings. Why, to ask a classic question, do we have homologies, even though they seem to serve no purpose? As we know, Darwinians have been aware of such questions. Indeed, we have seen how Darwin was able to explain homologies through common descent. But, nonadaptive features have never been the primary focus of Darwinians, as they were to Goethe and as they are to contemporary biologists such as Gould.

5. Problems of Classification

Whatever the nature of the fossil record, the world around us is not a smooth continuum, with one form grading imperceptibly into another. Rather, we see fairly distinct forms with breaks between them — fish, birds, mammals, dandelions, dogs, humans. In fact, not all of these types or kinds seem to be of quite the same order. I am a human. I am also a mammal and an animal. My dog Spencer is likewise a mammal and an animal. He is not a human. In the first reading of this section Ernst Mayr, the world's most distinguished living evolutionist, explains that classifiers ("taxonomists") follow the eighteenth-century thinker Carolus Linnaeus, in classifying all organisms hierarchically, with each individual being assigned to a sequence of groups (known as "taxa") at ever-higher more inclusive levels (known as "categories"). The lowest standard level is that of "species." There, humans belong to the taxon *Homo sapiens*. The next category up is "genus." There, we belong to *Homo*. We are now the only living representatives of *Homo*. It is believed that in the past there were other taxa, members of which also were included in *Homo*, namely *Homo habilis* and *Homo erectus*.

Biologists have a sense that there is something special about species, more so than about taxa of any other level. After all, Darwin's great work was not called *On the Origin of Genera*. There is a feeling that somehow species are uniquely "real" or "natural." The breaks between species are objective. Placing an organism in the taxon *Homo sapiens* tells you something about the organism that other assignments —for example, to the group "born on June 21, 1940"—do not. Even telling you that the organism belongs to the genus *Homo* does not have quite the same impact. Apart from anything else, we know that our ancestors were australopithecines, and some of the earliest members of *Homo* were very similar to some of the latest members of the coexisting *Australopithecus*.

Why are species different? What gives them their special status? To tackle the "species problem," we must first have some clear idea of what is meant by species. Mayr runs crisply through various proposals for characterizing species showing how, in the light of our commitment to evolutionism, some of the more obvious suggestions, resting on physical similarity and the like, just will not do. His own proposal, involving the reproductive barriers between members of different species, has been very influential. Yet, although a good definition is crucial, it is only part of the solution of the species problem. There still remains the question of "naturalness."

Much ink has been spilled in search of an answer to this question. Our second reading presents an argument by the philosopher David L. Hull who (with the biologist Michael Ghiselin) has proposed the most daring solution in recent years. He argues that no solution to the species problem can be forthcoming as long as we misidentify the true nature of species. Commonly, we think they are groups or classes, such as baseball teams. Just as Babe Ruth was a member of the Yankees, so Michael Ruse is a member of *Homo sapiens*. However, according to Hull, species are not classes at all. They are literally *individuals,* such as particular organisms. As my hand is part of me, so I am part of *Homo sapiens*. Once species are taken to be individuals the original problem is solved: they exist in their own right no less than particular organisms whose reality no one questions. Arthur L. Caplan, the author of the next essay in this section, flatly opposes the Hull/Ghiselin proposal. He argues that the traditional approach, taking species as classes, still has much to commend it.

The dispute between the champions and the opponents of the notion that species are literally individuals is by no means resolved. To this date, no one seems to have won over the other side. What I will draw your attention to is the fact that both sides seem to agree that this is not an issue that can be resolved by a simple appeal to common sense, whatever that might be. The status of species must be judged against the background of accepted biological thought. This point leads us back to the discussion of the last section. If you are an orthodox Darwinian, you may be quite happy with the species-as-a-class view. If, to the contrary, you are drawn toward the punctuated equilibria theory, you may well favor viewing species as individuals. In this latter case, you see species coming and going fairly distinctly, just as individuals, and when species are in existence, organisms within are integrated by being subject to shared constraints. Certainly, Gould embraces the Ghiselin/ Hull thesis in his article.

The species problem may be the most troublesome question about the foundations of taxonomy. However, in a way it is only the beginning of the problem, for you must then go on to ask questions about classification taken as a whole. In particular, what underlying theme or philosophy should guide you as you attempt to slot organisms into their various levels of the Linnean hierarchy? Why, for example, put whales with mammals rather than fish, or bats with mammals rather than birds, or humans with apes rather than a very social group such as ants? As the biologist Mark Ridley points out in the final reading of this section, essentially taxonomists face two options. On the one hand, they can try to classify entirely on raw physical similarity ("phenetically"). Or, they can bring their knowledge of evolution into play, insisting from the beginning that all classification must in some way reflect evolutionary principles. Ridley himself argues for this second option ("phylogenetic" classification), even though it may not seem the most intuitively attractive. Much of his discussion is given to defending this choice.

6. Teleology: Help or Hindrance?

For Paley, it made sense to ask "What's the eye for?" because he knew that it was for seeing. God had intended it or made it for that *purpose* or *end*. The purpose of the eye is to see, just as the purpose of the knife is to cut the bread. However, with the coming of natural selection, all of this design language becomes inappropriate. God may or may not have designed the world, but any such design is, at the least, remote. Now we see that organisms are produced directly by natural causes, no less than planets or other celestial bodies. Therefore, since it makes no sense to ask "What's the moon for?" — the moon is not for anything — it should make no sense to ask "What's the eye for?" The eye is not *for* anything. The eye is simply a product of nature, as is the moon.

Yet, look again at the selection from Darwin. Darwin talks about "adaptation" and about variations being "profitable" to their possessors ("useful to each being's welfare"). For all the difference it made to biological *language*, the *Origin of Species* might not have been written. Of course, you might point out that, notwithstanding his greatness, Darwin was a pioneer, and a certain careless carryover of thought and language was almost inevitable. Today, however, with the coming of genetics, we should find more linguistic purity and a decline in the use of old theological ways of thought.

In fact, we find nothing of the sort. Design language reigns triumphant in evolutionary biology. The first selection in this section, by George C. Williams, extracted from a highly praised and influential book on evolutionary biology, makes this point eloquently. Granted that evolutionists use design language, is this a good thing? Some biologists feel uncomfortable and urge us to drop all language with anthropomorphic connotations, whether it be talk of goals, strategies, or designs. One who takes this view is the botanist Paul J. Kramer, who believes that such language in this post-Darwinian age is inappropriate. He would therefore eliminate it all. People (and, presumably, Gods) have strategies and goals and ends; plants and animals and their parts do not. For Kramer, it is time biology grew up and took its place as a mature science, alongside physics and chemistry.

Yet, doubts remain. Apart from anything else, it may be much more difficult to eliminate the general anthropomorphism of biology than most reformers allow. Without comment, although having italicized certain interesting terms, I note that Kramer himself says that "plant structure and *function* are *compromises* to *conflicting pressures.*" And, in any case, perhaps there is a point to the language and implied understanding of biology that is worth acknowledging and preserving. Biologists face problems that are different from physicists — even the most hardened atheist must agree that plants and animals are design-like in appearance in a way that rocks and planets are not. Could it be that the very nature of biological phenomena forces a distinctive form of understanding on biologists?

This is certainly the feeling of many biologists and philosophers. They argue that in biology, as elsewhere in life, particularly when dealing with human actions and intentions, we can appropriately seek a special kind of understanding. Instead of the usual mode, where we try to explain things in terms of what has gone before — "Why was there a bang?" "Because the door slammed." — We can try to explain things in terms of what will happen — "Why is there this belt?" "To restrain you in an accident." These forward-looking "teleological" explanations include examples of the kind we have been considering — "Why is an arctic fox white?" "In order to conceal it against the snow." — and, thus laid out, we can readily see that it is not really the anthropomorphism of design that is a problem. After all, physicists use anthropomorphic terms such as "work" and "force" all the time, and no one objects.

Rather, it is the forward-looking aspect that troubles. In a normal explanation, that which explains is already history. We know it occurred (the slamming door). In a teleological explanation, it (the car accident, the concealed fox) is yet to come. But, what if it never comes? What if the accident never occurs? What if all the snow melts? Then you seem to be left with an explanation without a base. Unless, that is, you can cover yourself against the "missing goal object," as the philosopher Ernest Nagel called the problem.

It is this kind of issue that the biologist Francisco J. Ayala tackles in the final reading of this section. He is no mushy-thinking metaphysician, believing in mysterious forces. But Ayala does think that as a biologist, he can be and must be a teleologist. Clearly, in his understanding of "teleology," Ayala is far removed from Christian theists such as Paley and also from the champions of vital forces. Ayala argues that it would be wrong to expel teleology from biology the way it has been expelled from physics. In biology teleological hypotheses have a heuristic value they do not possess in physics and they frequently lead to new insights. Although there may not be a great Designer in the sky, because of natural selection organisms (unlike inanimate objects) are design-like — the eye is like a telescope and the heart is like a pump. Hence, asking function questions makes sense and advances biological understanding. (Referring back to a critic of ubiquitous adaptationism, such as Gould, it is clear that inasmuch as he thinks talk of adaptation appropriate, he would approve of design talk. It is just that he considers the Paley metaphor of design too seductive.)

7. Molecular Biology

We now move to a topic of pressing importance. I have already mentioned how our understanding of the mechanisms of heredity falls below the level of the "classical" understanding of genetics, locating the baseline in the functioning of long molecules of nucleic acid. The major breakthrough at the molecular level came in 1953, as a result of the brilliant work of two young researchers, James Watson and Francis Crick, who discovered that (in most organisms) genes are to be identified with lengths of deoxyribonucleic acid (DNA) twisted together in pairs, in a "double helix." This was the foundation of "molecular biology," which has had profound effects on traditional classical biology — as demonstrated by the amount of chemistry the average biology undergraduate has to take to complete a degree.

What is the effect of the molecular revolution, and is it a good thing? Let us start with the question of understanding and with the relationship between the older, nonmolecular biology and the newer, molecular biology. We are plunged again into the difficult question of "reduction." In this specific case, we are dealing with the connections between the physical sciences (where these are taken to include molecular biology) and (nonmolecular) biology itself.

To throw some light on the problem, let me refer to a three-fold division of Francisco J. Ayala, between "ontological," "methodological," and "theoretical" reduction. (As Ayala acknowledges, this is in spirit the same division he uses in his response to Gould.) By the first of these terms, ontological reduction, is meant the attempt to explain living things purely in terms of material particles — atoms, for instance. By the second term, methodological reduction, is meant the attempt to explain the larger always in terms of the smaller. Explain organisms in terms of cells, but even better, explain them in terms of genes and other cell components. By the third term, theoretical reduction, is meant the attempt to explain older theories in terms of newer theories. This sense of explanation usually centers on the notion of "deductive consequence." One must show that the older theory follows deductively from the newer theory. A paradigmatic example of this theoretical reduction occurred in the last century, when physicists showed how older claims about the behavior of gases, such as Boyle's law, were logical consequences of newer claims about the behavior of particles in groups. (Here was a clear case of methodological reduction as well as of theoretical reduction, showing how the various senses often overlap.)

The first essay of this section, by Kenneth F. Schaffner, will act as a foil, for in it he presents himself as a hard-line reductionist, in just about every respect. In the context of genetics, what does this mean and who would oppose him? At the ontological level, this means that Schaffner believes that organisms in general and their units of heredity in particular are no more than molecules. Mendelian genes are, ultimately, molecules — as we now know, DNA molecules. I doubt anyone today would deny this, so let us move at once to the next level. As one might expect, many biologists, especially those who endorse the species-as-individuals thesis, find methodological reductionism unduly restrictive. Yet, with respect to genetics, Schaffner believes that full understanding, even down to the origin of genetics itself, is to be found in our understanding of the entities of the physical sciences, particularly chemistry.

As Schaffner acknowledges, even in genetics there will be opposition to such methodological reductionism. The usual charge is that the physical sciences ignore the *organization* of living things. The DNA molecule is not just a number of smaller molecules strung together randomly. They are strung together in a particular order and can function only in this order. Schaffner, however, has a reply. He argues that physical scientists are likewise sensitive to organization — the order of the DNA molecule is part of chemistry as much as part of biology — and hence methodological reduction is still possible.

Perhaps, however, one can formulate the objection in another way. At least, this is the basis of the argument given by the philosopher John Beatty, in the second reading of this section. It is not so much that biological organization is peculiar in the sense of involving new phenomena, such as vital forces, but, rather that, as a product of evolution, biological organization raises different types of questions. Beatty argues that no matter how detailed our molecular understanding, there are still going to be questions about function and purpose in the overall evolutionary scheme of things.

As Beatty points out, Schaffner seems to agree with the claim that evolution demands and imposes new principles of understanding on biology. This being so, depending on one's perspective, one might conclude either that Schaffner answers the antireductionist completely, or that he gives most of the case away. One's conclusion will probably depend on how far one is prepared to allow Schaffner to bring in the whole evolutionary past of organisms as part of molecular biology! It is of some interest that since Schaffner's article was written, molecular biologists have themselves made major strides in this direction, suggesting that another nucleic acid (ribonucleic acid, RNA) was crucial in life's beginnings.

One thing on which everyone does now seem to agree — whether one labels oneself a "reductionist" or not — is that the organization of organisms is crucial, and that from an evolutionary perspective, new questions are asked in addition to the traditional ones of physics and chemistry. But does this then open the way to a full theoretical reduction, as Schaffner suggests? This is a strong request. What it would entail (among other things) is showing how Mendel's laws are direct deductive consequences of laws or claims about DNA molecules. Suffice it to say that, if John Beatty's argument shows us anything, it is that the actual demonstration of such a connection might be much more complex than Schaffner hints.

8. The Recombinant DNA Debate

Theoretical reduction as discussed in the last section is perhaps primarily a philosopher's problem. What is not such a problem is the fact that molecular biologists act as if methodological reductionism is not merely an ideal but a humanly achievable goal. And what this has meant is that, in recent years, the molecular advance has taken on a whole new meaning, as biologists have developed much more powerful techniques to blast apart the cell, recombining molecules in whole new forms, making fresh or radically revised organisms. (This technology is called "recombinant DNA" or "rDNA.") This raises all sorts of moral and practical questions, and the readings in this section address some of these issues. When, if ever, should one hold up the course of science, and what factors are pertinent in making such a decision? Jeremy Rifkin has had considerable success in the courts, preventing the application of the new biotechnology, and here he takes a very dim view of the supposed benefits (as compared to the risks) of rDNA. The philosopher Stephen P. Stich tries to present a more balanced position.

The question we must all answer is whether the risks are so great that any attempt at balance is simply naive optimism. What we can be sure of is that if we do not make the decisions, someone will do it for us. And that someone may not have our interests as the highest priority. Reductionism in practice strikes very close to home.

9. Human Sociobiology

Darwin was always convinced that his theory extended to our own species. Indeed, in his private notebooks the first clear statement showing that he had grasped the significance of natural selection is a passage applying it to human intelligence. However, for a number of fairly obvious reasons, the full application of Darwinism to humankind took much time and effort and indeed in some respects is still in its infancy. On the one hand, to make a convincing case, it was necessary to work out in some detail the actual path (as well as the processes) of human evolution. It was also necessary to extend and develop those areas of evolutionary theory that are particularly relevant to human nature. I refer here in particular to our understanding of social behavior, which is so very much part of what it is to be a human being.

In recent years, great progress has been made on both of these fronts. As mentioned earlier, we now have some good ideas about the break of the human line from that of other apes, and many of the details from the point of the break have been filled in. At the same time, students of social behavior have made impressive strides in understanding such phenomena, showing how natural selection is as much a factor in that realm as it is in other parts of biology. Indeed, this new subbranch of the Darwinian picture, "sociobiology," is one of the most intensively discussed topics in evolutionary studies today.

Yet, for all the advances in understanding and for all the realization that humans are part of the story of evolution, the development of the human branch of sociobiology has been clouded by controversy. In part, this is an accident of history. With the decline of the traditional Christian picture of mankind, various other claimants rushed to fill its place. Particular manifestations of this development have been Marxism and Freudianism, carrying their associated world pictures. Whether these pictures are right or wrong, and whether they are truly rivals to any biological picture, it has meant that the growing Darwinian account of humankind has had to fight for breathing space. And this has led to much discord.

Nevertheless, complaints about rivals aside, even its strongest supporters must agree that human sociobiology has been, in part, the cause of its own misfortunes. Its enthusiasts have rushed in with scant theory and scantier evidence to make bold pronouncements. As critics have rightly pointed out, too frequently human sociobiologists have assumed that contingent aspects of late twentieth-century western culture are manifestations of a deeper biological reality applicable to all people at all times. One thinks in particular of some of the sexual and social inequalities of our society that have been presented as ingrained in human nature.

Yet, having said this, one may still feel that there is a basis for optimism about the biological approach to humankind. This is the position of E. O. Wilson, chief spokesman for human sociobiology and author of *On Human Nature*, an extract from which opens our section. He does not argue that human beings are blind puppets, determined to strut through life at the mercy of their genes. He does not want to argue that culture and learning are powerless in affecting human lives.

However, Wilson does maintain that biology sets certain broad constraints, that these are founded in adaptive strategies in our evolutionary past, and that ignoring them (perhaps for the sake of some ideology) cannot serve any useful purpose.

There are various ways in which human sociobiology might be attacked. It might simply be dismissed on moral or political grounds, with the argument that it is only racist ideology brought fifty years forward. Even if sociobiology contained an element of truth, it is too dangerous to be explored. One may sympathize with this reaction — it is undeniable that genetic pictures of mankind have been used to justify reactionary and extremely vicious ends. However, to refuse absolutely to consider the questions of the sociobiologists is to allow Hitler power from beyond the grave. Alternatively, one might dismiss human sociobiology as bad science. It depends on "ultraadaptationism," suggesting, for instance, that differences between males and females may have a (biological) adaptive origin. As we know, Stephen Jay Gould — who is one of the foremost critics of human sociobiology — denies ultraadaptationism even in the animal world. To deny it in the human world is simply an extension of his general thesis, and it is precisely this that he does in the second selection of this section. He accuses people such as Wilson of following Rudyard Kipling in making up "just so" stories. Kipling suggested that the elephant's trunk grew because its nose was pulled. In the same way sociobiologists provide pseudo-historical accounts for various kinds of human behavior.

In the final selection of this section Arthur L. Caplan argues that it is unfair to reject sociobiology at this early stage of its development. If one applies the stringent criteria that Gould would impose on us, virtually no new science would ever get off the ground. To critics such as Gould Caplan replies that a new science should be shown tolerance if it is an outgrowth of an already-established science and that this holds in the present case precisely because human sociobiology is a part of orthodox Darwinism. This, of course, raises the twin questions of whether human sociobiology is truly a part of Darwinism or simply takes on a veneer to make it look respectable, and whether orthodox Darwinism is without blemish. Gould answers both of these questions differently from Caplan.

10. Extraterrestrials?

"Exobiology," the proper name for the study of the possibilities of life elsewhere in the universe, tends to get more coverage in the cinema than it does in the writings of philosophers. This was not always so. Indeed, some of the very greatest philosophers had thoughts on the subject. Aristotle claimed that the moon is inhabited. Expectedly, with the coming of Christianity, the focus of writers on exobiology turned primarily to theological issues. If there are beings — intelligent beings

—elsewhere in the universe, then what does this mean for our special relationship with God? Is Jesus being crucified on Friday—every Friday—somewhere in the universe? Conversely, if the universe is empty, then why on earth (or, rather, why in heaven) did God bother to create it in the first place?

Such theological worries no longer trouble most philosophers and scientists. The question itself nevertheless remains of great interest. The authors of our two selections reach very different conclusions. Like many people from the physical sciences, Robert Bieri is enthusiastic about intelligent beings living in outer space and like many biologists Ernst Mayr is skeptical. Mayr's negative answer stems from some deep-felt convictions about the nature of the evolutionary process. Before Darwin, it was natural to think that the world makes overall sense and that, even if not quite as described by Genesis, everything had led progressively up to humans. After Darwin, that faith in progress was gone, or, at least, it should have been. Evolution through natural selection is a blind random process going nowhere. The correct metaphor is a branching tree (Figure 4, p. 45), not an upwardly pointing escalator. Once this is realized, Mayr argues, naive hopes of intelligent extraterrestrials seem ill-founded.

11. Evolution and Ethics

Questions about right and wrong, about what we ought or ought not to do, are surely among the most important questions human beings can face. Traditional religions have answers to the questions ("Thou shalt not kill," "Love your neighbor as yourself") as do traditional philosophies ("Treat others as ends and not as means," "Promote the greatest happiness for the greatest number"). Evolutionists have also presented suggestions—after all, if we are modified monkeys this surely makes some difference to our thinking about right and wrong.

Even before the *Origin* appeared Herbert Spencer was promoting an evolutionary ethic, a philosophy taken up with enthusiasm by many in the late nineteenth century and nowhere more than in America, particularly by the Yale sociologist, William Graham Sumner, who opens the section with a passage from one of his best-known essays. Essentially, Sumner's moral philosophy is simple and straight forward. The evolutionary mechanism is natural selection brought on by the struggle for existence. This applies no less to humans, and therefore we ought to let the struggle have its head. Our moral obligations are to promote *laissez-faire* capitalism, in which the strong succeed and the weak fall by the wayside. That is "nature's law."

Sumner was no hypocrite. His personal motto was "root, hog, or die" and he challenged any university professor to oust him from his post by doing a better job than he. But, before we even get to the countercase, note how difficult it is to subscribe consistently to "Social Darwinism" as this particular philosophy, somewhat inaccurately, is known. Sumner would have us cherish monogamy and personal property. It is not readily apparent how this follows from natural selection. More plausibly, in a free-for-all, one might expect rape, pillage, and murder. We shall see in a moment that things are rather more complex, and such a moral nightmare is not necessarily the outcome of the evolutionary process. Sumner, though, does not show this.

Sumner's leading nineteenth-century opponent was Thomas Henry Huxley, Darwin's famous early supporter. Huxley has no time for the morality of the jungle, and he presents a forceful case for a traditional morality of caring and sensitivity. We must "combat" the evolutionary process, not acquiesce to it. It may be questioned whether Huxley is always consistent. At times he seems to argue that our moral sentiments no less than our immoral urges are the result of evolution. In that case we would hardly be opposing evolution in being kind and considerate. We would be using one aspect of evolution to fight another. Many biologists during the twentieth century have tried to develop this idea and attempt to use evolution to develop an acceptable moral code. One of the most influential attempts in this direction was made by T. H. Huxley's grandson, Julian, famous in his own right as one of the founders of neo-Darwinism. What he (and others) argued, pursuing a line of thought articulated most notably by the Russian anarchist Prince Peter Kropotkin, is that selection can work for the good of the group and hence a general ethical sentiment follows automatically.

However, today, as Gould notes (in "Darwinism and the Expansion of Evolutionary Theory"), such "group selectionist" ideas find little favor with evolutionists. They fear that individual interests will virtually always win out over such disinterested feelings. Yet, the sociobiologists have used this very point to breathe new life into evolutionary ethics. They argue that a kind of enlightened self-interest can promote group sentiments. For example, cooperation will often yield greater dividends than all-out attack, because cooperators get help in turn, unlike attackers. A similar line of argument is explored by the well-known Australian-born philosopher John L. Mackie in his contribution to this section. He develops the ideas of Richard Dawkins, who employs the powerful (and notorious) metaphor of a "selfish gene" — a concept implying that all of the adaptive benefits accrued during evolution must work ultimately for the individual actor. Following Dawkins, Mackie shows that this in no way implies that organisms (humans in particular) behave or think selfishly. To the contrary: the "law of the jungle" demands cooperation. Yet, Mackie would not return to an idealized Christian-type morality, in which we have an obligation to go on forgiving transgressors no matter what. Rather, Mackie believes biology supports a kind of common sense morality, in which we feel genuine obligations but are ever wary against being made "suckers."

Assume for the sake of argument that the Dawkins/Mackie thesis is well taken, something that a critic such as Gould would obviously not accept. Many would think that we are still only part of the way to an adequate ethical theory. Perhaps biology tells us what we do and why we think we should do what we do. But are we truly right in what we think and do? Are our moral sentiments justified? Surely one should draw a distinction between the existence and origin of beliefs and their justification. I may fervently believe that the Blue Jays will win the World Series next year, but (alas) that belief is far from justified.

In the opinion of most critics — a group that includes most professional philosophers — this is the point at which evolutionary ethics tends to come undone. Either one must go from claims about factual beliefs to claims about moral obligations, or one must suppose that evolution itself evinces values, perhaps through some supposedly progressive rise from molecules to men, a trend that is our moral obligation to sustain. But either way, critics note, we are in trouble. Going from fact to obligation, from an "is" statement to an "ought" statement, is to move illicitly from statements of one logical type to those of another type, committing the notorious "naturalistic fallacy." Finding value in life's upward progress is (as Mayr showed us in the last section and as Huxley hints in this) to read improperly into life's history precisely that which you would read out.

In the final essay of this section, Edward O. Wilson and I accept the critics' objections and try a different approach. In line with many of today's moral philosophers, we suggest that perhaps ethical claims have no ultimate objective foundations — any more than do (say) likes and dislikes about ice-cream flavors. This is not to say that moral obligations are mere likes and dislikes, and even less is it to say that our biology has left us free to make up our own minds about right and wrong. What we suggest is that even if there are no foundations, to get us to cooperate efficiently, evolution makes us *think* such foundations exist. Evolution tricks us into beliefs about objectivity, and therefore, in this sense, morality is truly a collective illusion of our species!

12. God and Biology

We come to our final topic: can one be both a believer in God and an evolutionist? Is the war between religion and science still raging, or has one side admitted defeat? Is Darwinism the death-knell of God?

Let us start with some history. Darwin published the *Origin* in 1859. This was some 1500 years after sincere Christians (notably St. Augustine) had started to grapple with the problems inherent in taking the Bible absolutely literally. By the third verse of the first chapter of Genesis God has created light, although the sun has to wait until the sixteenth verse. And the difficulties mount. The first two chapters of Genesis present two quite separate accounts of the Creation. Biblical scholars now believe that, in conception, the second came before the first. The story in Chapter 2 dates from the time of King David (he reigned 1000–961 BC), and was intended to illustrate the role of the king, in the form of Adam, as one who was both subject to God and yet had rule over the earth. The story in Chapter 1 dates from the Babylonian exile (post 587 BC), and, drawing on near-East myths, reassures the oppressed Jews of God's power and, through the status of Adam, their special place in His scheme.

None of this detail was known in Darwin's day, but, as mentioned in an earlier section, by the mid-nineteenth century educated Christians generally had moved beyond regarding Genesis as a manual of geology. They would have agreed with the present pope (John Paul II) when he said that the Bible tells us where we are going, not where we came from. Far more troublesome for the believer were the problems of natural theology, centering on design. The one sizable group who persisted in reading the Bible in an absolutely literal manner were the evangelical Christians in the Southern United States, who remained in the grip of an entirely indigenous form of extreme Protestantism. Uniquely in that part of the world, it was not long after the *Origin* that teachers began losing their jobs for telling their students that Adam and Eve might not have been the first humans.

In my contribution to this section, I tell how things came to a head in the 1920s, when a number of states began passing "monkey laws," actually banning the teaching of evolution in classrooms. I also relate how laws of this kind were blocked by ridicule as much as by law. However, this hardly halted matters. In recent years, rather than banning evolution, Biblical literalists have tried to get Genesis into the schools, along with evolution. The First Amendment to the Constitution separates Church and State. This makes it illegal to introduce Genesis as religion. Can it therefore be disguised as science, and slipped in surreptitiously? This has been the aim of the so-called "Creation Scientists," who have argued that everything in the Bible read absolutely literally can be given full scientific backing. Evolutionists underestimated the threat and eventually, in the states of Arkansas and Louisiana, the Creationists succeeded in having bills passed mandating the teaching of Creation–Science along with evolution in state-supported public schools.

The laws based on these bills have since been declared unconstitutional. However, they raise important questions about the nature of biological science and its relationship to religion. I was an expert witness in the Arkansas case. The kinds of things I had to say are included in my article. This collection also reproduces the relevant chapters of Genesis and the passages about Creation–Science in the Arkansas bill. The two are identical, as indeed they are intended to be, for Creation–Science is simply a means of circumventing the First Amendment. The question is surely irresistible whether a loving Creator would have given us powers and sense and reason and then let us be so grievously misled about evolution. How could He have placed the animals inhabiting the Galapagos archipelago in such a grotesquely deceitful arrangement?

Anybody rejecting Creationism as an adequate form of Christianity, let alone science, has the obligation to think further about the true relationship between science and religion. Some theologians solve the problem by simply declaring that science and religion speak to different realms of experience—physical and moral—and the two never can meet and be in conflict. Others, like Arthur Peacocke, who is both an Anglican priest and a physical chemist, do not believe that matters can or should be solved so readily. They see more contact between the physical and the spiritual, and in the final reading of this section we see one man's attempt to bring evolutionary thought to bear on his faith. For Peacocke, evolution is no threat to be denied or ignored or trivialized. It is, rather, one of the glories of the universe, and our ability to understand it attests to our special place in this creation.

13. Cloning

In February of 1997, a British veterinary surgeon, Ian Wilmut, working in Edinburgh, successfully cloned a lamb, "Dolly," from an adult sheep. He and his team took DNA from the cell of the adult's udder, introduced it into an ovum, and grew this into a new organism. Immediately people saw that this was more than just a good method for propagating farm animals and does indeed have profound implications for us humans. Does this mean now that millionaires can have themselves replicated *ad infinitum*? Could a madman like Hitler be reproduced many times? Should all of our future physicists be clones from Newton or Einstein? Could we now get all of our boring and dirty jobs done by cloning today's most efficient garbage collectors and cleaning ladies? Worries like this have hovered over us since Aldous Huxley (the grandson of Thomas Henry Huxley) published his futuristic novel *Brave New World,* between the Wars. In this newly added final section, we have expressions from holders of two very different points of view. Philip Hefner, a liberal Lutheran theologian, would have us take very great care in how we proceed. He is certainly not blindly against cloning, but he would have us pay full attention to what one might call the spiritual dimension of life. This demands extreme care and caution when dealing with something like cloning. One might violate or destroy the essentially human. Ronald A. Lindsay, to the contrary, takes a resolutely secular approach. He would have us treat cloning in a moral perspective, certainly, but he sees no reason for invoking any special norms or principles. In particular, he argues that there is no need at all for a theological or spiritual approach. Essentially, he sees cloning as a good thing and thinks there is no need for special emphasis on dangers.

WHAT IS LIFE?

1

ARISTOTLE*

The Generation of Animals

THERE IS A considerable difficulty in understanding how the plant is formed out of the seed or any animal out of the semen. Everything that comes into being or is made must be made out of something, be made by the agency of something, and must become something. Now that out of which it is made is the material; this some animals have in its first form within themselves, taking it from the female parent, as all those which are not only born alive but produced as a grub or an egg; others receive it from the mother for a long time by sucking, as the young of all those which are not only externally but also internally viviparous. Such, then, is the material out of which things come into being, but we now are inquiring not out of what the parts of an animal are made, but by what agency. Either it is something external which makes them, or else something existing in the seminal fluid and the semen; and this must either be soul or a part of soul, or something containing soul.

Now it would appear irrational to suppose that any of either the internal organs or the other parts is made by something external, since one thing cannot set up a motion in another without touching it, nor can a thing be affected in any way by anything that does not set up a motion in it. Something then of the sort we require exists in the embryo itself, being either a part of it or separate from it. To suppose that it

*From Jonathan Barnes, ed., *Complete Works of Aristotle: The Revised Oxford Translation,* Bollingen Series 71, pp. 38–41. Copyright © 1984 by Princeton University Press. Reprinted by permission of Princeton University Press.

should be something else separate from it is irrational. For after the animal has been produced does this something perish or does it remain in it? But nothing of the kind appears to be in it, nothing which is not a part of the whole plant or animal. Yet, on the other hand, it is absurd to say that it perishes after making either all the parts or only some of them. If it makes some of the parts and then perishes, what is to make the rest of them? Suppose this something makes the heart and then perishes, and the heart makes another organ, by the same argument either all the parts must perish or all must remain. Therefore it is preserved. Therefore it is a part of the embryo itself which exists in the semen from the beginning; and if indeed there is no part of the soul which does not exist in some part of the body, it would also be a part containing soul in it from the beginning.

How, then, does it make the other parts? Either all the parts, as heart, lung, liver, eye, and all the rest, come into being together or in succession, as is said in the verse ascribed to Orpheus, for there he says that an animal comes into being in the same way as the knitting of a net. That the former is not the fact is plain even to the senses, for some of the parts are clearly visible as already existing in the embryo while others are not; that it is not because of their being too small that they are not visible is clear, for the lung is of greater size than the heart, and yet appears later than the heart in the original development. Since, then, one is earlier and another later, does the one make the other, and does the later part exist on account of the part which is next to it, or rather does the one come into being only *after* the other? I mean, for instance, that it is not the fact that the heart, having come into being first, then makes the liver, and the liver again another organ, but that the liver only comes into being *after* the heart, and not by the agency of the heart, as a man becomes a man *after* being a boy, not by his agency. An explanation of this is that, in all the productions of nature or of art, what already exists potentially is brought into being only by what exists actually; therefore if one organ formed another the form and the character of the later organ would have to exist in the earlier, e.g. the form of the liver in the heart. And otherwise also the theory is strange and fictitious.

Yet again, if the whole animal or plant is formed from semen or seed, it is impossible that any part of it should exist ready made in the semen or seed, whether that part be able to make the other parts or no. For it is plain that, if it exists in it from the first, it was made by that which made the semen. But semen must be made first, and that is the function of the generating parent. So, then, it is not possible that any part should exist in it, and therefore it has not within itself that which makes the parts.

But neither can this agent be external, and yet it must needs be one or other of the two. We must try, then, to solve this difficulty, for perhaps some one of the statements made cannot be made without qualification, e.g. the statement that the parts cannot be made by what is external to the semen. For if in a certain sense they cannot, yet in another sense they can. (Now it makes no difference whether we say 'the semen' or 'that from which the semen comes', in so far as the semen has in itself the movement initiated by the other.) It is possible, then, that A should move B, and B move C; that, in fact, the case should be the same as with the automatic puppets. For the parts of such puppets while at rest have a sort of potentiality of motion in them, and when any external force puts the first of them in motion, immediately the next is moved in actuality. As, then, in these automatic puppets the external force moves the parts in a certain sense (not by touching any part at the moment, but by having touched one previously), in like manner also that from which the semen comes, or in other words that which made the semen, sets up the movement in the embryo and makes the parts of it by having first touched something though not continuing to touch it. In a way it is the innate motion that does this, as the act of building builds the house. Plainly, then, while there is something which makes the parts, this does not exist as a definite object, nor does it exist in the semen at the first as a complete part.

But how is each part formed? We must answer this by starting in the first instance from the principle that, in all products of nature or art, a thing is made by something actually existing out of that which is potentially such as the finished product. Now the semen is of such a nature, and has in it such a principle of motion, that when the motion is ceasing each of the parts comes into being, and that as a part having life or soul. For there is no such thing as face or flesh without soul in it; it is only homonymously that they will be called face or flesh if the life has gone out of them, just as if they had been made of stone or wood. And the homogeneous parts and the organic come into being together. And just as we should not say that an axe or other instrument or organ was made by the fire alone, so neither shall we say that foot or hand were made by heat alone. The same applies also to flesh, for this too has a function. While, then, we may allow that hardness and softness, stickiness and brittleness, and whatever other qualities are found in the parts that have life and soul, may be caused by mere heat and cold, yet, when we come to the principle in virtue of which flesh is flesh and bone is bone, that is no longer so; what makes them is the movement set up by the male parent, who is in actuality what that out of which the offspring is made is in potentiality. This is what we find in the products of art; heat and cold may make the iron soft and hard, but what makes a sword is the movement of the tools employed, this movement containing the principle of the art. For the art is the starting-point and form of the product; only it exists in something else, whereas the movement of nature exists in the product itself, issuing from another nature which has the form in actuality.

Has the semen soul, or not? The same argument applies here as in the question concerning the parts. As no part, if it participate not in soul, will be a part except homonymously (as the eye of a dead man is still called an eye), so no soul will exist in anything except that of which it is soul; it is plain therefore that semen both has soul, and is soul, potentially.

But a thing existing potentially may be nearer or further from its realization in actuality, just as a sleeping geometer is further away than one awake and the latter than one actually studying. Accordingly it is not any part that is the cause of the soul's coming into being, but it is the first moving cause from outside. (For nothing generates itself, though when it has come into being it thenceforward increases itself.) Hence it is that only one part comes into being first and not all of them together. But that must first come into being which has a principle of increase (for this nutritive power exists in all alike, whether animals or plants, and this is the same as the power that enables an animal or plant to generate another like itself, that being the function of them all if naturally perfect). And this is necessary for the reason that whenever a living thing is produced it must grow. It is produced, then, by something else of the same name, as e.g. man is produced by man, but it is increased by means of itself. There is, then, something which increases it. If this is a single part, this must come into being first. Therefore if the heart is first made in some animals, and what is analogous to the heart in the others which have no heart, it is from this or its analogue that the first principle of movement would arise.

2

J. B. S. HALDANE*

What Is Life?

I AM NOT going to answer this question. In fact, I doubt if it will ever be possible to give a full answer, because we know what it feels like to be alive, just as we know what redness, or pain, or effort are. So we cannot describe them in terms of anything else. But it is not a foolish question to ask, because we often want to know whether a man is alive or not, and when we are dealing with the microscopic agents of disease, it is clear enough that bacteria are alive, but far from clear whether viruses, such as those which cause measles and smallpox, are so.

So we have to try to describe life in terms of something else, even if the description is quite incomplete. We might try some such expression as "the influence of spirit on matter." But this would be of little use for several reasons. For one thing, even if you are sure that man, and even dogs, have spirits, it needs a lot of faith to find a spirit in an oyster or a potato. For another thing, such a definition would certainly cover great works of art, or books which clearly show their author's mind, and go on influencing readers long after he is dead. Similarly, it is no good trying to define life in terms of a life force. George Bernard Shaw and Professor C. E. M. Joad think there is a life force in living things. But if this has any meaning, which I doubt, you can only detect the life force in an animal or plant by its effects on matter. So we should have to define life in terms of matter. In ordinary life we recognise living things partly by their shape and texture. But these do not change for some

*From J. B. S. Haldane, *What is Life?* (London: Alcuin Press, 1949), pp. 58–62.

hours after death. In the case of mammals and birds we are sure they are dead if they are cold.

This test will not work on a frog or a snail. We take it that they are dead if they will not move when touched. But in the case of a plant the only obvious test is whether it will grow, and this may take months to find out. All these tests agree in using some kind of motion or change as the criterion of life, for heat is only irregular motion of atoms. They also agree in being physical rather than chemical tests. There is no doubt, I think, that we can learn a lot more about life from a chemical than from a physical approach. This does not mean that life has been fully explained in terms of chemistry. It does mean that it is a pattern of chemical rather than physical events. Perhaps I can make this clear by an example.

Suppose a blind man and a deaf man both go to performances of *Macbeth* and of *Alexander Nevsky*. The deaf man will understand little of the play. He will not know Duncan was murdered, let alone who did it. The blind man will miss far less. The essential part of Shakespeare's plays are the words. But with the film it will be the other way round.

What is common to all life is the chemical events. And these are extraordinarily similar in very different organisms. We may say that life is essentially a pattern of chemical happenings, and that in addition there is some building of a characteristic shape in almost all living things, characteristic motion in most animals, and feeling and purpose in some of them. The chemical make-up of different living things is very different. A tree consists largely of wood, which is not very like any of the constituents of a man, though rather like a stuff called glycogen which is part of most, if not all, of our organs. But the chemical changes which go on in the leaves, bark, and roots of a tree, particularly the roots, are surprisingly like those which go on in human organs. The roots need oxygen just as a man does, and you can see whether a root is alive, just as you can see whether a dog is alive, by measuring the amount of oxygen which it consumes per minute. And the oxygen is used in the same kinds of chemical processes, which may roughly be described as controlled burning of foodstuffs at a low temperature. Under ordinary circumstances oxygen does not combine with sugar unless both are heated. It does so in almost all living things through the agency of what are called enzymes. Most of the oxygen which we use has first to unite with an enzyme consisting mainly of protein, but containing a little iron. Warburg discovered this in yeast in 1924. In 1926 I did some rather rough experiments which showed the same, or very nearly the same, enzyme in green plants, moths, and rats. Since then it has been found in a great variety of living things.

Just the same is true for other processes. A potato makes sugar into starch and your liver makes it into glycogen by substantially the same process. Most of the steps by which sugar is broken down in alcoholic fermentation and muscular contraction are the same. And so on. The end results of these processes are, of course, very different. A factory may switch over from making bren guns to making sewing machines or bicycles without very great changes. Similarly the chemical processes by which an insect makes its skin and a snail its slime are very similar, though the products differ greatly.

In fact, all life is characterised by a fundamentally similar set of chemical processes arranged in very different patterns. Thus, animals use up foodstuffs, while most plants make them. But in both plants and animals the building up and breaking down are both going on all the time. The balance is different.

Engels said that life was the mode of existence of proteins (the word which he used is often translated as "albuminous substances"). This is true in so far as all enzymes seem to be proteins. And it is true in so far as the fundamental similarity of all living things is a chemical one. But enzymes and other proteins can be purified and will carry on their characteristic activities in glass bottles. And no biochemist would say they were alive.

In the same way Shakespeare's plays consist of words, whereas words are a very small part of Eisenstein's films. It is important to know this, as it is important to know that life consists of chemical processes. But the arrangement of the words is even more important than the words themselves. And in the same way life is a pattern of chemical processes.

This pattern has special properties. It begets a similar pattern, as a flame does, but it regulates itself as a flame does not except to a slight extent. And, of course, it has many other peculiarities. So when we have said that life is a pattern of chemical processes, we have said something true and important. It is practically important because we are at last learning how to control some of them, and the first fruits of this knowledge are practical inventions like the use of sulphonamides, penicillin, and streptomycin.

But to suppose that one can describe life fully on these lines is to attempt to reduce it to mechanism, which I believe to be impossible. On the other hand, to say that life does not consist of chemical processes is to my mind as futile and untrue as to say that poetry does not consist of words.

EXPLAINING DESIGN

3

WILLIAM PALEY*

Natural Theology

State of the Argument

IN CROSSING A heath, suppose I pitched my foot against a *stone*, and were asked how the stone came to be there, I might possibly answer, that for any thing I knew to the contrary it had lain there for ever; nor would it, perhaps, be very easy to show the absurdity of this answer. But suppose I had found a *watch* upon the ground, and it should be inquired how the watch happened to be in that place, I should hardly think of the answer which I had before given, that for any thing I knew the watch might have always been there. Yet why should not this answer serve for the watch as well as for the stone; why is it not as admissible in the second case as in the first? For this reason, and for no other, namely, that when we come to inspect the watch, we perceive—what we could not discover in the stone—that its several parts are framed and put together for a purpose, *e.g.* that they are so formed and adjusted as to produce motion, and that motion so regulated as to point out the hour of the day; that if the different parts had been differently shaped from what they are, or placed after any other manner or in any other order than that in which they are placed, either no motion at all would have been carried on in the machine, or none which would have answered the use that is now served by it. To reckon up a few of the plainest of these parts and of their offices, all tending to one result: We see a cylindrical

*From William Paley, *Natural Theology* (London: Rivington, 1802).

box containing a coiled elastic spring, which, by its endeavor to relax itself, turns round the box. We next observe a flexible chain — artificially wrought for the sake of flexure — communicating the action of the spring from the box to the fusee. We then find a series of wheels, the teeth of which catch in and apply to each other, conducting the motion from the fusee to the balance and from the balance to the pointer, and at the same time, by the size and shape of those wheels, so regulating that motion as to terminate in causing an index, by an equable and measured progression, to pass over a given space in a given time. We take notice that the wheels are made of brass, in order to keep them from rust; the springs of steel, no other metal being so elastic; that over the face of the watch there is placed a glass, a material employed in no other part of the work, but in the room of which, if there had been any other than a transparent substance, the hour could not be seen without opening the case. This mechanism being observed — it requires indeed an examination of the instrument, and perhaps some previous knowledge of the subject, to perceive and understand it; but being once, as we have said, observed and understood, the inference we think is inevitable, that the watch must have had a maker — that there must have existed, at some time and at some place or other, an artificer or artificers who formed it for the purpose which we find it actually to answer, who comprehended its construction and designed its use. . . .

Suppose, in the next place, that the person who found the watch should after some time discover, that in addition to all the properties which he had hitherto observed in it, it possessed the unexpected property of producing in the course of its movement another watch like itself — the thing is conceivable; that it contained within it a mechanism, a system of parts — a mould, for instance, or a complex adjustment of lathes, files, and other tools — evidently and separately calculated for this purpose; let us inquire what effect ought such a discovery to have upon his former conclusion.

I. The first effect would be to increase his admiration of the contrivance, and his conviction of the consummate skill of the contriver. Whether he regarded the object of the contrivance, the distinct apparatus, the intricate, yet in many parts intelligible mechanism by which it was carried on, he would perceive in this new observation nothing but an additional reason for doing what he had already done — for referring the construction of the watch to design and to supreme art. If that construction *without* this property, or which is the same thing, before this property had been noticed, proved intention and art to have been employed about it, still more strong would the proof appear when he came to the knowledge of this further property, the crown and perfection of all the rest.

II. He would reflect, that though the watch before him were *in some sense* the maker of the watch which was fabricated in the course of its movements, yet it was in a very different sense from that in which a carpenter, for instance, is the maker of a chair—the author of its contrivance, the cause of the relation of its parts to their use. With respect to these, the first watch was no cause at all to the second; in no such sense as this was it the author of the constitution and order, either of the parts which the new watch contained, or of the parts by the aid and instrumentality of which it was produced. We might possibly say, but with great latitude of expression, that a stream of water ground corn; but no latitude of expression would allow us to say, no stretch of conjecture could lead us to think, that the stream of water built the mill, though it were too ancient for us to know who the builder was. What the stream of water does in the affair is neither more nor less than this: by the application of an unintelligent impulse to a mechanism previously arranged, arranged independently of it and arranged by intelligence, an effect is produced, namely, the corn is ground. But the effect results from the arrangement. The force of the stream cannot be said to be the cause or the author of the effect, still less of the arrangement. Understanding and plan in the formation of the mill were not the less necessary for any share which the water has in grinding the corn; yet is this share the same as that which the watch would have contributed to the production of the new watch, upon the supposition assumed in the last section. Therefore,

III. Though it be now no longer probable that the individual watch which our observer had found was made immediately by the hand of an artificer, yet doth not this alteration in anywise affect the inference, that an artificer had been originally employed and concerned in the production. The argument from design remains as it was. Marks of design and contrivance are no more accounted for now than they were before. In the same thing, we may ask for the cause of different properties. We may ask for the cause of the color of a body, of its hardness, of its heat; and these causes may be all different. We are now asking for the cause of that subserviency to a use, that relation to an end, which we have remarked in the watch before us. No answer is given to this question, by telling us that a preceding watch produced it. There cannot be design without a designer; contrivance, without a contriver; order, without choice; arrangement, without any thing capable of arranging; subserviency and relation to a purpose, without that which could intend a purpose; means suitable to an end, and executing their office in accomplishing that end, without the end ever having been contemplated, or the means accommodated to it. Arrangement, disposition of parts, subserviency of means to an end, relation of instruments to a use, imply the presence of intelligence and mind. No one, therefore, can rationally believe that the insensible, inanimate watch, from which the watch before us issued, was the proper cause of the mechanism we so much admire in it—could be truly said to have constructed the instrument, disposed its parts, assigned their office, determined their order, action, and mutual dependency, combined their several motions into one result, and that also a result connected with the utilities of other beings. All these properties, therefore, are as much unaccounted for as they were before. . . .

The conclusion which the *first* examination of the watch, of its works, construction, and movement, suggested, was, that it must have had, for cause and author of that construction, an artificer who understood its mechanism and designed its use. This conclusion is invincible. A *second* examination presents us with a new discovery. The watch is found, in the course of its movement, to produce another watch similar to itself; and not only so, but we perceive in it a system or organization separately calculated for that purpose. What effect would this discovery have, or ought it to have, upon our former inference? What, as hath already been said, but to increase beyond measure our admiration of the skill which had been employed in the formation of such a machine? Or shall it, instead of this, all at once turn us round to an opposite conclusion, namely, that no art or skill whatever has been concerned in the business, although all other evidences of art and skill remain as they were, and this last and supreme piece of art be now added to the rest? Can this be maintained without absurdity? Yet this is atheism.

Application of the Argument

This is atheism; for every indication of contrivance, every manifestation of design which existed in the watch, exists in the works of nature, with the difference on the side of nature of being greater and more, and that in a degree which exceeds all computation. I mean, that the contrivances of nature surpass the contrivances of art, in the complexity, subtilty, and curiosity of the mechanism; and still more, if possible, do they go beyond them in number and variety; yet, in a multitude of cases, are not less evidently mechanical, not less evidently contrivances, not less evidently accommodated to their end or suited to their office, than are the most perfect productions of human ingenuity.

I know no better method of introducing so large a subject, than that of comparing a single thing with a single thing: an eye, for example, with a telescope. As far as the examination of the instrument goes, there is precisely the same proof that the eye was made for vision, as there is that the telescope was made for assisting it. They are made upon the same principles; both being adjusted to the laws by which the transmission and refraction of rays of light are regulated. I speak not of the origin of the laws themselves; but such laws being fixed, the construction in both cases is adapted to them. For instance, these laws require, in order to produce the same effect, that the rays of light, in passing from water into the eye, should be refracted by a more convex surface than when it passes out of air into the eye. Accordingly we find that the eye of a fish, in that part of it called the crystalline lens, is much rounder than the eye of terrestrial animals. What plainer manifestation of design can there be than this difference? What could a mathematical instrument maker have done more to show his knowledge of his principle, his application of that knowledge, his suiting of his means to his end—I will not say to display the compass or excellence of his skill and art, for in these all comparison is indecorous, but to testify counsel, choice, consideration, purpose?

4

CHARLES DARWIN*

Origin of Species

Struggle for Existence

BEFORE ENTERING ON the subject of this chapter, I must make a few preliminary remarks, to show how the struggle for existence bears on Natural Selection. It has been seen in the last chapter that amongst organic beings in a state of nature there is some individual variability; indeed I am not aware that this has ever been disputed. It is immaterial for us whether a multitude of doubtful forms be called species or sub-species or varieties; what rank, for instance, the two or three hundred doubtful forms of British plants are entitled to hold, if the existence of any well-marked varieties be admitted. But the mere existence of individual variability and of some few well-marked varieties, though necessary as the foundation for the work, helps us but little in understanding how species arise in nature. How have all those exquisite adaptations of one part of the organisation to another part, and to the conditions of life, and of one distinct organic being to another being, been perfected? We see these beautiful co-adaptations most plainly in the woodpecker and missletoe; and only a little less plainly in the humblest parasite which clings to the hairs of a quadruped or feathers of a bird; in the structure of the beetle which dives through the water; in the plumed seed which is wafted by the gentlest breeze; in short, we

*From Charles Darwin, *Origin of Species* (London: John Murray, 1859), extracts from Chapters 3 and 4, pp. 60–150.

see beautiful adaptations everywhere and in every part of the organic world.

Again, it may be asked, how is it that varieties, which I have called incipient species, become ultimately converted into good and distinct species, which in most cases obviously differ from each other far more than do the varieties of the same species? How do those groups of species, which constitute what are called distinct genera, and which differ from each other more than do the species of the same genus, arise? All these results, as we shall more fully see in the next chapter, follow inevitably from the struggle for life. Owing to this struggle for life, any variation, however slight and from whatever cause proceeding, if it be in any degree profitable to an individual of any species, in its infinitely complex relations to other organic beings and to external nature, will tend to the preservation of that individual, and will generally be inherited by its offspring. The offspring, also, will thus have a better chance of surviving, for, of the many individuals of any species which are periodically born, but a small number can survive. I have called this principle, by which each slight variation, if useful, is preserved, by the term of Natural Selection, in order to mark its relation to man's power of selection. We have seen that man by selection can certainly produce great results, and can adapt organic beings to his own uses, through the accumulation of slight but useful variations, given to him by the hand of Nature. But Natural Selection, as we shall hereafter see, is a power incessantly ready for action, and is as immeasurably superior to man's feeble efforts, as the works of Nature are to those of Art.

We will now discuss in a little more detail the struggle for existence. In my future work this subject shall be treated, as it well deserves, at much greater length. The elder De Candolle and Lyell have largely and philosophically shown that all organic beings are exposed to severe competition. In regard to plants, no one has treated this subject with more spirit and ability than W. Herbert, Dean of Manchester, evidently the result of his great horticultural knowledge. Nothing is easier than to admit in words the truth of the universal struggle for life, or more difficult—at least I have found it so—than constantly to bear this conclusion in mind. Yet unless it be thoroughly engrained in the mind, I am convinced that the whole economy of nature, with every fact on distribution, rarity, abundance, extinction, and variation, will be dimly seen or quite misunderstood. We behold the face of nature bright with gladness, we often see superabundance of food; we do not see, or we forget, that the birds which are idly singing round us mostly live on insects or seeds, and are thus constantly destroying life; or we forget how largely these songsters, or their eggs, or their nestlings, are destroyed by birds and beasts of prey; we do not always bear in mind, that though food may be now superabundant, it is not so at all seasons of each recurring year.

I should premise that I use the term Struggle for Existence in a large and metaphorical sense, including dependence of one being on another, and including (which is more important) not only the life of the individual, but success in leaving progeny. Two canine animals in a time of dearth, may be truly said to struggle with each other which shall get food and live. But a plant on the edge of a desert is said to struggle for life against the drought, though more properly it should be said to be dependent on the moisture. A plant which annually produces a thousand seeds, of which on an average only one comes to maturity, may be more truly said to struggle with the plants of the same and other kinds which already clothe the ground. The missletoe is dependent on the apple and a few other trees, but can only in a far-fetched sense be said to struggle with these trees, for if too many of these parasites grow on the same tree, it will languish and die. But several seedling missletoes, growing close together on the same branch, may more truly be said to struggle with each other. As the missletoe is disseminated by birds, its existence depends on birds; and it may metaphorically be said to struggle with other fruit-bearing plants, in order to tempt birds to devour and thus disseminate its seeds rather than those of other plants. In these several senses, which pass into each other, I use for convenience sake the general term of struggle for existence.

A struggle for existence inevitably follows from the high rate at which all organic beings tend to increase. Every being, which during its natural lifetime produces several eggs or seeds, must suffer destruction during some period of its life, and during some season or occasional year, otherwise, on the principle of geometrical increase, its numbers would quickly become so inordinately great that no country could support the product. Hence, as more individuals are produced than can possibly survive, there must in every case be a struggle for existence, either one individual with another of the same species, or with the individuals of distinct species, or with the physical conditions of life. It is the doctrine of Malthus applied with manifold force to the whole animal and vegetable kingdoms; for in this case there can be no artificial increase of food, and no prudential restraint from marriage. Although some species may be now increasing, more or less rapidly, in numbers, all cannot do so, for the world would not hold them. . . .

Summary of Chapter. — If during the long course of ages and under varying conditions of life, organic beings vary at all in the several parts of their organisation, and I think this cannot be disputed; if there be, owing to the high geometrical powers of increase of each species, at some age, season, or year, a severe struggle for life, and this certainly cannot be disputed; then, considering the infinite complexity of the relations of all organic beings to each other and to their conditions of existence, causing an infinite diversity in structure, constitution, and habits, to be advantageous to them, I think it would be a most extraordinary fact if no variation ever had occurred useful to each being's own welfare, in the same way as so many variations have occurred useful to man. But if variations useful to any organic being do occur, assuredly individuals thus characterised will have the best chance of being preserved in the struggle for life; and from the strong principle of inheritance they will tend to produce offspring similarly characterised. This principle of preservation, I have called, for the sake of brevity, Natural Selection. Natural selection, on the principle of qualities being inherited at corresponding ages, can modify the egg, seed, or young, as easily as the adult. Amongst many animals, sexual selection will give its aid to ordinary selection, by assuring to the most vigorous and best adapted males the greatest number of offspring. Sexual selection will also give characters useful to the males alone, in their struggles with other males.

Whether natural selection has really thus acted in nature, in modifying and adapting the various forms of life to their several conditions and stations, must be judged of by the general tenour and balance of evidence given in the following chapters. But we already see how it entails extinction; and how largely extinction has acted in the world's history, geology plainly declares. Natural selection, also, leads to divergence of character; for more living beings can be supported on the same area the more they diverge in structure, habits, and constitution, of which we see proof by looking at the inhabitants of any small spot or at naturalised productions. Therefore during the modification of the descendants of any one species, and during the incessant struggle of all species to increase in numbers, the more diversified these descendants become, the better will be their chance of succeeding in the battle of life. Thus the small differences distinguishing varieties of the same species, will steadily tend to increase till they come to equal the greater differences between species of the same genus, or even of distinct genera.

We have seen that it is the common, the widely-diffused, and widely-ranging species, belonging to the larger genera, which vary most; and these will tend to transmit to their modified offspring that superiority which now makes them dominant in their own countries. Natural selection, as has just been remarked, leads to divergence of character and to much extinction of the less improved and intermediate forms of life. On these principles, I believe, the nature of the affinities of all organic beings may be explained. It is a truly wonderful fact — the wonder of which we are apt to overlook from familiarity — that all animals and all plants throughout all time and space should be related to each other in group subordinate to group, in the manner which we everywhere behold — namely, varieties of the same species most closely related together, species of the same genus less closely and unequally related together, forming sections and sub-genera, species of distinct genera much less closely related, and genera related in different degrees, forming sub-families, families, orders, sub-classes, and classes. The several subordinate groups in any class cannot be ranked in a single file,

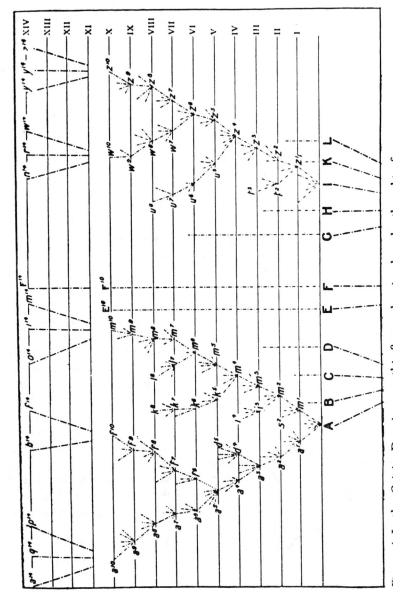

Figure 4. In the *Origin*, Darwin gave this figure showing how he thought of evolution as a (branching) tree of life.

but seem rather to be clustered round points, and these round other points, and so on in almost endless cycles. On the view that each species has been independently created, I can see no explanation of this great fact in the classification of all organic beings; but, to the best of my judgment, it is explained through inheritance and the complex action of natural selection, entailing extinction and divergence of character, as we have seen illustrated in the diagram.

The affinities of all the beings of the same class have sometimes been represented by a great tree. I believe this simile largely speaks the truth. The green and budding twigs may represent existing species; and those produced during each former year may represent the long succession of extinct species. At each period of growth all the growing twigs have tried to branch out on all sides, and to overtop and kill the surrounding twigs and branches, in the same manner as species and groups of species have tried to overmaster other species in the great battle for life. The limbs divided into great branches, and these into lesser and lesser branches, were themselves once, when the tree was small, budding twigs; and this connexion of the former and present buds by ramifying branches may well represent the classification of all extinct and living species in groups subordinate to groups. Of the many twigs which flourished when the tree was a mere bush, only two or three, now grown into great branches, yet survive and bear all the other branches; so with the species which lived during long-past geological periods, very few now have living and modified descendants. From the first growth of the tree, many a limb and branch has decayed and dropped off; and these lost branches of various sizes may represent those whole orders, families, and genera which have now no living representatives, and which are known to us only from having been found in a fossil state. As we here and there see a thin straggling branch springing from a fork low down in a tree, and which by some chance has been favoured and is still alive on its summit, so we occasionally see an animal like the Ornithorhynchus or Lepidosiren, which in some small degree connects by its affinities two large branches of life, and which has apparently been saved from fatal competition by having inhabited a protected station. As buds give rise by growth to fresh buds, and these, if vigorous, branch out and overtop on all sides many a feebler branch, so by generation I believe it has been with the great Tree of Life, which fills with its dead and broken branches the crust of the earth, and covers the surface with its ever branching and beautiful ramifications.

5

MICHAEL DENTON*

Beyond the Reach of Chance

He who believes that some ancient form was transformed suddenly . . . will further be compelled to believe that many structures beautifully adapted to all the other parts of the same creature and to the surrounding conditions, have been suddenly produced . . . To admit all this is, as it seems to me, to enter into the realms of miracle, and to leave those of Science.

ACCORDING TO THE central axiom of Darwinian theory, the initial elementary mutational changes upon which natural selection acts are entirely random, completely blind to whatever effect they may have on the function or structure of the organism in which they occur, "drawn," in Monod's words,[1] from "the realm of pure chance." It is only after an innovation has been disclosed by chance that it can be seen by natural selection and conserved.

Thus it follows that every adaptive advance, big or small, discovered during the course of evolution along every phylogenetic line must have been found as a result of what is in effect a purely random search strategy. The essential problem with this "gigantic lottery" conception of evolution is that all experience teaches that searching for solutions by purely random search procedures is hopelessly inefficient.

Consider first the difficulty of finding by chance English words within the infinite space of all possible combinations of letters. A section of this space would resemble the following block of letters:

*From Michael Denton, *Evolution: A Theory in Crisis*, Chapter 13 (London: Burnett Books, 1985).

FLNWCYTQONMCDFEUTYLDWPQXCZNMIPQZXHGOT
IRJSALXMZVTNCTDHEKBUZRLHAJCFPTQOZPNOTJXD
WHYGCBZUDKGTWIBMZGPGLAOTDJZKXUEMWBCNX
YTKGHSBQJVUCPDLWKSMYJVGXUZIEMTJBYGLMPSJS
KFURYEBWNQPCLXKZUFMTYBUDISTABWNCPDORIS
MXKALQJAUWNSPDYSHXMCKFLQHAVCPDYRTSIZSJR
YFMAHZLVPRITMGYGBFMDLEPE

Within the total letter space would occur every single English word and every single English sentence and indeed every single English book that has been or will ever be written. But most of the space would consist of an infinity of pure gibberish.

Simple three letter English words would be relatively common. There are $26^3 = 17,000$ combinations three letters long, and, as there are about five hundred three letter words in English, then about one in thirty combinations will be a three letter word. All other three letter combinations are nonsense. To find by chance three letter words, eg "not," "bud," "hut," would be a relatively simple task necessitating a search through a string of only about thirty or so letters.

Because three letter words are so probable it is very easy to go from one three letter word to another by making random changes to the letter string. In the case of the word "hat" for example, by randomly substituting letters in the position occupied by h in the word we soon hit on a new three letter word:

hat
aat
bat
cat
dat
eat
fat

Thus not only is it possible to find three letter English words by chance but because the probability gaps between them are small, it is easy to transform any word we find into a quite different word through a sequence of probable intermediates:

hat → cat → can → tan → tin

However, to find by chance longer words, say seven letters long such as "English" or "require," would necessitate a vastly increased search. There are 26^7 or 10^9, that is, one thousand million combinations of letters seven letters long. As there are certainly less than ten thousand English words seven letters long, then to find one by chance we would have to search through letter strings in the order of one hundred thousand units long. Twelve letter words such as "construction" or "unreasonable" are so rare that they occur only once by chance in strings of letters 10^{14} units long; as there are about 10^{14} minutes in one thousand million years one can imagine how long a monkey at a typewriter would take to type out by chance one English word twelve letters long. Intuitively it seems unbelievable that such apparently simple entities as twelve letter English words could be so rare, so inaccessible to a random search.

The problem of finding words by chance arises essentially because the space of all possible letter combinations is immense and the overwhelming majority are complete nonsense; consequently meaningful sequences are very rare and the probability of hitting one by chance is exceedingly small.

Moreover, even if by some lucky fluke we were to find, say, one twelve letter word by chance, because each word is so utterly isolated in a vast ocean of nonsense strings it is very difficult to get another meaningful letter string by randomly substituting letters and testing each new string to see if it forms an English word. Take the word "unreasonable." There are a few closely related words such as "reasonable," "reason," "season," "treason," or "able," which can be reached by making changes to the letter sequence but the necessary letter changes are unfortunately highly specific and finding them by chance involves a far longer and more difficult search than was the case with three letter words.

Sentences, of course, even short ones, are even rarer and long sentences rare almost beyond imagination. Linguists have estimated a total of 10^{25} possible English sentences one hundred letters long, but as there are a total of 26^{100} or 10^{130} possible sequences one hundred letters long, then less than one in about 10^{100} will be an English sentence. The figure of 10^{100} is beyond comprehension — some idea of the immensity it represents can be grasped by recalling that there are only 10^{70} atoms in the entire observable universe.

Each English sentence is a complex system of letters which are integrated together in highly specific ways: firstly into words, then into word phrases, and finally into sentences. If the subsystems are all to be combined in such a way that they will form a grammatical English sentence than their integration must follow rigorously the *a priori* rules of English grammar. For example, one of the rules is that the letters in the sentence must be combined in such a way that they form words belonging to the lexicon of the English language.

However, random strings of English words, eg "horse," "cog," "blue," "fly," "extraordinary," do not form sentences because there exists a further set of rules — the rules of syntax which dictate, among other things, that a sentence must possess a subjective and a verbal clause.

On top of this there exists a further set of rules which governs the semantic relationship of the components of a sentence. Obviously, not all strings of English words which are arranged correctly according to the rules of English syntax are meaningful. For example: "The raid (subject) ate (verb) the sky (object)." Each word is from the English lexicon and their arrangement satisfies the rules of syntax. However, the sequence disobeys the rules of semantics and is as nonsensical as a completely random string of letters.

The rules of English grammar are so stringent that only highly specialized letter combinations can form grammatical sentences and consequently, because of the immensity of the space of all possible letter combinations, such highly specialized strings are utterly lost within it, infinitely rare and isolated, absolutely beyond the reach of any sort of random search that could be conceivably carried out in a finite time even with the most advanced computers on earth. Moreover, because sentences are so rare and isolated, even if one was discovered by chance the probability gap between it and the nearest related sentence is so immeasurable that no conceivable sort of random change to the letter or word sequence will ever carry us across the gap.

Consider the sentence: "Because of the complexity of the rules of English grammar most English sentences are completely isolated." If we set out to reach another sentence by randomly substituting a new word in place of an existing word and then testing the newly created sequence of words to see if it made a grammatical sentence, we would find very few substitutions were grammatically acceptable and even to find *one* grammatical substitution would take an unbelievably long time if we searched by pure chance. Some of the few grammatical substitutions are shown below:

because	English
of	grammar
the	most → all → some
complexity → nature	english
of	sentences
the	are
rules → algorithms	completely → totally → invariably → always
of	isolated → immutable

There are about 10^5 words in the lexicon of the English language and, as there are sixteen words in the above sentence, we would have 1.6×10^6 possibilities to test. If there are, say, two hundred individual words out of the 1.6×100^6 which can be substituted grammatically, we would have to test about eight thousand words on average before we found a grammatical substitution.

Testing one new word per minute, it would take us five days working day and night to find by chance our first grammatical substitution, and to test all the possible words in every position in the sentence would take about three years, and after three years of searching all we would have achieved would be a handful of sentences closely related to the one with which we started.

Sentences are not the only complex systems which are beyond the reach of chance. The same principles apply, for example, to watches, which are also highly improbable, and where consequently each different functional watch is intensely isolated by immense probability gaps from its nearest neighbours.

To see why, we must begin by trying to envisage a universe of mechanical objects containing all possible combinations of watch components: springs, gears, levers, cogwheels, each of every conceivable size and shape. Such a universe would contain every functional watch that has ever existed on earth and every functional watch that could possibly exist at any time in the future. Although we cannot in this case calculate the rarity of functional combinations (watches that work) as we could in the case of words and sentences, common sense tells us that they would be exceedingly rare. Our imaginary universe would mostly consist of combinations of gears and cogwheels which would be entirely useless; each functional watch would, like a meaningful sentence, be an isolated island separated from all other islands of function by a surrounding infinity of junk composed only of incoherent and functionless combinations.

Again, as with sentences, because the total number of incoherent nonsense combinations of components vastly exceeds, by an almost inconceivable amount, the tiny fraction which can form coherent combinations, function is exceedingly rare. If we were to look by chance for a functional watch we would have to search for an eternity amid an infinity of combinations until we hit upon a functional watch.

The basic reason why functional watches are so exceedingly improbable is because, to be functional, a combination of watch components must satisfy a number of very stringent criteria (equivalent to the rules of grammar), and these can only be satisfied by highly specialized unique combinations of components which are coadapted to function together. One rule might be that all cogwheels must possess perfect regularly-shaped cogs; another rule might be that all the cogs must fit together to allow rotation of one wheel to be transmitted throughout the system.

It is obviously impossible to contemplate using a random search to find combinations which will satisfy the stringent criteria which govern functionality in watches. Yet, just as a speaker of a language cognizant with the rules of grammar can generate a functional sentence with great ease, so too a watchmaker has little trouble in assembling a watch by following the rules which govern functionality in combinations of watch components.

What is true of sentences and watches is also true of computer programs, airplane engines, and in fact of all known complex systems. Almost invariably, function is restricted to unique and fantastically improbable combinations of subsystems, tiny islands of meaning lost in an infinite sea of incoherence. Because the number of nonsense combinations of component subsystems vastly exceeds by unimaginable orders of magnitude the infinitesimal fraction of combinations in which the components are capable of undergoing coherent or meaningful interactions. Whether we are searching for a functional sentence or a functional watch or the best move in a game of chess, the goals of our search are in each case so far lost in an infinite space of possibilities that, unless we guide our search by the use of algorithms which direct us to very specific regions of the space, there is no realistic possibility of success.

Discussing a well known checker-playing program, Professor Marvin Minsky of the Massachusetts Institute of Technology comments:[2]

> This game exemplifies the fact that many problems can in principle be solved by trying all possibilities — in this case exploring all possible moves, all the opponent's replies all the player's possible replies to the opponent's replies and so on. If this could be done, the player could see which move has the best chance of winning. In practice, however, this approach is out of the question, even for a computer; the tracking down of every possible line of play would involve some 10^{40} different board positions. A similar analysis for the game of chess would call for some 10^{120} positions. Most interesting problems present far too many possibilities for complete trial and error analysis.

Nevertheless, as he continues, a computer can play checkers if it is capable of making intelligent limited searches:

> Instead of tracking down every possible line of play the program uses a partial analysis (a "static evaluation") of a relatively small number of carefully selected features of a board position — how many men there are on each side, how advanced they are and certain other simple relations. This incomplete analysis is not in itself adequate for choosing the best move for a player in a current position. By combining the partial analysis with a limited search for some of the consequences of the possible moves from the current position, however, the program selects its move as if on the basis of a much deeper analysis. The program contains a collection of rules for deciding when to continue the search and when to stop. When it stops, it assesses the merits of the "terminal position" in terms of the static evaluation. If the computer finds by this search that a given move leads to an advantage for the player in all the likely positions that may occur a few moves later, whatever the opponent does, it can select this move with confidence.

The inability of unguided trial and error to reach anything but the most trivial of ends in almost every field of interest obviously raises doubts as to its validity in the biological realm. Such doubts were recently raised by a number of mathematicians and engineers at an international symposium entitled "Mathematical Challenges to the Neo-Darwinian Interpretation of Evolution,"[3] a meeting which also included many leading evolutionary biologists. The major argument presented was that Darwinian evolution by natural selection is merely a special case of the general procedure of problem solving by trial and error. Unfortunately, as the mathematicians present at the symposium such as Schutzenberger and Professor Eden from MIT pointed out, trial and error is totally inadequate as a problem solving technique without the guidance of specific algorithms, which has led to the consequent failure to simulate Darwinian evolution by computer analogues. For similar reasons, the biophysicist Pattee has voiced scepticism over natural selection at many leading symposia over the past two decades. At one meeting entitled "Natural Automata and Useful Simulations," he made the point:[4]

> Even some of the simplest artificial adaptive problems and learning games appear practically insolvable even by multistage evolutionary strategies.

Living organisms are complex systems, analogous in many ways to non-living complex systems. Their design is stored and specified in a linear sequence of symbols, analogous to coded information in a computer programme. Like any other system, organisms consist of a number of subsystems which are all coadapted to interact together in a coherent manner: molecules are assembled into multimolecular systems, multimolecular assemblies are combined into cells, cells into organs and organ systems finally into the complete organism. It is hard to believe that the fraction of meaningless combinations of molecules, of cells, of organ systems, would not vastly exceed the tiny fraction that can be combined to form assemblages capable of exhibiting coherent interactions. Is it really possible that the criteria for function which must be satisfied in the design of living systems are at every level far less stringent than those which must be satisfied in the design of functional watches, sentences or computer programs? Is it possible to design an automaton to construct an object like the human brain, laying down billions of specific connections, without having to satisfy criteria every bit as exacting and restricting as those which must be met in other areas of engineering?

Given the close analogy between living systems and machines, particularly at a molecular level, there cannot be any objective basis to the assumption that functional organic systems are likely to be less isolated or any easier to find by chance. Surely it is far more likely that functional combinations in the space of all organic possibilities are just as isolated, just as rare and improbable, just as inaccessible to a random search and just as functionally immutable by any sort of random process. The only warrant for believing that functional living systems are probable, capable of undergoing functional transformation by random mechanisms, is belief in evolution by the natural selection of purely random changes in the structure of living things. But this is precisely the question at issue.

If complex computer programs cannot be changed by random mechanisms, then surely the same must apply to the genetic programmes of living organisms. The fact that systems in every way analogous to living organisms cannot undergo evolution by pure trial and error and that their functional distribution invariably conforms to an improbable discontinuum comes, in my opinion, very close to a formal disproof of the whole Darwinian paradigm of nature. By what strange capacity do living organisms defy the laws of chance which are apparently obeyed by all analogous complex systems?

We now have machines which exhibit many properties of living systems. Work on artificial intelligence has advanced and the possibility of constructing a self-reproducing machine was discussed by the mathematician von Neumann in his now famous *Theory of Self-Reproducing Automata*.[5] Although some advanced machines can solve simple problems, none of them can undergo evolution by the selection of random changes in their structure without the guidance of already existing programmes. The only sort of machine that might, at some future date, undergo some sort of evolution would be one exhibiting artificial intelligence. Such a machine would be capable of altering its own organization in an intelligent way. However evolution of this sort would be more akin to Lamarckian, but by no stretch of the imagination could it be considered Darwinian. The construction of a self-evolving intelligent machine would only serve to underline the insufficiency of unguided trial and error as a causal mechanism of evolution.

It was the close analogy between living systems and complex machines and the impossibility of envisaging how objects could have been assembled by chance that led the natural theologians of the eighteenth and early nineteenth centuries to reject as inconceivable the possibility that chance would have played any role in the origin of the complex adaptations of living things. William Paley, in his classic analogy between an organism and a watch, makes precisely this point:[6]

> Nor would any man in his senses think the existence of the watch, with its various machinery, accounted for, by being told that it was one out of possible combinations of material forms; that whatever he had found in the place where he found the watch, must have contained some internal configuration or other; and that this configuration might be the structure now exhibited, viz. of the works of a watch, as well as a different structure.

It is true that some authorities have seen an analogy to evolution by natural selection in gradual technological advances. Jukes, for instance, in a recent letter to *Nature* drew an analogy between the evolution of the Boeing 747 from Bleriots' 1909 monoplane through the Boeing Clippers in the 1930s to the first Boeing jet airliner, the 707, which started in service in 1959 and which was the immediate predecessors of the 747s, and biological evolution. In his words:[7]

> The brief history of aircraft technology is filled with branching processes, phylogeny and extinctions that are a striking counterpart of three billion years of biological evolution.

Unfortunately, the analogy is false. At no stage during the history of the aviation industry was the design of any flying machine achieved by chance, but only by the most rigorous applications of all the rules which govern function in the field of aerodynamics. It is true, as Jukes states, that "wide-bodied jets evolved from small contraptions made in bicycle shops, or in junkyards," but they did not evolve by chance.

There is no way that a purely random search could ever have discovered the design of an aerodynamically feasible flying machine from a random assortment of mechanical components — again, the space of all possibilities is inconceivably large. All such analogies are false because in *all* such cases the search for function is intelligently guided. It cannot be stressed enough that evolution by natural selection is analogous to problem solving without any intelligent guidance, without any intelligent input whatsoever. No activity which involves an intelligent input can possibly be analogous to evolution by natural selection.

The above discussion highlights one of the fundamental flaws in many of the arguments put forward by defenders of the role of chance in evolution. Most of the classic arguments put forward by leading Darwinists, such as the geneticist H. J. Muller and many other authorities including G. G. Simpson, in defence of natural selection make the implicit assumption that islands of function are common, easily found by chance in the first place, and that it is easy to go from island to island through functional intermediates.

This is how Simpson, for example, envisages evolution by natural selection:[8]

> How natural selection works as a creative process can perhaps best be explained by a very much oversimplified analogy. Suppose that from a pool of all the letters of the alphabet in large, equal abundance you tried to draw simultaneously the letters c, a, and t, in order to achieve a purposeful combination of these into the word "cat." Drawing out three letters at a time and then discarding them if they did not form this useful combination, you obviously would have very little chance of achieving your purpose. You might spend days, weeks, or even years at your task before you finally succeeded. The possible number of combinations of three letters is very large and only one of these is suitable for your purpose. Indeed, you might well never succeed, because you might have drawn all the c's, a's, or t's in wrong combinations and have discarded them before you succeeded in drawing all three together. But now suppose that every time you draw a c, an a, or a t in a wrong combination, you are allowed to put these desirable letters back in the pool and to discard the undesirable letters. Now you are sure of obtaining your result, and your chances of obtaining it quickly are much improved. In time there will be only c's, a's, and t's in the pool, but you probably will have succeeded long before that. Now suppose that in addition to returning c's, a's, and t's to the pool and discarding all other letters, you are allowed to clip together any two of the desirable letters when you happen to draw them at the same time. You will shortly have in the pool a large number of clipped ca, ct, and at combinations plus an also large number of the t's, a's, and c's needed to complete one of these if it is drawn again. Your chances of quickly obtaining the desired result are improved still more, and by these processes you have "generated a high degree of improbability" — you have made it probable that you will quickly achieve the combi-

nation *cat*, which was so improbable at the outset. Moreover, you have created something. You did not create the letters *c*, *a*, and *t*, but you have created the word "cat," which did not exist when you started.

The obvious difficulty with the whole scheme is that Simpson assumes that finding islands of function in the first place (the individual letters *c*, *a*, and *t*) is highly probable and that the functional island "cat" is connected to the individual letter islands by intermediate functional islands *ca*, *ac*, *ct*, *at*, *ta*, *fc*, so that we can cross from letters to islands by natural selection in unit mutational steps. In other words, Simpson has assumed that islands of function are very probable, but this is the very assumption which must be proved to show that natural selection would work.

Obviously, if islands of function in the space of all organic possibilities are common, like three or four letter words, then of course functional biological systems will be within the reach of chance; and because the probability gaps will be small, random mutations will easily find a way across. However, as is evident from the above discussion, Simpson's scheme, and indeed the whole Darwinian framework, collapse completely if islands of function are like twelve letter words or English sentences.

These considerations of the likely rarity and isolation of functional systems within their respective total combinatorial spaces also reveal the fallacy of the current fashion of turning to saltational models of evolution to escape the impasse of gradualism. For as we have seen, in the case of every kind of complex functional system the total space of all combinatorial possibilities is so nearly infinite and the isolation of meaningful systems so intense, that it would truly be a miracle to find one by chance. Darwin's rejection of chance saltations as a route to new adaptive innovations is surely right. For the combinatorial space of all organic possibilities is bound to be so great, that the probability of a sudden macromutational event transforming some existing structure or converting *de novo* some redundant feature into a novel adaptation exhibiting, that "perfection of coadaptation" in all its component parts so obvious in systems like the feather, the eye or the genetic code and which is necessarily ubiquitous in the design of all complex functional systems biological or artifactual, is bound to be vanishingly small. Ironically, in any combinatorial space, it is the very same restrictive criteria of function which prevent gradual functional change which also isolate all functional systems to vastly improbable and inaccessible regions of the space.

To determine, finally, whether the distribution of islands of function in organic nature conforms to a probable continuum or an improbable discontinuum and to assess definitively the relevance of chance in evolution would be a colossal task. Just as in the case of the sentences and watches, we would have to begin by constructing a multi-dimensional universe filled with all possible combinations of organic chemicals. Within this space of all possibilities there would exist every conceivable functional biological system, including not only those which exist on earth, but all other functional biological systems which could possibly work elsewhere in the universe. The functional systems would range from simple protein molecules capable of particular catalytic functions right up to immensely complex systems such as the human brain. Within this universe of all possibilities we would find many strange biological systems, such as enzymes capable of transforming unique substrates not found on earth, and perhaps nervous systems resembling those found among vertebrates on earth, but far more advanced. We would also find many sorts of complex aggregates, the function of which may not be clear but which we could dimly conceive as being of some value on some alien planet.

Such a space would of course consist mainly of combinations which would have no conceivable biological function — merely junk aggregates ranging from functionless proteins to entirely disordered nervous systems reminiscent of Cuvier's incompatibilities. From the space we would be able to calculate exactly how probable functional biological systems are and how easy it is to go from one functional system to another, Darwinian fashion, in a series of unit mutational steps through functional intermediates. Of course if analogy is any guide then the space would in all probability conjure up a vision of nature more in harmony with the thinking of Cuvier and the early 19th-century typology than modern evolutionary thought in which each island of meaning is intensely isolated unlinked by transitional forms and quite beyond the reach of chance.

At present we are very far from being able to construct such a space of all organic possibilities and to calculate the probability of functional living systems. Nevertheless, for some of the lower order functional systems, such as individual proteins, their rarity in the space can be at least tentatively assessed.

A protein . . . is fundamentally a long chain-like molecule built up out of twenty different kinds of amino acids. After its assembly the long amino acid chain automatically folds into a specific stable 3D configuration. Particular protein functions depend on highly specific 3D shapes and, in the case of proteins which possess catalytic functions, depend on the protein possessing a particular active site, again of highly specific 3D configuration.

Although the exact degree of isolation and rarity of functional proteins is controversial it is now generally conceded by protein chemists that most functional proteins would be difficult to reach or to interconvert through a series of successive individual amino acid mutations. Zuckerkandl comments:[9]

> Although, abstractly speaking, any polypeptide chain can be transformed into any other by successive amino acid substitutions and other mutational events, in concrete situations the pathways between a

poorly and a highly adapted molecule will be mostly impracticable. Any such pathway, whether the theoretically shortest, or whether a longer one, will perforce include stages of favorable change as well as hurdles. Of the latter some will be surmountable and some not. Some of the latter will presumably be present along the pathways of adaptive change in a very large majority of ill adapted de novo polypeptide chains.

Consequently, when a protein molecule is selected for its weak enzymatic activity and in spite of limited substrate specificity, it will most often represent a dead end road.

The impossibility of gradual functional transformation is virtually self-evident in the case of proteins: mere casual observation reveals that a protein is an interacting whole, the function of every amino acid being more or less (like letters in a sentence or cogwheels in a watch) essential to the function of the entire system. To change, for example, the shape and function of the active site (like changing the verb in a sentence or an important cogwheel in a watch) in isolation would be bound to disrupt all the complex intramolecular bonds throughout the molecule, destabilizing the whole system and rendering it useless. Recent experimental studies of enzyme evolution largely support this view, revealing that proteins are indeed like sentences, and are only capable of undergoing limited degrees of functional change through a succession of individual amino acid replacements. The general consensus of opinion in this field is that significant functional modification of a protein would require several simultaneous amino acid replacements of a relatively improbable nature. The likely impossibility of major functional transformation through individual amino acid steps was raised by Brian Hartley, a specialist in this area, in an article in the journal *Nature* in 1974. From consideration of the atomic structure of a family of closely related proteins which, however, have different amino acid arrangements in the central region of the molecule, he concluded that their functional interconversion would be impossible:[10]

> It is hard to see how these alternative arrangements could have evolved without going through an intermediate that could not fold correctly (i.e. would be non functional).

Here then, is at least one functional subset of the space of all organic possibilities which almost certainly conforms to the general discontinuous pattern observed in the case of other complex systems. But how discontinuous is the pattern of the distribution of proteins within the space of all organic possibilities? Might functional proteins be beyond the reach of chance?

In attempting to answer the question—how rare are functional proteins?—we must first decide what general restrictions must be imposed on a sequence of amino acids before it can form a biologically functional protein. In other words, what are the rules or criteria which govern functionality in an amino acid sequence?

First, a protein must be a stable structure so that it can hold a particular 3D shape for a sufficiently long period to allow it to undergo a specific interaction with some other entity in the cell. Second, a protein must be able to fold into its proper shape. Third, if a protein is to possess catalytic properties it must have an active site which necessitates a highly specific arrangement of atoms in some region of its surface to form this site.

From the tremendous advances that have been made over the past two decades in our knowledge of protein structure and function, there are compelling reasons for believing that these criteria for function would inevitably impose severe limitations on the choice of amino acids. It is very difficult to believe that the criteria for stability and for a folding algorithm would not require a relatively severe restriction of choice in at least twenty per cent of the amino acid chain. To get the precise atomic 3D shape or active sites may well require an absolute restriction in between one and five per cent of the amino acid sequence.

There is a considerable amount of empirical evidence for believing that the criteria for function must be relatively stringent. One line of evidence, for example, is the very strict conservation of overall shape and the exact preservation of the configuration of active sites in homologous proteins such as the cytochromes in very diverse species. Further evidence suggests that most mutations which cause changes in the amino acid sequence of proteins tend to damage function to a greater or lesser degree. The effects of such mutations have been carefully documented in the case of haemoglobin, and some of them were described in an excellent article in *Nature* by Max Perutz,[11] who himself pioneered the X-ray crystallographic work which first revealed the detailed 3D structure of proteins. As Perutz shows, although many of the amino acids occupying positions on the surface of the molecule can be changed with little effect on function, most of the amino acids in the centre of the protein cannot be changed without having drastic deleterious effects on the stability and function of the molecule.

There are, in fact, both theoretical and empirical grounds for believing that the *a priori* rules which govern function in an amino acid sequence are relatively stringent. If this is the case, and all the evidence points to this direction, it would mean that functional proteins could well be exceedingly rare. The space of all possible amino acid sequences (as with letter sequences) is unimaginably large and consequently sequences which must obey particular restrictions which can be defined, like the rules of grammar, are bound to be fantastically rare. Even short unique sequences just ten amino acids long only occur once by chance in about 10^{13} average-sized proteins; unique sequences twenty amino acids long once in about 10^{26} proteins; and unique sequences thirty amino acids long once in about 10^{39} proteins!

As it can easily be shown that no more than 10^{40} possible proteins could have ever existed on earth since its formation, this means that, if proteins, functions reside in sequences any less probable than 10^{-40}, it becomes increasingly unlikely that any functional proteins could ever have been discovered by chance on earth.

We have seen . . . that envisaging how a living cell could have gradually evolved through a sequence of simple protocells seems to pose almost insuperable problems. If the estimates above are anywhere near the truth then this would undoubtedly mean that the alternative scenario — the possibility of life arising suddenly on earth by chance — is infinitely small. To get a cell by chance would require at least one hundred functional proteins to appear simultaneously in one place. That is one hundred simultaneous events each of an independent probability which could hardly be more than 10^{-20} giving a maximum combined probability of 10^{-2000}. Recently, Hoyle and Wickramasinghe in *Evolution from Space* provided a similar estimate of the chance of life originating, assuming functional proteins to have a probability of 10^{-20}:[12]

> By itself, this small probability could be faced, because one must contemplate not just a single shot at obtaining the enzyme, but a very large number of trials such as are supposed to have occurred in an organic soup early in the history of the Earth. The trouble is that there are about two thousand enzymes, and the chance of obtaining them all in a random trial is only one part in $(10^{20})^{2000} = 10^{40,000}$ an outrageously small probability that could not be faced even if the whole universe consisted of organic soup.

Although at present we still have insufficient knowledge of the rules which govern function in amino acid sequences to calculate with any degree of certainty the actual rarity of functional proteins, it may be that before long quite rigorous estimates may be possible. Over the next few decades advances in molecular biology are inevitably going to reveal in great detail many more of the principles and rules which govern the function and structure of protein molecules. In fact, by the end of the century, molecular engineers may be capable of specifying quite new types of functional proteins. From the first tentative steps in this direction it already seems that, in the design of new functional proteins, chance will play as peripheral a role as it does in any other area of engineering.[13]

The Darwinian claim that all the adaptive design of nature has resulted from a random search, a mechanism unable to find the best solution in a game of checkers, is one of the most daring claims in the history of science. But it is also one of the least substantiated. No evolutionary biologist has ever produced any quantitive proof that the designs of nature are in fact within the reach of chance. There is not the slightest justification for claiming, as did Richard Dawkins recently:[14]

> . . . Charles Darwin showed how it is possible for blind physical forces to mimic the effects of conscious design, and, by operating as a cumulative filter of chance variations, to lead eventually to organised and adaptive complexity, to mosquitoes and mammoths, to humans and therefore, indirectly, to books and computers.

Neither Darwin, Dawkins nor any other biologist has ever calculated the probability of a random search finding in the finite time available the sorts of complex systems which are so ubiquitous in nature. Even today we have no way of rigorously estimating the probability or degree of isolation of even one functional protein. It is surely a little premature to claim that random processes could have assembled mosquitoes and elephants when we still have to determine the actual probability of the discovery by chance of one single functional protein molecule!

NOTES

1. Monod, J. (1972) *Chance and Necessity*, Collins, London, p 114.
2. Minsky, M. (1966) "Artificial Intelligence," *Scientific American*, 215(3) September, p 246–60, see pp 247–48.
3. Moorhead, P. S. and Kaplan, M. M., eds (1967) *Mathematical Challenges to the Darwinian Interpretation of Evolution*, Wistar Institute Symposium Monograph.
4. Pattee, H. H. (1966) "Introduction to Session One" in *Natural Automata and Useful Simulations*, eds H. H. Pattee et al, Spartan Books, Washington, pp 1–2.
5. Von Neumann, J. (1966) *Theory of Self-Reproducing Automata*, University of Illinois Press, Urbana.
6. Paley, W. (1818) *Natural Theology on Evidence and Attributes of Deity*, 18th ed, Lackington, Allen and Co, and James Sawers, Edinburgh, p 13.
7. Jukes, T. H. (1982) "Aircraft Evolution," *Nature*, 295, p 548.
8. Simpson, G. G. (1947) "The Problem of Plan and Purpose in Nature," *Scientific Monthly*, 64: 481–495, see p 493.
9. Zuckerkandl, E. (1975) "The Appearance of New Structures in Proteins During Evolution," *J. Mol. Evol.*, 7:1–57, see p 21.
10. Rigby, P. W. J., Burleigh, B. D. Jnr, and Hartley, B. S. (1974) "Gene Duplication in Experimental Enzyme Evolution," *Nature*, 251:200–204, see p 200.
11. Perutz, M. F. and Lehmann, H. (1968) "Molecular Pathology of Human Haemoglobin," *Nature*, 219: 902–09.
12. Hoyle, F. and Wickramasinghe, C. (1981) *Evolution from Space*, J. M. Dent and Sons, London, p 24.
13. Paba, C. (1983) "Designing Proteins and Peptides," *Nature*, 301:200.
14. Dawkins, R. (1982) "The Necessity of Darwinism," *New Scientist*, 94, (1301) 15 April, pp 130–132, see p 130.

Editor's note: The reference on p. 57 is to the early nineteenth century French biologist Georges Curvier, who denied the very possibility of evolution on the grounds that transitional organisms would simply be functionally inadequate. A reptile-bird intermediate, for instance, would supposedly fail both as a reptile and as a bird.

6

RICHARD DAWKINS*

Accumulating Small Change

WE HAVE SEEN that living things are too improbable and too beautifully "designed" to have come into existence by chance. How, then, did they come into existence? The answer, Darwin's answer, is by gradual, step-by-step transformations from simple beginnings, from primordial entities sufficiently simple to have come into existence by chance. Each successive change in the gradual evolutionary process was simple enough, *relative to its predecessor*, to have arisen by chance. But the whole sequence of cumulative steps constitutes anything but a chance process, when you consider the complexity of the final end-product relative to the original starting point. The cumulative process is directed by nonrandom survival. The purpose of this chapter is to demonstrate the power of this *cumulative selection* as a fundamentally nonrandom process.

If you walk up and down a pebbly beach, you will notice that the pebbles are not arranged at random. The smaller pebbles typically tend to be found in segregated zones running along the length of the beach, the larger ones in different zones or stripes. The pebbles have been sorted, arranged, selected. A tribe living near the shore might wonder at this evidence of sorting or arrangement in the world, and might develop a myth to account for it, perhaps attributing it to a Great Spirit in the sky with a tidy mind and a sense of order. We might give a

*Reprinted from *The Blind Watchmaker: Why the Evidence of Evolution Reveals a Universe without a Design*. Copyright © 1996, 1987, 1986 by Richard Dawkins. Reprinted by permission of W. W. Norton & Company, Inc., and Sterling Lord Literistic, Inc.

superior smile at such a superstitious notion, and explain that the arranging was really done by the blind forces of physics, in this case the action of waves. The waves have no purposes and no intentions, no tidy mind, no mind at all. They just energetically throw the pebbles around, and big pebbles and small pebbles respond differently to this treatment so they end up at different levels of the beach. A small amount of order has come out of disorder, and no mind planned it.

The waves and the pebbles together constitute a simple example of a system that automatically generates non-randomness. The world is full of such systems. The simplest example I can think of is a hole. Only objects smaller than the hole can pass through it. This means that if you start with a random collection of objects above the hole, and some force shakes and jostles them about at random, after a while the objects above and below the hole will come to be nonrandomly sorted. The space below the hole will tend to contain objects smaller than the hole, and the space above will tend to contain objects larger than the hole. Mankind has, of course, long exploited this simple principle for generating non-randomness, in the useful device known as the sieve.

The Solar System is a stable arrangement of planets, comets and debris orbiting the sun, and it is presumably one of many such orbiting systems in the universe. The nearer a satellite is to its sun, the faster it has to travel if it is to counter the sun's gravity and remain in stable orbit. For any given orbit, there is only one speed at which a satellite can travel and remain in that orbit. If it were travelling at any other velocity, it would either move out into deep space, or crash into the Sun, or move into another orbit. And if we look at the planets of our solar system, lo and behold, every single one of them is travelling at exactly the right velocity to keep it in its stable orbit around the Sun. A blessed miracle of provident design? No, just another natural "sieve". Obviously all the planets that we see orbiting the sun must be travelling at exactly the right speed to keep them in their orbits, or we wouldn't see them there because they wouldn't be there! But equally obviously this is not evidence for conscious design. It is just another kind of sieve.

Sieving of this order of simplicity is not, on its own, enough to account for the massive amounts of nonrandom order that we see in living things. Nowhere near enough. [Think of] the analogy of the combination lock. The kind of non-randomness that can be generated by simple sieving is roughly equivalent to opening a combination lock with one dial: it is easy to open it by sheer luck. The kind of non-randomness that we see in living systems, on the other hand, is equivalent to a gigantic combination lock with an almost uncountable number of dials. To generate a biological molecule like haemoglobin, the red pigment in blood, by simple sieving would be equivalent to taking all the amino-acid building blocks of haemoglobin, jumbling them up at random, and hoping that the haemoglobin molecule would reconstitute itself by sheer luck. The amount of luck that would be required for this feat is unthinkable, and has been used as a telling mind-boggler by Isaac Asimov and others.

A haemoglobin molecule consists of four chains of amino acids twisted together. Let us think about just one of these four chains. It consists of 146 amino acids. There are 20 different kinds of amino acids commonly found in living things. The number of possible ways of arranging 20 kinds of things in chains 146 links long is an inconceivably large number, which Asimov calls the "haemoglobin number." It is easy to calculate, but impossible to visualize the answer. The first link in the 146-long chain could be any one of the 20 possible amino acids. The second link could also be any one of the 20, so the number of possible 2-link chains is 20×20, or 400. The number of possible 3-link chains is $20 \times 20 \times 20$, or 8,000. The number of possible 146-link chains is 20 times itself 146 times. This is a staggeringly large number. A million is a 1 with 6 noughts after it. A billion (1,000 million) is a 1 with 9 noughts after it. The number we seek, the "haemoglobin number," is (near enough) a 1 with 190 noughts after it! This is the chance against happening to hit upon haemoglobin by luck. And a haemoglobin molecule has only a minute fraction of the complexity of a living body. Simple sieving, on its own, is obviously nowhere near capable of generating the amount of order in a living thing. Sieving is an essential ingredient in the generation of living order, but it is very far from being the whole story. Something else is needed. To explain the point, I shall need to make a distinction between "single-step" selection and "cumulative" selection. The simple sieves we have been considering so far in this chapter are all examples of single-step selection. Living organization is the product of cumulative selection.

The essential difference between single-step selection and cumulative selection is this. In single-step selection the entities selected or sorted, pebbles or whatever they are, are sorted once and for all. In cumulative selection, on the other hand, they "reproduce"; or in some other way the results of one sieving process are fed into a subsequent sieving, which is fed into . . ., and so on. The entities are subjected to selection of sorting over many "generations" in succession. The end-product of one generation of selection is the starting point for the next generation of selection, and so on for many generations. It is natural to borrow such words as "reproduce" and "generation," which have associations with living things, because living things are the main examples we know of things that participate in cumulative selection. They may in practice be the only things that do. But for the moment I don't want to beg that question by saying so outright.

Sometimes clouds, through the random kneading and carving of the winds, come to look like familiar objects. There is a much published photograph, taken by the pilot of a small aeroplane, of what looks a bit like the face of Jesus, staring out of the sky. We have all seen clouds that reminded us of something—a sea horse, say, or a smiling face.

These resemblances come about by single-step selection, that is to say by a single coincidence. They are, consequently, not very impressive. The resemblance of the signs of the zodiac to the animals after which they are named, Scorpio, Leo, and so on, is as unimpressive as the predictions of astrologers. We don't feel overwhelmed by the resemblance, as we are by biological adaptations — the products of cumulative selection. We describe as weird, uncanny or spectacular, the resemblance of, say, a leaf insect to a leaf or a praying mantis to a cluster of pink flowers. The resemblance of a cloud to a weasel is only mildly diverting, barely worth calling to the attention of our companion. Moreover, we are quite likely to change our mind about exactly what the cloud most resembles.

Hamlet.   Do you see yonder cloud that's almost in shape of a camel?
Polonius.   By the mass, and 'tis like a camel, indeed.
Hamlet.   Methinks it is like a weasel.
Polonius.   It is backed like a weasel.
Hamlet.   Or like a whale?
Polonius.   Very like a whale.

I don't know who it was first pointed out that, given enough time, a monkey bashing away at random on a typewriter could produce all the works of Shakespeare. The operative phrase is, of course, given enough time. Let us limit the task facing our monkey somewhat. Suppose that he has to produce, not the complete works of Shakespeare but just the short sentence "Methinks it is like a weasel," and we shall make it relatively easy by giving him a typewriter with a restricted keyboard, one with just the 26 (capital) letters, and a space bar. How long will he take to write this one little sentence?

The sentence has 28 characters in it, so let us assume that the monkey has a series of discrete "tries," each consisting of 28 bashes at the keyboard. If he types the phrase correctly, that is the end of the experiment. If not, we allow him another "try" of 28 characters. I don't know any monkeys, but fortunately my 11-month old daughter is an experienced randomizing device, and she proved only too eager to step into the role of monkey typist. Here is what she typed on the computer:

UMMK JK CDZZ F ZD DSDSKSM
S SS FMCV PU I DDRGLKDXRRDO
RDTE QDWFDVIOY UDSKZWDCCVYT
H CHVY NMGNBAYTDFCCVD D
RCDFYYYRM N DFSKD LD K WDWK
JJKAUIZMZI UXDKIDISFUMDKUDXI

She has other important calls on her time, so I was obliged to program the computer to simulate a randomly typing baby or monkey:

WDLDMNLT DTJBKWIRZREZLMQCO P
Y YVMQKZPGJXWVHGLAWFVCHQYOPY
MWR SWTNUXMLCDLEUBXTQHNZVJQF

FU OVAODVYKDGXDEKYVMOGGS VT
HZQZDSFZIHIVPHZPETPWVOVPMZGF
GEWRGZRPBCTPGQMCKHFDBGW ZCCF

And so on and on. It isn't difficult to calculate how long we should reasonably expect to wait for the random computer (or baby or monkey) to type METHINKS IT IS LIKE A WEASEL. Think about the total number of *possible* phrases of the right length that the monkey or baby or random computer *could* type. It is the same kind of calculation as we did for haemoglobin, and it produces a similarly large result. There are 27 possible letters (counting "space" as one letter) in the first position. The chance of the monkey happening to get the first letter — M — right is therefore 1 in 27. The chance of it getting the first two letters — ME — right is the chance of it getting the second letter — E — right (1 in 27) *given that* it has also got the first letter — M — right, therefore $1/27 \times 1/27$, which equals 1/729. The chance of it getting the first word — METHINKS — right is 1/27 for each of the 8 letters, therefore $(1/27) \times (1/27) \times (1/27)$ . . ., etc. 8 times, or (1/27) to the power 8. The chance of it getting the entire phrase of 28 characters right is (1/27) to the power 28, i.e. (1/27) multiplied by itself 28 times. These are very small odds, about 1 in 10,000 million million million million million million. To put it mildly, the phrase we seek would be a long time coming, to say nothing of the complete works of Shakespeare.

So much for single-step selection of random variation. What about cumulative selection; how much more effective should this be? Very very much more effective, perhaps more so than we at first realize, although it is almost obvious when we reflect further. We again use our computer monkey, but with a crucial difference in its program. It again begins by choosing a random sequence of 28 letters, just as before:

WDLMNLT DTJBKWIRZREZLMQCO P

It now "breeds from" this random phrase. It duplicates it repeatedly, but with a certain chance of random error — "mutation" — in the copying. The computer examines the mutant nonsense phrases, the "progeny" of the original phrase, and chooses the one which, *however slightly*, most resembles the target phrase, METHINKS IT IS LIKE A WEASEL. In this instance the winning phrase of the next "generation" happened to be:

WDLTMNLT DTJBSWIRZREZLMQCO P

Not an obvious improvement! But the procedure is repeated, again mutant "progeny" are "bred from" the phrase, and a new "winner" is chosen. This goes on, generation after generation. After 10 generations, the phrase chosen for "breeding" was:

MDLDMNLS ITJISWHRZREZ MECS P

After 20 generations it was:

MELDINLS IT ISWPRKE Z WECSEL

By now, the eye of faith fancies that it can see a resemblance to the target phrase. By 30 generations there can be no doubt:

METHINGS IT ISWLIKE B WECSEL

Generation 40 takes us to within one letter of the target:

METHINKS IT IS LIKE I WEASEL

And the target was finally reached in generation 43. A second run of the computer began with the phrase:

Y YVMQKZPFJXWVHGLAWFVCHQXYOPY,

passed through (again reporting only every tenth generation):

Y YVMQKSPFTXWSHLIKEFV HQYSPY
YETHINKSPITXISHLIKEFA WQYSEY
METHINKS IT ISSLIKE A WEFSEY
METHINKS IT ISBLIKE A WEASES
METHINKS IT ISJLIKE A WEASEO
METHINKS IT IS LIKE A WEASEP

and reached the target phrase in generation 64. In a third run the computer started with:

GEWRGZRPBCTPGQMCKHFDBGW ZCCF

and reached METHINKS IT IS LIKE A WEASEL in 41 generations of selective "breeding."

The exact time taken by the computer to reach the target doesn't matter. If you want to know, it completed the whole exercise for me, the first time, while I was out to lunch. It took about half an hour. (Computer enthusiasts may think this unduly slow. The reason is that the program was written in BASIC, a sort of computer baby-talk. When I rewrote it in Pascal, it took 11 seconds.) Computers are a bit faster at this kind of thing than monkeys, but the difference really isn't significant. What matters is the difference between the time taken by *cumulative* selection, and the time which the same computer, working flat out at the same rate, would take to reach the target phrase if it were forced to use the other procedure of *single-step selection*: about a million million million million million years. This is more than a million million million times as long as the universe has so far existed. Actually it would be fairer just to say that, in comparison with the time it would take either a monkey or a randomly programmed computer to type our target phrase, the total age of the universe so far is a negligibly small quantity, so small as to be well within the margin of error for this sort of back-of-an-envelope calculation. Whereas the time taken for a computer working randomly but with the constraint of *cumulative selection* to perform the same task is of the same order as humans ordinarily can understand, between 11 seconds and the time it takes to have lunch.

There is a big difference, then, between cumulative selection (in which each improvement, however slight, is used as a basis for future building), and single-step selection (in which each new "try" is a fresh one). If evolutionary progress had had to rely on single-step selection, it would never have got anywhere. If, however, there was any way in which the necessary conditions for *cumulative* selection could have been set up by the blind forces of nature, strange and wonderful might have been the consequences. As a matter of fact that is exactly what happened on this planet, and we ourselves are among the most recent, if not the strangest and most wonderful, of those consequences.

It is amazing that you can still read calculations like my haemoglobin calculation, used as though they constituted arguments *against* Darwin's theory. The people who do this, often expert in their own field, astronomy or whatever it may be, seem sincerely to believe that Darwinism explains living organization in terms of chance—"single-step selection"—alone. This belief, that Darwinian evolution is "random," is not merely false. It is the exact opposite of the truth. Chance is a minor ingredient in the Darwinian recipe, but the most important ingredient is cumulative selection which is quintessentially *non*random.

Clouds are not capable of entering into cumulative selection. There is no mechanism whereby clouds of particular shapes can spawn daughter clouds resembling themselves. If there were such a mechanism, if a cloud resembling a weasel or a camel could give rise to a lineage of other clouds of roughly the same shape, cumulative selection would have the opportunity to get going. Of course, clouds do break up and form "daughter" clouds sometimes, but this isn't enough for cumulative selection. It is also necessary that the "progeny" of any given cloud should resemble its "parent" *more* than it resembles any old "parent" in the "population." This vitally important point is apparently misunderstood by some of the philosophers who have, in recent years, taken an interest in the theory of natural selection. It is further necessary that the chances of a given cloud's surviving and spawning copies should depend upon its shape. Maybe in some distant galaxy these conditions did arise, and the result, if enough millions of years have gone by, is an ethereal, wispy form of life. This might make a good science fiction story—*The White Cloud*, it could be called—but for our purposes a computer model like the monkey/Shakespeare model is easier to grasp.

DARWINISM AND THE
TAUTOLOGY PROBLEM

7

A. G. N. FLEW*

❧

The Structure of Darwinism

❧

THOSE WHO DO philosophy of science tend to equate science with physics.[1] The same thing also seems often to be true of those who try to present scientific thought to non-scientists. Yet one of the most important of all scientific theories is that developed by Darwin in the *Origin of Species*.[2] Covering the entire range of biological phenomena its scope is enormous. While if any scientific theory is interesting philosophically this one is.[3] It happens also to be exceptionally suitable as an elementary example for teaching. No outlandish or elaborate concepts are involved. The nature of most of the supporting evidence is familiar, and its relevance fairly easy to grasp. The deductive moves made are short and simple. The essential material can be studied in a single original source—conveniently brief, interesting, and written well. Darwin introduces no "theoretical entities" (photons, electro-magnetic waves, the libido, or what have you)[4] and employs no mathematics either within his theory or in presenting his case for it. The lack of these two generally essential features of modern science is at an elementary stage a positive advantage. Especially when, as here, these defects in the example may be made good afterwards by considering the later development of genetics; which both provides a hypothetical entity, the gene, and applies mathematics abundantly. The object of the

*From A. G. N. Flew, "The Structure of Darwinism," *New Biology,* Vol. 28 (London: Penguin Books Ltd., 1959), pp. 25–44. Reprinted by permission of the author.

present paper is not to say anything strikingly novel, but simply to suggest some of the general morals which may be drawn from an examination of this particular conceptual structure.

The Deductive Core of Darwinism

Darwin himself remarked in the last chapter: "this whole volume is one long argument" (p. 413). Again in the *Autobiography*, in a typically modest and engaging passage, he claims: "*The Origin of Species* is one long argument from the beginning to the end."[5] A recent interpreter goes further: "The old arguments for evolution were only based on circumstantial evidence. . . . But the core of Darwin's argument was of a different kind. It did not make it more probable—it made it a certainty. Given his facts his conclusion *must* follow: like a proposition in geometry. You do not show that any two sides of a triangle are very *probably* greater than the third. You show they *must* be so. Darwin's argument was a *de*ductive one—whereas an argument based on circumstantial evidence is *in*ductive."[6]

This is a challenging contention. For surely Darwin was a great empirical naturalist, concerned to discover what as a matter of contingent fact *is* the case, though it might not have been? What business had he with deductive *a priori* arguments purporting to demonstrate as in a theorem in geometry that some things *must* be so? We must consider what precisely this deductive core is: how much it does prove and how; and how much it leaves open to be settled by further research.

That Darwin's argument does indeed contain such a deductive core is suggested by a passage in the "Introduction": "As many more individuals of each species are born than can possibly survive; and as, *consequently*, there is a frequently recurring struggle for existence, *it follows that* any being, if it vary however slightly in any manner profitable to itself, under the complex and sometimes varying conditions of life will have a better chance of surviving and thus be naturally selected. From the strong principle of inheritance, any selected variety will tend to propagate its new and modified form" (p. 4: my italics). He promises that in the chapter "Struggle for Existence" he will treat this struggle "amongst all organic beings throughout the world, *which inevitably follows from* the high geometrical ratio of their increase" (p. 4: my italics). In that chapter he develops the argument: "A struggle for existence *inevitably follows* from the high rate at which all organic beings tend to increase . . . as more individuals are produced than can possibly survive, *there must in every case be* a struggle for existence, either one individual with another of the same species, or with the individuals of distinct species, or with the physical conditions of life. It is the doctrine of Malthus applied with manifold force to the whole animal and vegetable kingdoms, for in this case there can be no artificial increase of food, and no prudential restraint from marriage" (p. 59:

my italics). Just as the struggle for existence is derived as a consequence of the combination of a geometrical ratio of increase with the finite possibilities of survival: so in the chapter "Natural Selection" this in turn is derived as a consequence of the combination of the struggle for existence with variation. Darwin summarizes his argument here: "If . . . organic beings vary at all in the several parts of their organization, and I think this cannot be disputed; if there be . . . a severe struggle for life . . . and this certainly cannot be disputed; then . . . I think it would be a most extraordinary fact if no variation had ever occurred useful to each being's own welfare, in the same manner as so many variations have occurred useful to man. But if variations useful to any organic being do occur, assuredly individuals thus characterized will have the best chance of being preserved in the struggle for life: and from the strong principle of inheritance they will tend to produce offspring similarly characterized. This principle of . . . Natural Selection . . . leads to the improvement of each creature in relation to its organic and inorganic conditions of life" (p. 115).

Since Darwin's argument does indeed contain this deductive core the pedagogic device used by Julian Huxley in his latest exposition of evolutionary theory is exceptionally apt. It can, he claims, "be stated in the form of two general evolutionary equations. The first is that reproduction plus mutation produces natural selection; and the second that natural selection plus time produces the various degrees of biological improvement that we find in nature."[7] The idea is excellent, but the execution here is curiously slapdash. For the first equation, as Huxley gives it ($R + M \rightarrow NS$), is not valid. Reproduction plus mutation would not necessarily lead to natural selection. It is necessary also to add the struggle for existence: and that in turn has to be derived from the sum of the geometrical ratio of increase plus the limited resources for living — limited *Lebensraum* as Hitler's Germans might have called it. So to represent the core of Darwin's argument we need something more like: $GRI + LR \rightarrow SE$; $SE + V \rightarrow NS$; and, not on all fours with the first two, $NS + T \rightarrow BI$. To avoid anachronism V (for heritable variation) must be substituted for M (mutation). The third equation represents part of the argument which does not properly belong to what we are calling the deductive core; and so from now on we shall neglect it.

Of course to make this core, and the equations used to represent it schematically, ideally rigorous one would have to construct for all the crucial terms definitions to include explicitly every necessary assumption. There are in fact several, many of which when uncovered and noticed may seem too obvious to have been worth stating. Take for instance one to which Darwin himself refers, rather obliquely, in a passage already quoted: "I think it would be a most extraordinary fact if no variation had ever occurred useful to each being's own welfare" (p. 115). It would indeed. He offers a very powerful reason for believing that this has not in fact been the case, adding after the phrase just quoted; "in the same manner as so many variations have occurred useful to man" (p. 115). Nevertheless, in this particular case bringing out the assumption may have a value other than that of rigorization for its own sake. For it may suggest, what has in fact proved to be the case, that one of the main effects of natural selection is to eliminate unfavourable variations. It not only helps to generate biological improvement. It is essential to prevent biological degeneration.

We shall not attempt here to develop Darwin's argument quite rigorously or to formalize the result. This is an exercise which might be instructive. But it would of course involve transforming, and hence in a way misrepresenting, what Darwin actually said. It is perfectly possible to point the main morals without this.

(i) Though the argument itself proceeds *a priori*, because the premises are empirical it can yield conclusions which are also empirical. That living organisms all tend to reproduce themselves at a geometrical ratio of increase; that the resources they need to sustain life are limited; and that while each usually reproduces after its kind sometimes there are variations which in their turn usually reproduce after their kind: all these propositions are nonetheless contingent and empirical for being manifestly and incontestably true. That there is a struggle for existence; and that through this struggle for existence natural selection occurs: both these propositions equally are nonetheless contingent and empirical for the fact that it follows, necessarily as a matter of logic *a priori*, that wherever the first three hold the second two must hold also.

(ii) The premises are matters of obvious fact. The deductive steps are short and simple. The conclusions are enormously important. Yet these conclusions, and that they were implied by these premises, was before Darwin very far indeed from being obvious to able men already sufficiently familiar with the necessary premises. This should give us a greater respect for the power of simple logic working on the obvious. Of course, he did not just have to make some short deductive moves from a few very wide-ranging empirical premises already provided as such. He had first to recognize that these propositions did constitute essential premises; and then, after making the deductions from them, to appreciate that these premises and these conclusions contained and linked together concepts crucial for understanding the problem of the origin of species. The premises, the concepts, the deductions, the conclusions, all are simple. To bring them together and to see the importance of the theoretical scheme so constructed was a simple matter too. But this simplicity is the simplicity of genius.

(iii) The conclusions of the deductive argument are proved beyond dispute: for though the premises are as empirical generalizations in principle open to revision, in fact, as Darwin urged in a passage quoted already, they cannot reasonably be questioned. It is therefore all the more important to appreciate what it does not prove; and hence what, at least as far as this argument is concerned, is left open to be settled by further inquiry.

It certainly does not prove that all "the various degrees of biological improvement that we find in nature"[8] can be accounted for in these terms. It proves at most only that some "biological improvement" must occur. Darwin needed, and provided, other facts and considerations to support his far wider and more revolutionary conclusion: "that species are not immutable; but that those belonging to what are called the same genera are lineal descendants of some other and generally extinct species, in the same manner as the acknowledged varieties of any one species are the descendants of that species . . . I am convinced that Natural Selection has been the main but not exclusive means of modification" (pp. 5–6). Though he built up a mighty case for this sweeping conclusion the case is one which could not in principle be complete unless the whole science of evolutionary biology were complete. Here the field for inquiry is open and without limit.

Again, this deductive argument has nothing to say about the causes and mechanisms of variation. Indeed the *Origin of Species* as a whole has on these little to say. Darwin begins his summary of the chapter "Laws of Variation": "Our ignorance of the laws of variation is profound" (p. 151). While in "Recapitulation and Conclusion" one of the advantages of accepting his theory urged is: "A grand and almost untrodden field of inquiry will be opened, on the causes and laws of variation, on correlation of growth, on the effects of use and disuse, on the direct action of external conditions, and so forth" (p. 437). Thus Lamarckism is quite compatible with Darwin's theory; and Darwin indeed does himself accept in the chapter on variation the inheritance of acquired characters. There would not, of course, be this elasticity had Darwin, like Julian Huxley, held a theory in which V (variation) would be replaced in the equations by M (mutation); and "mutation" would be so defined as to exclude Lamarckism—the infamous thing! Thus, again, it would be strictly compatible with this deductive core of Darwinism to maintain that some or all favourable variations were the results of special interventions by the Management. Though any such arbitrary and anti-scientific postulations would be entirely out of harmony with Darwin's own thoroughly naturalistic spirit, and his Lyellian insistence on continuity of development: ". . . species are produced and exterminated by slowly acting and still existing causes, and not by miraculous acts of creation and by catastrophes . . ." (p. 439). Notwithstanding the fact that in the *Origin of Species* he always concedes a special creation for the first life: ". . . life, with its several powers, having been breathed by the Creator into a few forms or into one . . ." (p. 441); yet "I should infer from analogy that probably all the organic beings which have ever lived on this earth have descended from some one primordial form, into which life was first breathed by the Creator" (p. 436).

(iv) Darwinism provides an outstanding example to show how a good theory guides and stimulates inquiry, setting whole new ranges of fruitful questions. This Darwin sees clearly. He rightly claims as a great advantage attending the acceptance of his theory, that: "A grand and almost untrodden field of inquiry will be opened, on the causes and laws of variation. . . . A new variety raised by man will be a more important and interesting subject of study than one more species added to the infinitude of already recorded species. Our classifications will come to be, as far as they can be so made, genealogies . . . we have to discover and trace the many diverging lines of descent in our natural genealogies. . . . Rudimentary organs will speak infallibly with respect to the nature of long lost structures. . . . Embryology will reveal to us the structure, in some degree obscured, of the prototypes of each great class" (pp. 437–8). Ranging further: "In the distant future I see open fields for far more important researches. Psychology will be based on a new foundation, that of the necessary acquirement of each mental power and capacity by gradation" (p. 439).[9]

(v) Again, Darwinism makes an excellent text-book example to show how a theory explains: by showing that the elements to be explained are not after all just a lot of separate brute facts — just one damn thing after or along with another — but rather what, granted the assumptions of the theory, is to be expected.[10]

Whereas before Darwin the general opinion even among biologists was that species constituted natural kinds, created separately:[11] he tried to show that "Although much remains obscure, and will long remain obscure . . . the view which most naturalists entertain, and which formerly I entertained — namely, that each species has been independently created — is erroneous" (p. 5); and that, on the contrary, all species have their places on one single family tree — or at most on four or five — and are thus in the most strict and literal sense related. In what we have called the deductive core of his theory he shows how, on certain assumptions, a struggle for existence, natural selection, and some biological improvement must be expected. In applying this conceptual framework in detail in the attempt to account for the origin of species he shows how not only the existence of an enormous number of species but also many other very general biological facts previously isolated and brute are just what, granted its premises and some other equally plausible assumptions, is to be expected.

In his "Recapitulation" (pp. 422–32) Darwin reviews some of these very general facts which he has considered earlier in more detail. For instance: "As natural selection acts solely by accumulating slight, successive, favourable variations, it can produce no great or sudden modification. . . . Hence the canon of *"Natura non facit saltum"* [Literally: Nature does not take a jump — A.F.], which every fresh addition to our knowledge tends to make truer, is on this theory simply intelligible. We can see why nature is prodigal in variety, though niggard in innovation. But why this should be a law of nature if each species has been independently created, no man can explain" (p. 424: italics mine). Then:

"Looking to geographical distribution, if we admit that there has been during the long course of ages much migration . . . then we can understand, on the theory of descent with modification, most of the great leading facts in Distribution . . . we can understand, by the aid of the Glacial period, the identity of some few plants, and the close alliance of many others, on the most distant mountains, and likewise the close alliance of some of the inhabitants of the sea in the northern and southern temperate zones, though separated by the whole intertropical ocean" (pp. 428–9). And so on.

If we ask how Darwinism explains, or indeed how any other theory in the natural sciences explains, the answer seems to lie in its powers: to provide connexions between elements which without it would be unconnected — just a lot of loose and separate facts; and to show how the phenomena to be explained are, on certain assumptions, exactly what is to be expected.[12]

(vi) Darwin, always conscientious and generous in acknowledging his debts, refers several times to Malthus: ". . . the Struggle for Existence amongst all organic beings throughout the world, which inevitably follows from the high geometrical ratio of their increase, . . . This is the doctrine of Malthus, applied to the whole animal and vegetable kingdoms" (p. 4: cf. passage on p. 9 quoted above).[13] It is therefore perhaps not surprising that the logical skeleton of theory which provided the organizing and supporting framework for all Malthus' inquiries and recommendations about population resembles in almost every respect so far considered the theoretical framework of the *Origin of Species*.

Even for those who are not antecedently inclined, either by their political and religious ideology or by their deep emotional drives, to eschew Malthusian ideas, these similarities may be obscured by the fact that Malthus' theory as he presents it himself contains several logical faults, albeit easily remediable ones. Also, unfortunately and unnecessarily, it has built into it various always controversial and now generally obnoxious value commitments. Nevertheless Malthus, like Darwin, in his theory does proceed *a priori* from very general, scarcely contestable, empirical premises. From the manifest power of the human animal to reproduce in a geometrical ratio of increase; and from an assumption, too precisely formulated as an arithmetical ratio, which expresses the fact that the possibilities of increasing the resources necessary for living are, though elastic, finite, Malthus would deduce some very important empirical conclusions (cf. (i) above). The deductive moves once the premises are assembled are not elaborate (cf. (ii) above). The limitations of the purely deductive argument are important (cf. (iii) above). For neither in Malthus' original form nor in a slightly amended and corrected form will the premises yield the conclusion that the power of increase must be checked *everywhere and always* absolutely. The same is true of the premises from which Darwin derives the conclusion that there must be a struggle for existence. These alone are not sufficient to prove that no species ever has or ever could have enjoyed *even for a short time* an environment in which its possibilities of increase were not checked by competition either from other species or from other members of the same species. They prove only that such a condition if it did ever occur must necessarily be very short lived. But whereas this qualification is on the evolutionary time scale with which Darwin was dealing insignificant: in the field of human affairs with which Malthus was concerned it may sometimes be very important indeed. For here a generation is a lifetime; and in determining practical policy some weight may properly be given to Keynes' reminder that in the long run we are all dead.

Again, the theory of Malthus, like that of Darwin, has the power to guide and stimulate inquiry, setting ranges of fruitful new questions (cf. (iv) above). Recognition of the enormous animal power of reproductive increase, and of the inescapable fact that this power must always in the fairly short run be checked at the finite though elastic limits of the possibility of increasing the resources necessary for living — if it is not checked earlier by something else: generates the master question "How in fact is this enormous power checked?" And it was precisely this question which provoked all the empirical inquiries the results of which Malthus embodied in the second version of the *Essay on Population*.[14]

Finally, Malthus' theory too makes an excellent text-book example. It can be used to show how a theory picks out and explains by showing connexions between the vital elements in the situation. It can be used to show how granted certain conditions a theory can explain why certain other conditions also hold; by revealing that in these circumstances and on this theory these things are what is to be expected (cf. (v) above). Furthermore, it has the advantage of belonging to the field of social studies, still even more neglected by philosophers of science than biology. It can illustrate excellently the temptation confusingly and implicitly to incorporate controversial value commitments into the very structure of a theory, thereby determining the directions to be taken by any policy founded thereon. At this point the analogy between Malthus and Darwin begins to break down. It would be illuminating to develop a full Plutarchian comparison, to distinguish and catalogue the similarities and dissimilarities between the two theories. But here it is sufficient simply to mention a few of the similarities between Darwinism and a system of ideas which, partly because it is uncongenial both to the two most powerful ideological groups in the world[15] and to certain psychologically more elemental sources of prejudice,[16] is still very generally misunderstood and underestimated.[17]

The Conceptual Changes Required by Darwinism

Darwinism, as again Darwin himself clearly saw, implies that we must abandon assumptions implicit in the previous use of certain categorial terms such as *genus* and *species*: "When the views advanced by me in this volume . . . are generally admitted . . . there will be a considerable revolution in natural history. Systematists will be able to pursue their labours at present; but they will not be incessantly haunted by the shadowy doubt whether this or that form be in essence a species" (p. 436). Which, he remarks, feelingly: "I feel sure, and I speak after experience, will be no slight relief" (p. 436). Hence: "I look at the term *species* as one arbitrarily given for the sake of convenience to a set of individuals closely resembling each other, and that it does not essentially differ from the term *variety*, which is given to less distinct and more fluctuating forms. The term *variety*, again, in comparison with mere individual differences, is also applied arbitrarily, and for mere convenience's sake" (p. 49: italics mine). "Hereafter we shall be compelled to acknowledge that the only distinction between species and well-marked varieties is, that the latter are known, or believed, to be connected at the present day by intermediate gradations, whereas species were formerly thus connected" (p. 436). "In short, we shall have to treat species in the same manner as those naturalists treat genera, who admit that genera are merely artificial combinations made for convenience" (p. 437).

The assumptions which Darwin's theory commits him to challenge are more particular cases of those very general prejudices about language and classification which Locke had begun to uncover and to question in his *Essay concerning Human Understanding* (1690). These are the assumptions: that all things belong to certain natural kinds, in virtue of their "essential natures"; that there are no marginal cases falling outside and between these sharply delimited collections of individuals; that there must always be straight yes or no answers to the question "Is this individual a so and so or not?"; and that men have only to uncover and, as it were, write the labels for, the classes to which nature has antecedently allocated every individual thing. To such assumptions in the biological field the pictures appropriate are: that of *Genesis*, of all creatures created after their sharply different kinds; and that of the tradition, illustrated by William Blake, of Adam naming the beasts. They would justify the approach to classification epitomized in the citation of the Swedish Academy in honouring Linnaeus: "he discovered the essential nature of insects." Or that of his own famous but dark saying, quoted by Darwin: ". . . the characters do not make the genus, but . . . the genus gives the characters . . ." (p. 372).

In one aspect Darwin's work can be seen as a continuation and application of that of Locke, particularly of the third book of the *Essay* "Of Words"; though there seems to be no evidence that he had ever read the philosopher. For he is insisting on "that old canon in natural history, '*Natura non facit saltum*'" (p. 185; cf. pp. 174–5 and p. 424); and arguing that on his theory "we shall at least be freed from the vain search for the undiscovered and undiscoverable essence of the term *species*" (p. 437: italics mine in this and the previous quotation). Sir Arthur Keith in his Plutarchian comparison of Locke with Darwin fails surprisingly to notice this continuity.[18]

As an example of philosophy within scientific theory this analysis and reinterpretation of the concepts of *species, variety,* and *genus,* might be compared with Einstein's analysis of *motion* and *simultaneity* in relativity theory. In both cases the analysis is required by the theory. In both cases it had been to a greater or lesser extent anticipated by a philosopher before it was reworked and put to use by a scientist. But whereas Darwin seems never to have read Locke, Einstein certainly did read Mach; and acknowledged indebtedness to him.

The Model Contained and "Deployed" in Darwinism

Recently the useful notion of the "deployment" of a model has been introduced into the philosophy of the physical sciences. "At the stage at which a new model is introduced the data that we have to go on, the phenomena which it is used to explain, do not justify us in prejudging, either way, which of the questions which normally make sense when asked of things which, say, travel will eventually be given a meaning on the new theory also. . . . One might speak of models in physics as more or less 'deployed.'"[19] How far a given model is said to be deployed is a matter of how far the analogy, between that model and the phenomena to which it is applied, is believed to hold. This idea is relevant also elsewhere than in the physical sciences. The acceptance of Darwin's theory made possible a massive deployment of one model which had been curiously boxed up and impotent for an extraordinarily long time. This was the model of the family: with which was associated the method of representation employed to express both familial relationships ("the family tree") and one system of classification ("the tree of Porphyry").[20] Once again Darwin saw what he was doing: "The terms used by naturalists of affinity, relationship, community of type, paternity . . . will cease to be metaphorical and will have a plain signification"; while "Our classifications will come to be, as far as they can be made, genealogies . . ." (p. 437). The remarkable points about this particular model are: first, that it does not seem originally to have been introduced as any part of an attempt at scientific explanation; and, second, that its suggestive force does not seem to have had much influence towards the production of the theory which made possible its deployment. The various terms appropriate to this model were, apparently, introduced because naturalists noticed analogies which made the idea of family relationship seem apt as a metaphor. But the suggestion that the metaphor might be considerably more than a mere metaphor, that the model could be deployed, seems almost always to have been blocked by the strong resistance of the accepted doctrine of the fixity of species.

We can find in Darwin's letters and other papers indications of the strength of this resistance: though, of course, we with the advantage of hindsight see before Darwin the occasional deviations to a doctrine of descent standing out. It is a mistake to interpret this resistance simply as a matter of religion. Certainly belief in the evolution of species would be incompatible with the acceptance of the creation myths of *Genesis*, interpreted literally: (to say nothing of the incompatibility between the two myths themselves). But the evolutionary geology of Lyell is equally irreconcilable with a literal reading of the Pentateuch; and that was already becoming generally accepted among scientific men when Darwin began to work on the origin of species. Quite apart from any considerations of ideology it *is* an inescapable biological fact that living things as we now see them around us do seem, with only comparatively infrequent exceptions, to belong to various natural kinds; which mate with and reproduce, again with only comparatively minor exceptions, after their kinds. It was precisely this massive and stubborn fact which the aetiological myth makers of *Genesis* were trying to account for with their theory of the special creation of fixed species. Even after the evolutionary geologists had opened up the vast time scale needed for a smooth and uniformitarian process of biological evolution, the way of this concept was still blocked by the difficulty of suggesting any mechanisms which could possibly have brought about such an enormous development. To this problem Darwin found the clue in the concept of natural selection, suggested to him by his reading of Malthus.

Another bulwark of the doctrine of fixed species would presumably be the unevolutionary view of language referred to above. For while the general fact of the apparent stability and separateness of seeming "natural kinds" is a main source and stay for certain false assumptions about language: those assumptions, usually hidden, would in turn support the idea of the fixity of species. For to the extent that we assume that, because a question of the "Is this an X?" form can usually be answered by a straight "Yes" or "No," such a question of identification must always be susceptible of a straight yes or no answer: to this extent we are mistaking it that all the things in the world must without exception either definitely fit or definitely not fit into the classificatory pigeonholes provided by our language, labelled with the words presently available in the vocabulary of that language. This is an assumption which has only to be revealed and recognized to be challenged. Nevertheless it is one which has been, and remains, protean, powerful, and pervasive.

The "Philosophical Implications" of Darwinism

Darwin revolutionized biological studies. But in addition, rightly or wrongly, his work has had considerable effects in shaping world outlooks and determining the general climate of opinion. This influence has been felt in two spheres. First, Darwin succeeded in indicating, at least in outline, how the appearance of design in living things might perhaps be accounted for without any appeal to actual interventions by some Supernatural Designer. Second, his ideas have been taken, or mistaken, to justify various moral and political policies.

Now it may well be that David Hume in his masterpiece, the *Dialogues concerning Natural Religion* (First edition, 1779), provided all the instruments needed to dismantle the Argument to Design. But Hume's subtle speculative arguments, presented as they are in the discreet and curiously difficult form of a philosophical dialogue, have never made much impact directly on popular thinking. Alternatively, it may well be that some version of this hardiest and most perennial of all the arguments of natural theology can be salvaged and refined for reuse, even if you have rejected absolutely the idea that supernatural interventions may be postulated to explain the particular course of nature. Nevertheless, the popular version of this argument, developed perhaps most powerfully in the *Natural Theology: or Evidences of the Existence and Attributes of the Deity collected from the Appearances of Nature* (First edition, 1802), though not of course originated by Paley, urges that just as from the observation of a watch we may infer the existence of a watchmaker, so by parity of reasoning, from the existence of mechanisms so marvellous as the human eye we must infer the existence of a Designer. Paley explicitly repudiates as an alternative any suggestion "that the eye, the animal to which it belongs, every other animal, every plant . . . are only so many out of the possible varieties and combinations of being which the lapse of infinite ages has brought into existence: that the present world is the relic of that variety; millions of other bodily forms and other species having perished, being by the defect of their constitutions incapable of preservation. . . . Now there is no foundation whatever for this conjecture in anything which we observe in the works of nature . . ." (*Paley's Works*, 1838 edition, Vol. I, p. 32). It is this version of the Argument to Design to which the *Origin of Species* is crucially relevant. For in it Darwin offers a demonstration, backed by a mass of empirical material, of how adaptation however remarkable might be the result not of supernatural artifice but of natural selection.

Again, the doctrine of Darwinism has been taken to justify the most extraordinarily diverse, and often mutually incompatible, moral and political policies. As a purely scientific theory by itself it could not entail any normative conclusions (conclusions, that is, about what *ought* to be); because it would not, so long as it remained a purely scientific theory, contain any normative premises. Coming, however, from this important truism in the abstract to our particular case it is perhaps just worth mentioning that one or two of Darwin's scientific ideas are peculiarly open to ethical misinterpretation; though he himself is usually careful to avoid and to discourage such misconstructions.

Thus the concept of the survival of the fittest in the struggle for existence is easily mistaken to imply that Nature favours the survival of the most admirable and best. Whereas of course "fittest" here is to be defined neutrally as "having whatever as a matter of fact it may take to survive." Again, because natural selection in a struggle for existence enables the fittest (in this neutral sense) to win out against their competitors, it has seemed that evolutionary biology provides a ready-made justification for unrestricted competition, and extinction take the hindmost. So it has been found necessary to compile books on mutual aid, to show that co-operation within and between species has also sometimes paid off in the biological rat race. While, on the other side, the descriptive law of the jungle has been accepted as the prescriptive law of a natural order of Nature: ". . . one general law, leading to the advancement of all organic beings, namely, multiply, vary, let the strongest live and the weakest die" (p. 219). In fact Darwinism provides no reason: either for saying that what will survive in unrestricted competition will be the most excellent and worthy; or for denying that some co-operation may pay off within the general struggle for existence. One dramatic turn of phrase does not commit Darwin to the error of regarding the laws of nature which *describe* what does go on, as laws *prescribing* what ought to go on. Again, the not particularly Darwinian notion of higher animals, combined with the Darwinian idea of their evolution by natural selection, may raise a hope that Nature is somehow in favour of progress. Of this Darwin himself was not altogether innocent. While in considering "Geological Succession" he remarks noncommittally: "The inhabitants of each successive period in the world's history have beaten their predecessors in the race for life, and are, insofar, higher in the scale of nature; and this may account for that vague yet ill-defined sentiment, felt by many palaeontologists, that organization on the whole has progressed" (p. 309). In his final peroratory paragraph he goes further into the world of evaluation, claiming: ". . . from the war of nature, from famine and death, the most exalted object we are capable of conceiving, the higher animal, directly follows. There is grandeur in this view of life . . ." (p. 441). Yet in pure evolutionary theory nothing is valuable and nothing is without value. By it nothing is justified, and no values are guaranteed.[21] Things just happen.

ACKNOWLEDGMENT

I should like to express here my thanks to my colleague at Keele, Dr. R. G. Evans for his helpful criticism of an earlier draft of this paper. He is in no way responsible for the biological and other errors which no doubt it still contains.

NOTES

1. See for example Stephen Toulmin's *The Philosophy of Science* (Hutchinson, 1953).
2. All references are given to the World's Classics edition first published in 1902 by the Oxford University Press.

3. Contrast Wittgenstein in the *Tractatus Logico-Philosophicus* (Kegan Paul, 1922): "The Darwinian theory has no more to do with philosophy than has any other hypothesis of natural science" (4.1122).
4. I particularly do *not* want to suggest that all these, and all the others classed sometimes as "scientific" or "theoretical" entities, enjoy exactly the same ontological status.
5. *Autobiography of Charles Darwin* (Watts, 1929), p. 75.
6. C. F. A. Pantin in *The History of Science* (Cohen and West, 1953), p. 137: italics in original.
7. Julian Huxley *The Process of Evolution* (Chatto and Windus, 1953), p. 38.
8. *Ibid.*
9. "About thirty years ago there was much talk that geologists ought only to observe and not to theorize; and I well remember someone saying that at this rate a man might as well go into a gravel pit and count the pebbles and describe the colours. How odd it is that anyone should not see that all observation must be for or against some view if it is to be of any service." (*More Letters of Charles Darwin*, ed. Francis Darwin and A. Seward, Vol. I, p. 195.)
10. The best short study of explanation which I know is by John Hospers in *Essays in Conceptual Analysis* (Ed. Flew, Macmillan, 1956). With this may be compared Norman Campbell *What is Science?* Ch. V. (Dover Publications, New York, 1952: the original 1921 U.K. edition has been out of print for years). Campbell's thinking is oriented towards physics and chemistry: so it may not be obvious immediately how or how far his emphatic distinction between laws (which do not explain) and theories (which do) can be applied to evolutionary biology. It is also unfortunate that he fails to make any distinction between: explaining *why* something occurs; and explaining *to* some particular person the meaning of some notion. He thus commits the howler: "Explanation in general is the expression of an assertion in a more acceptable and satisfactory form" (p. 77). If this were so, no explanation could ever give us any new information.
11. I do not want here or elsewhere in this paper to enter into the difficult question of where exactly Darwin's originality lay. See on this J. Arthur Thomson "Darwin's Predecessors" in *Darwin and Modern Science* (ed. A. C. Seward, C.U.P., 1909).
12. I insert the qualification "in the natural sciences" simply in order to bypass the recent attempts to show that the second clause does not hold good of explanations in history.
13. The well-known passage in the *Autobiography* relevant here is particularly worth quoting for its timely moral to specialists inclined to neglect their general studies: "In October 1838, that is, fifteen months after I had begun my systematic enquiry, I happened to read for amusement Malthus on *Population*, and being well prepared to appreciate the struggle for existence which everywhere goes on from long-continued observation of the habits of animals and plants, it at once struck me that under these circumstances favourable variations would tend to be preserved, and unfavourable ones to be destroyed. The result of this would be the formation of new species. Here, then, I had at last got a theory by which to work . . ." (p. 57). Compare A. R. Wallace *My Life, A Record of Events and Opinions* (London, 1905), Vol. I, p. 232, for his account of the operation of the same stimulus in his parallel case.
14. The second and all later editions are so substantially different from the first as to constitute different books. It would be good if the practice of distinguishing them as *First Essay* and *Second Essay* became general. Since the Appendices are important, the *Second Essay* is best studied in one of the later editions in which these are included: e.g. the sixth edition of 1826.

15. At the UNO World Population Conference in Rome in 1954 it was notable how often Roman Catholic and Communist delegates were found standing side by side in rejecting Malthusian ideas. For a very sober comment on the opposition of this Holy-Unholy Alliance see *World Population and Resources* (P.E.P. Reports, 1955), pp. 307 ff.
16. See J. C. Flügel *Population, Psychology, and Peace* (Watts, Thinkers' Library).
17. For an exposition and an attempt at reconstructive criticism of Malthus' theory I may perhaps be allowed to refer to my "The Structure of Malthus' Population Theory" in *Australasian Journal of Philosophy*, 1957.
18. "Darwin's Place among Philosophers" in *Rationalist Annual 1955*.
19. Stephen Toulmin: *loc. cit.* p. 37.
20. On the importance of "methods of representation" we may again refer to Stephen Toulmin *loc cit., passim*. The "tree of Porphyry" was the name given both to a certain method of classification and to a way of representing this method by a sort of family tree.
21. On the subject of this last section see: Julian Huxley's "Progress" in *Essays of a Biologist* (Chatto and Windus, 1923) and *Evolution and Ethics* (Pilot Press, 1947). This last contains a reprint of T. H. Huxley's Romanes Lecture. For a rather unsympathetic criticism see Stephen Toulmin "World Stuff and Nonsense" in *Cambridge Journal*, Vol. I.

EDITORIAL NOTE

Flew makes particular reference to two men whose ideas much influenced Darwin. The first is the Reverend Thomas Robert Malthus who argued in his *Essay on a Principle of Population* (6th. ed., 1826) that state welfare systems only compound the problem of the poor because (human) population numbers have a tendency to increase geometrically, thus outstripping food supplies which at most can only increase arithmetically. There is therefore an invariable struggle for existence, and no change in the human state is possible. Darwin took Malthus' premises but turned his conclusion on its head, arguing that the struggle is precisely the key to change. Darwin made similar use of the ideas of Charles Lyell, who argued in his *Principles of Geology* (1830–33) for a "uniformitarian" perspective, namely, that of a very old world cycling on indefinitely with the only causes of change being natural, like rain and frost and earthquakes and volcanoes. Darwin's biological uniformitarianism, which likewise eschewed non-natural causes, translated into evolutionism!

Flew refers also to the ideas of the early nineteenth-century evolutionist Jean Baptiste de Lamarck. "Lamarckism" sees evolutionary change as coming through the inheritance of acquired characters, as in the long neck of the giraffe through stretching. This idea, rejected by today's evolutionists, was accepted by Darwin, and he may well have been influenced by Lamarck, for he read his work.

8

TOM BETHELL*

Darwin's Mistake

All Change Is Not Progress

HOW DO WE COME to have horses and tigers and things? There are at least a million species in existence today, according to the paleontologist George Gaylord Simpson, and for every one extant, perhaps 100 are extinct. Such profusion! Such variety! How did it come about? The old answer was that they are created by God. But with the increasingly scientific temper of the eighteenth and nineteenth centuries, this explanation began to look insufficient. God was invisible, and so could not be part of any scientific explanation.

So an alternative explanation was proposed by a number of savants, among them Jean Baptiste Lamarck and Erasmus Darwin: the various forms of life did not just appear (as at the tip of a magician's wand), but evolved by a process of gradual transformation. Horses came from something slightly less horselike, tigers from something slightly less tigerlike, and so on back, until finally, if you went back far enough in time, you would come to a primitive blob of life which itself got started (perhaps) by lightning striking the primeval soup.

"Either each species of crocodile has been specially created," said Thomas Henry Huxley, "or it has arisen out of some pre-existing form by the operation of natural causes. Choose your hypothesis; I have chosen mine."

That's all very well, replied more conservative thinkers. If all of this life got here by evolution from more primitive life, then how did evolution occur? No answer was immediately forthcoming. Genesis prevailed. Then Charles Darwin (grandson of Erasmus) furnished what looked like the solution. He proposed the machinery of evolution, and claimed that it existed in nature. Natural selection, he called it.

His idea was accepted with great rapidity. Once stated it seemed only too obvious. The survival of the fittest — of course! Some types are fitter than others, and given the competition — the "struggle for existence" — the fitter ones will survive to propagate their kind. And so animals, plants, all life in fact, will tend to get better and better. They would have to, with the fitter ones inevitably replacing those that are less fit. Nature itself, then, had evolving machinery built into it. "How extremely stupid not to have thought of that!" Huxley commented, after reading the *Origin of Species*. Huxley had coined the term *agnostic*, and he remained one. Meanwhile, the Genesis version didn't entirely fade away, but it inevitably took on a slightly superfluous air.

The Evolution Debate

That was a little over 100 years ago. By the time of the Darwin Centennial Celebrations at the University of Chicago in 1959, Darwinism was triumphant. At a panel discussion Sir Julian Huxley (grandson of Thomas Henry) affirmed that "the evolution of life is no longer a theory; it is a fact." He added sternly: "We do not intend to get bogged down in semantics and definitions." At about the same time, Sir Gavin de Beer of the British Museum remarked that if a layman sought to "impugn" Darwin's conclusions, it must be the result of "ignorance or effrontery." Garrett Hardin of the California Institute of Technology asserted that anyone who did not honor Darwin "inevitably attracts the speculative psychiatric eye to himself." Sir Julian Huxley saw the need for "true belief."

So that was it, then. The whole matter was settled — as I assumed, and as I imagined most people must. Darwin had won. No doubt there were backward folk tucked away in the remoter valleys of Appalachia who still clung to their comforting beliefs, but they, of course, lacked education. Not everyone was enlightened — goodness knows the Scopes trial had proved that, if nothing else. And some of them still wouldn't let up, apparently — they were trying to change the textbooks and get the Bible back into biology. Well, there are always diehards.

So it was only casually, about a year ago, that I picked up a copy of *Darwin Retried*, a slim volume by one Norman Macbeth, a Harvard-trained lawyer. An odd field for a lawyer, certainly. But an endorsement on the cover by Karl Popper caught my eye. "I regard the book as . . . a really important contribution to the debate," Popper had written.

The debate? What debate? This interested me. I had studied philosophy, and in my undergraduate days Popper was regarded as one of the top philosophers—especially important for having set forth "rules" for discriminating between genuine and pseudo science. And Popper evidently thought there had been a "debate" worth mentioning. In his bibliography Macbeth listed a few articles that had appeared in academic philosophy journals in recent years and evidently were a part of this debate.

That was, as I say, a year ago, and by now I have read these articles and a good many others. In fact, I have spent a good portion of the last year familiarizing myself with this debate. It is surprising that so little word of it has leaked out, because it seems to have been one of the most important academic debates of the 1960s, and as I see it the conclusion is pretty staggering: Darwin's theory, I believe, is on the verge of collapse. In his famous book, *On the Origin of Species by Means of Natural Selection, or The Preservation of Favored Races in the Struggle for Life*, Darwin made a mistake sufficiently serious to undermine his theory. And that mistake has only recently been recognized as such. The machinery of evolution that he supposedly discovered has been challenged, and it is beginning to look as though what he really discovered was nothing more than the Victorian propensity to believe in progress. At one point in his argument, Darwin was misled. I shall try to elucidate here precisely where Darwin went wrong.

What was it, then, that Darwin discovered? What was this mechanism of natural selection? Here it comes as a slight shock to learn that Darwin really didn't "discover" anything at all, certainly not in the same way that Kepler, for example, discovered the laws of planetary motion. The *Origin of Species* was not a demonstration but an argument—"one long argument," Darwin himself said at the end of the book—and natural selection was an idea, not a discovery. It was an idea that occurred to him in London in the late 1830s which he then pondered in the Home Counties over the next twenty years. As we now know, several other thinkers came up with the same or a very similar idea at about the same time. The most famous of these was Alfred Russel Wallace, but there were several others.

The British philosopher Herbert Spencer was one who came within a hair's breadth of the idea of natural selection, in an essay called "The Theory of Population" published in the *Westminster Review* seven years before the *Origin of Species* came out. In this article Spencer used the phrase "the survival of the fittest" for the first time. Darwin then appropriated the phrase in the fifth edition of the *Origin of Species*, considering it an admirable summation of his argument. This argument was in fact an analogy, as follows:

While in his country retreat Darwin spent a good deal of time with pigeon fanciers and animal breeders. He even bred pigeons himself. Of particular relevance to him was that breeders bred for certain characteristics (length of feather, length of wool, coloring), and that the offspring of the selected mates often tended to have the desired characteristic more abundantly, or more noticeably, than its parents. Thus, it could perhaps be said, a small amount of "evolution" had occurred between one generation and the next.

By analogy, then, the same process occurred in nature, Darwin thought. As he wrote in the *Origin of Species*: "How fleeting are the wishes of man! how short his time! and consequently how poor will his productions be, compared with those accumulated by nature during whole geological periods. Can we wonder, then, that nature's productions should be far 'truer' in character than man's productions?"

Just as the breeders selected those individuals best suited to the breeders' needs to be the parents of the next generation, so, Darwin argued, nature selected those organisms that were best fitted to survive the struggle for existence. In that way evolution would inevitably occur. And so there it was: a sort of improving machine inevitably at work in nature, "daily and hourly scrutinizing," Darwin wrote, "silently and insensibly working . . . at the improvement of each organic being." In this way, Darwin thought, one type of organism could be transformed into another—for instance, he suggested, bears into whales. So that was how we came to have horses and tigers and things —by natural selection.

The Great Tautology

For quite some time Darwin's mechanism was not seriously examined, until the renowned geneticist T. H. Morgan, winner of the Nobel Prize for his work in mapping the chromosomes of fruit flies, suggested that the whole thing looked suspiciously like a tautology. "For, it may appear little more than a truism," he wrote, "to state that the individuals that are the best adapted to survive have a better chance of surviving than those not so well adapted to survive."

The philosophical debate of the past ten to fifteen years has focused on precisely this point. The survival of the fittest? Any way of identifying the fittest other than by looking at the survivors? The preservation of "favored" races? Any way of identifying them other than by looking at the preserved ones? If not, then Darwin's theory is reduced from the status of scientific theory to that of tautology.

Philosophers have ranged on both sides of this critical question: are there criteria of fitness that are independent of survival? In one corner we have Darwin himself, who assumed that the answer was yes, and his supporters, prominent among them David Hull of the University of Wisconsin. In the other corner are those who say no, among whom may be listed A. G. N. Flew, A. R. Manser, and A. D. Barker. In a nutshell here is how the debate has gone:

Darwin, as I say, just assumed that there really were independent criteria of fitness. For instance, it seemed obvious to him that extra speed would be useful for a wolf in an environment where prey was scarce, and only those wolves first on the scene of a kill would get enough to eat and, therefore, survive. David Hull has supported this line of reasoning, giving the analogous example of a creature that was better able than its mates to withstand desiccation in an arid environment.

The riposte has been as follows: a mutation that enables a wolf to run faster than the pack only enables the wolf to survive better if it does, in fact, survive better. But such a mutation could also result in the wolf outrunning the pack a couple of times and getting first crack at the food, and then abruptly dropping dead of a heart attack, because the extra power in its legs placed an extra strain on its heart. Fitness must be identified with survival, because it is the overall animal that survives, or does not survive, not individual parts of it.

However, we don't have to worry too much about umpiring this dispute, because a look at the biology books shows us that the evolutionary biologists themselves, perhaps in anticipation of this criticism, retreated to a fortified position some time ago, and conceded that "the survival of the fittest" was in truth a tautology. Here is C. H. Waddington, a prominent geneticist, speaking at the aforementioned Darwin Centennial in Chicago:

"Natural selection, which was at first considered as though it were a hypothesis that was in need of experimental or observational confirmation turns out on closer inspection to be a tautology, a statement of an inevitable although previously unrecognized relation. It states that the fittest individuals in a population (defined as those which leave most offspring) will leave most offspring."

The admission that Darwin's theory of natural selection was tautological did not greatly bother the evolutionary theorists, however, because they had already taken the precaution of redefining natural selection to mean something quite different from what Darwin had in mind. Like the philosophical debate of the past decade, this remarkable development went largely unnoticed. In its new form, natural selection meant nothing more than that some organisms have more offspring than others: in the argot, differential reproduction. This indeed was an empirical fact about the world, not just something true by definition, as was the case with the claim that the fittest survive.

The bold act of redefining selection was made by the British statistician and geneticist R. A. Fisher in a widely heralded book called *The Genetical Theory of Natural Selection*. Moreover, by making certain assumptions about birth and death rates, and combining them with Mendelian genetics, Fisher was able to qualify the resulting rates at which population ratios changed. This was called population genetics, and it brought great happiness to the hearts of many biologists, because the mathematical formulae looked so deliciously scientific and seemed to enhance the status of biology, making it more like physics. But here is what Waddington recently said about *this* development:

"The theory of neo-Darwinism is a theory of the evolution of the population in respect to leaving offspring and not in respect to anything else. . . . Everybody has it in the back of his mind that the animals that leave the largest number of offspring are going to be those best adapted also for eating peculiar vegetation, or something of this sort, but this is not explicit in the theory. . . . There you do come to what is, in effect, a vacuous statement: Natural selection is that some things leave more offspring than others; and, you ask, which leave more offspring than others; and it is those that leave more offspring, and there is nothing more to it than that. *The whole real guts of evolution — which is how do you come to have horses and tigers and things — is outside the mathematical theory* [my italics]."

Here, then, was the problem. Darwin's theory was supposed to have answered this question about horses and tigers. They had gradually developed, bit by bit, as it were, over the eons, through the good offices of an agency called natural selection. But now, in its new incarnation, natural selection was only able to explain how horses and tigers became more (or less) numerous — that is, by "differential reproduction." This failed to solve the question of how they came into existence in the first place.

This was no good at all. As T. H. Morgan had remarked, with great clarity: "Selection, then, has not produced anything new, but only more of certain kinds of individuals. Evolution, however, means producing new things, not more of what already exists."

One more quotation should be enough to convince most people that Darwin's idea of natural selection was quietly abandoned, even by his most ardent supporters, some years ago. The following comment, by the geneticist H. J. Muller, another Nobel Prize winner, appeared in the Proceedings of the American Philosophical Society in 1949. It represents a direct admission by one of Darwin's greatest admirers that, however we come to have horses and tigers and things, it is not by natural selection. "We have just seen," Muller wrote, "that if selection could be somehow dispensed with, so that all variants survived and multiplied, the higher forms would nevertheless have arisen."

I think it should now be abundantly clear that Darwin made a mistake in proposing his natural-selection theory, and it is fairly easy to detect the mistake. We have seen that what the theory so grievously lacks is a criterion of fitness that is independent of survival. If only there were some way of identifying the fittest beforehand, without always having to wait and see which ones survive, Darwin's theory would be testable rather than tautological.

But as almost everyone now seems to agree, fittest inevitably means "those that survive best." Why, then, did Darwin assume that there were independent criteria? And the answer is, because in the case of artificial selection, from which he worked by analogy, *there really are independent criteria*. Darwin went wrong in thinking that this aspect of his analogy was valid. In our sheep example, remember, long wool was the "desirable" feature—the independent criterion. The lambs of woolly parental sheep may possess this feature even more than their parents, and so be "more evolved"—more in the desired direction.

In nature, on the other hand, the offspring may differ from their parents in any direction whatsoever and be considered "more evolved" than their parents, provided only that they survive and leave offspring themselves. There is, then, no "selection" by nature at all. Nor does nature "act," as it is so often said to do in biology books. One organism may indeed be "fitter" than another from an evolutionary point of view, but the only event that determines this fitness is death (or infertility). This, of course, is not something which helps *create* the organism, but is something that terminates it. It occurs at the end, not the beginning of life.

Onward and Upward

Darwin seems to have made the mistake of just assuming that there were independent criteria of fitness because he lived in a society in which change was nearly always perceived as being for the good. R. C. Lewontin, Agassiz Professor of Zoology at Harvard, has written on this point: "The bourgeois revolution not only established change as the characteristic element of the cosmos, but added direction and progress as well. A world in which a man could rise from humble origins must have seemed, to him at least, a good world. Change per se was a moral quality. In this light, Spencer's assertion that change *is* progress is not surprising." One may note also James D. Watson's remark in *The Double Helix* that "cultural traditions play major roles" in the development of science.

Lewontin goes on to point out that "the bourgeois revolution gave way to a period of consolidation, a period in which we find ourselves now." Perhaps that is why only relatively recently has the concept of natural selection come under strong attack.

There is, in a way, a remarkable conclusion to this brief history of natural selection. The idea started out as a way of explaining how one type of animal gradually changed into another, but then it was redefined to be an explanation of how a given type of animal became more numerous. But wasn't natural selection supposed to have a *creative* role? the evolutionary theorists were asked. Darwin had thought so, after all. Now watch how they responded to this:

The geneticist Theodosius Dobzhansky compared natural selection to "a human activity such as performing or composing music." Sir Gavin de Beer described it as a "master of ceremonies." George Gaylord Simpson at one point likened selection to a poet, at another to a builder. Ernst Mayr, Lewontin's predecessor at Harvard, compared selection to a sculptor. Sir Julian Huxley topped them all, however, by comparing natural selection to William Shakespeare.

Life on Earth, initially thought to constitute a sort of prima facie case for a creator, was, as a result of Darwin's idea, envisioned merely as being the outcome of a process and a process that was, according to Dobzhansky, "blind, mechanical, automatic, impersonal," and, according to de Beer, was "wasteful, blind, and blundering." But as soon as these criticisms were leveled at natural selection, the "blind process" itself was compared to a poet, a composer, a sculptor, Shakespeare — to the very notion of creativity that the idea of natural selection had originally replaced. It is clear, I think, that there was something very, very wrong with such an idea.

I have not been surprised to read, therefore, in Lewontin's recent book, *The Genetic Basis of Evolutionary Change* (1974), that in some of the latest evolutionary theories "natural selection plays no role at all." Darwin, I suggest, is in the process of being discarded, but perhaps in deference to the venerable old gentleman, resting comfortably in Westminster Abbey next to Sir Isaac Newton, it is being done as discreetly and gently as possible, with a minimum of publicity.

9

STEPHEN JAY GOULD*

Darwin's Untimely Burial

IN ONE OF the numerous movie versions of *A Christmas Carol*, Ebenezer Scrooge encounters a dignified gentleman sitting on a landing, as he mounts the steps to visit his dying partner, Jacob Marley, "Are you the doctor?" Scrooge inquires. "No," replies the man, "I'm the undertaker; ours is a very competitive business." The cutthroat world of intellectuals must rank a close second, and few events attract more notice than a proclamation that popular ideas have died. Darwin's theory of natural selection has been a perennial candidate for burial. Tom Bethell held the most recent wake in a piece called "Darwin's Mistake" (*Harper's*, February 1976): "Darwin's theory, I believe, is on the verge of collapse. . . . Natural selection was quietly abandoned, even by his most ardent supporters, some years ago." News to me, and I, although I wear the Darwinian label with some pride, am not among the most ardent defenders of natural selection. I recall Mark Twain's famous response to a premature obituary: "The reports of my death are greatly exaggerated."

Bethell's argument has a curious ring for most practicing scientists. We are always ready to watch a theory fall under the impact of new data, but we do not expect a great and influential theory to collapse from a logical error in its formulation. Virtually every empirical scientist has a touch of the Philistine. Scientists tend to ignore academic

From *Natural History*, Vol. 85, no. 8. By permission of the author.

philosophy as an empty pursuit. Surely, any intelligent person can think straight by intuition. Yet Bethell cites no data at all in sealing the coffin of natural selection, only an error in Darwin's reasoning: "Darwin made a mistake sufficiently serious to undermine his theory. And that mistake has only recently been recognized as such. . . . At one point in his argument, Darwin was misled."

Although I will try to refute Bethell, I also deplore the unwillingness of scientists to explore seriously the logical structure of arguments. Much of what passes for evolutionary theory is as vacuous as Bethell claims. Many great theories are held together by chains of dubious metaphor and analogy. Bethell has correctly identified the hogwash surrounding evolutionary theory. But we differ in one fundamental way: for Bethell, Darwinian theory is rotten to the core; I find a pearl of great price at the center.

Natural selection is the central concept of Darwinian theory —the fittest survive and spread their favored traits through populations. Natural selection is defined by Spencer's phrase "survival of the fittest," but what does this famous bit of jargon really mean? Who are the fittest? And how is "fitness" defined? We often read that fitness involves no more than "differential reproductive success"—the production of more surviving offspring than other competing members of the population. Whoa! cries Bethell, as many others have before him. This formulation defines fitness in terms of survival only. The crucial phrase of natural selection means no more than "the survival of those who survive"—a vacuous tautology. (A tautology is a phrase—like "my father is a man"—containing no information in the predicate ("a man") not inherent in the subject ("my father"). Tautologies are fine as definitions, but not as testable scientific statements—there can be nothing to test in a statement true by definition.)

But how could Darwin have made such a monumental, two-bit mistake? Even his severest critics have never accused him of crass stupidity. Obviously, Darwin must have tried to define fitness differently— to find a criterion for fitness independent of mere survival. Darwin did propose an independent criterion, but Bethell argues quite correctly that he relied upon analogy to establish it, a dangerous and slippery strategy. One might think that the first chapter of such a revolutionary book as *Origin of Species* would deal with cosmic questions and general concerns. It doesn't. It's about pigeons. Darwin devotes most of his first forty pages to "artificial selection" of favored traits by animal breeders. For here an independent criterion surely operates. The pigeon fancier knows what he wants. The fittest are not defined by their survival. They are, rather, allowed to survive because they possess desired traits.

The principle of natural selection depends upon the validity of an analogy with artificial selection. We must be able, like the pigeon fancier, to identify the fittest beforehand, not only by their subsequent survival. But nature is not an animal breeder; no preordained purpose regulates the history of life. In nature, any traits possessed by survivors must be counted as "more evolved"; in artificial selection, "superior" traits are defined before breeding even begins. Later evolutionists, Bethell argues, recognized the failure of Darwin's analogy and redefined "fitness" as mere survival. But they did not realize that they had undermined the logical structure of Darwin's central postulate. Nature provides no independent criterion of fitness; thus, natural selection is tautological.

Bethel then moves to two important corollaries of his major argument. First, if fitness only means survival, then how can natural selection be a "creative" force, as Darwinians insist. Natural selection can only tell us how "a given type of animal became more numerous"; it cannot explain "how one type of animal gradually changed into another." Secondly, why were Darwin and other eminent Victorians so sure that mindless nature could be compared with conscious selection by breeders. Bethell argues that the cultural climate of triumphant industrial capitalism had defined any change as inherently progressive. Mere survival in nature could only be for the good: "It is beginning to look as though what Darwin really discovered was nothing more than the Victorian propensity to believe in progress."

I believe that Darwin was right and that Bethell and his colleagues are mistaken: criteria of fitness independent of survival can be applied to nature and have been used consistently by evolutionists. But let me first admit that Bethell's criticism applies to much of the technical literature in evolutionary theory, especially to the abstract mathematical treatments that consider evolution only as an alteration in numbers, not as a change in quality. These studies do assess fitness only in terms of differential survival. What else can be done with abstract models that trace the relative successes of hypothetical genes A and B in populations that exist only on computer tape? Nature, however, is not limited by the calculations of theoretical geneticists. In nature, A's "superiority" over B will be *expressed* as differential survival, but it is not *defined* by it — or, at least, it better not be so defined, lest Bethell et al. triumph and Darwin surrender.

My defense of Darwin is neither startling, novel, nor profound. I merely assert that Darwin was justified in analogizing natural selection with animal breeding. In artificial selection, a breeder's desire represents a "change of environment" for a population. In this new environment, certain traits are superior a priori; (they survive and spread by our breeder's choice, but this is a *result* of their fitness, not a definition of it). In nature, Darwinian evolution is also a response to changing environments. Now, the key point: certain morphological, physiological, and behavioral traits should be superior a priori as designs for living in new environments. These traits confer fitness by an engineer's criterion of good design, not by the empirical fact of their survival and spread. It got colder before the woolly mammoth evolved its shaggy coat.

Why does this issue agitate evolutionists so much? OK, Darwin was right: superior design in changed environments is an independent criterion of fitness. So what? Did anyone ever seriously propose that the poorly designed shall triumph? Yes, in fact, many did. In Darwin's day, many rival evolutionary theories asserted that the fittest (best designed) must perish. One popular notion—the theory of racial life cycles—was championed by a former inhabitant of the office I now occupy, the great American paleontologist Alpheus Hyatt. Hyatt claimed that evolutionary lineages, like individuals, had cycles of youth, maturity, old age, and death (extinction). Decline and extinction are programmed into the history of species. As maturity yields to old age, the best-designed individuals die and the hobbled, deformed creatures of phyletic senility take over. Another anti-Darwinian notion, the theory of orthogenesis, held that certain trends, once initiated, could not be halted, even though they must lead to extinction caused by increasingly inferior design. Many nineteenth-century evolutionists (perhaps a majority) held that Irish elks became extinct because they could not halt their evolutionary increase in antler size; thus, they died—caught in trees or bowed (literally) in the mire. Likewise, the demise of saber-toothed "tigers" was often attributed to canine teeth grown so long that the poor cats couldn't open their jaws wide enough to use them.

Thus, it is not true, as Bethell claims, that any traits possessed by survivors must be designated as fitter. "Survival of the fittest" is not a tautology. It is also not the only imaginable or reasonable reading of the evolutionary record. It is testable. It had rivals that failed under the weight of contrary evidence and changing attitudes about the nature of life. It has rivals that may succeed, at least in limiting its scope.

If I am right, how can Bethell claim, "Darwin, I suggest, is in the process of being discarded, but perhaps in deference to the venerable old gentleman, resting comfortably in Westminster Abbey next to Sir Isaac Newton, it is being done as discreetly and gently as possible with a minimum of publicity." I'm afraid I must say that Bethell has not been quite fair in his report of prevailing opinion. He cites the gadflies C. H. Waddington and H. J. Muller as though they epitomized a consensus. He never mentions the leading selectionists of our present generation —E. O. Wilson or D. Janzen, for example. And he quotes the architects of neo-Darwinism—Dobzhansky, Simpson, Mayr, and J. Huxley— only to ridicule their metaphors on the "creativity" of natural selection. (I am not claiming that Darwinism should be cherished because it is still popular; I am enough of a gadfly to believe that uncriticized consensus is a sure sign of impending trouble. I merely report that, for better or for worse, Darwinism is alive and thriving, despite Bethell's obituary.)

But why was natural selection compared to a composer by Dobzhansky; to a poet by Simpson; to a sculptor by Mayr; and to, of all people, Mr. Shakespeare by Julian Huxley? I won't defend the choice of metaphors, but I will uphold the intent, namely, to illustrate the essence of Darwinism — the creativity of natural selection. Natural selection has a place in all anti-Darwinian theories that I know. It is cast in a negative role as an executioner, a headsman for the unfit (while the fit arise by such non-Darwinian mechanisms as the inheritance of acquired characters or direct induction of favorable variation by the environment). The essence of Darwinism lies in its claim that natural selection creates the fit. Variation is ubiquitous and random in direction. It supplies the raw material only. Natural selection directs the course of evolutionary change. It preserves favorable variants and builds fitness gradually. In fact, since artists fashion their creations from the raw material of notes, words, and stone, the metaphors do not strike me as inappropriate. Since Bethell does not accept a criterion of fitness independent of mere survival, he can hardly grant a creative role to natural selection.

According to Bethell, Darwin's concept of natural selection as a creative force can be no more than an illusion encouraged by the social and political climate of his times. In the throes of Victorian optimism in imperial Britain, change seemed to be inherently progressive; why not equate survival in nature with increasing fitness in the nontautological sense of improved design.

I am a strong advocate of the general argument that "truth" as preached by scientists often turns out to be no more than prejudice inspired by prevailing social and political beliefs. I have devoted several essays to this theme because I believe that it helps to "demystify" the practice of science by showing its similarity to all creative human activity. But the truth of a general argument does not validate any specific application, and I maintain that Bethell's application is badly misinformed.

Darwin did two very separate things: he convinced the scientific world that evolution had occurred and he proposed the theory of natural selection as its mechanism. I am quite willing to admit that the common equation of evolution with progress made Darwin's first claim more palatable to his contemporaries. But Darwin failed in his second quest during his own lifetime. The theory of natural selection did not triumph until the 1940s. Its Victorian unpopularity, in my view, lay primarily in its denial of general progress as inherent in the workings of evolution. Natural selection is a theory of *local* adaptation to changing environments. It proposes no perfecting principles, no guarantee of general improvement; in short, no reason for general approbation in a political climate favoring innate progress in nature.

Darwin's independent criterion of fitness is, indeed, "improved design," but not "improved" in the cosmic sense that contemporary Britain favored. To Darwin, improved meant only "better designed for an immediate, local environment." Local environments change constantly: they get colder or hotter, wetter or drier, more grassy or more forested. Evolution by natural selection is no more than a tracking of these changing environments by differential preservation of organisms better designed to live in them: hair on a mammoth is not progressive in any cosmic sense. Natural selection can produce a trend that tempts us to think of more general progress — increase in brain size does characterize the evolution of group after group of mammals. But big brains have their uses in local environments; they do not mark intrinsic trends to higher states. And Darwin delighted in showing that local adaptation often produced "degeneration" in design — anatomical simplification in parasites, for example.

If natural selection is not a doctrine of progress, then its popularity cannot reflect the politics that Bethell invokes. If the theory of natural selection contains an independent criterion of fitness, then it is not tautological. I maintain, perhaps naïvely, that its current, unabated popularity must have something to do with its success in explaining the admittedly imperfect information we now possess about evolution. I rather suspect that we'll have Charles Darwin to kick around for some time.

THE CHALLENGE OF
PUNCTUATED EQUILIBRIUM

10

STEPHEN JAY GOULD*

Darwinism and the Expansion of Evolutionary Theory

BEN SIRA, AUTHOR of the apocryphal book of Ecclesiasticus, paid homage to the heroes of Israel in a noted passage beginning, "let us now praise famous men." He glorified great teachers above all others, for their fame shall eclipse the immediate triumphs of kings and conquerors. And he argued that the corporeal death of teachers counts for nothing —indeed, it should be celebrated—since great ideas must live forever: "His name will be more glorious than a thousand others, and if he dies, that will satisfy him just as well." These sentiments express the compulsion we feel to commemorate the deaths of great thinkers; for their ideas still direct us today. Charles Darwin died 100 years ago, on 19 April 1882, but his name still causes fundamentalists to shudder and scientists to draw battle lines amidst their accolades.

What Is Darwinism?

Darwin often stated that his biological work had embodied two different goals (1): to establish the fact of evolution, and to propose natural selection as its primary mechanism. "I had," he wrote, "two distinct objects in view; firstly to show that species had not been separately created, and secondly, that natural selection had been the chief agent of change" (2).

*Reprinted with permission from Stephen Jay Gould, "Darwinism and the Expansion of Evolutionary Theory," *Science*, Vol. 216, no. 23 (April 1982), pp. 380–87. Copyright © 1982 American Association for the Advancement of Science.

Although "Darwinism" has often been equated with evolution itself in popular literature, the term should be restricted to the body of thought allied with Darwin's own theory of mechanism, his second goal. This decision does not provide an unambiguous definition, if only because Darwin himself was a pluralist who granted pride of place to natural selection, but also advocated an important role for Lamarckian and other nonselectionist factors. Thus, as the 19th century drew to a close, G. J. Romanes and A. Weismann squared off in a terminological battle for rights to the name "Darwinian"—Romanes claiming it for his eclectic pluralism, Weismann for his strict selectionism (3).

If we agree, as our century generally has, that "Darwinism" should be restricted to the world view encompassed by the theory of natural selection itself, the problem of definition is still not easily resolved. Darwinism must be more than the bare bones of the mechanics: the principles of superfecundity and inherited variation, and the deduction of natural selection therefrom. It must, fundamentally, make a claim for wide scope and dominant frequency; natural selection must represent the primary directing force of evolutionary change.

I believe that Darwinism, under these guidelines, can best be defined as embodying two central claims and a variety of peripheral and supporting statements more or less strongly tied to the central postulates; Darwinism is not a mathematical formula or a set of statements, deductively arranged.

1) The creativity of natural selection. Darwinians cannot simply claim that natural selection operates since everyone, including Paley and the natural theologians, advocated selection as a device for removing unfit individuals at both extremes and preserving, intact and forever, the created type (4). The essence of Darwinism lies in a claim that natural selection is the primary directing force of evolution, in that it creates fitter phenotypes by differentially preserving, generation by generation, the best adapted organisms from a pool of random variants (5) that supply raw material only, not direction itself. Natural selection is a creator; it builds adaptation step by step.

Darwin's contemporaries understood that natural selection hinged on the argument for creativity. Natural selection can only eliminate the unfit, his opponents proclaimed; something else must create the fit. Thus, the American Neo-Lamarckian E. D. Cope wrote a book with the sardonic title *The Origin of the Fittest* (6), and Charles Lyell complained to Darwin that he could understand how selection might operate like two members of the "Hindoo triad"—Vishnu the preserver and Siva the destroyer—but not like Brahma the creator (7).

The claim for creativity has important consequences and prerequisites that also become part of the Darwinian corpus. Most prominently, three constraints are imposed on the nature of genetic variation (or at least the evolutionarily significant portion of it). (i) It must be copious since selection makes nothing directly and requires a large pool of raw material. (ii) It must be small in scope. If new species characteristically arise all at once, then the fit are formed by the process of variation itself, and natural selection only plays the negative role of executioner for the unfit. True saltationist theories have always been considered anti-Darwinian on this basis. (iii) It must be undirected. If new environments can elicit heritable, adaptive variation, then creativity lies in the process of variation, and selection only eliminates the unfit. Lamarckism is an anti-Darwinian theory because it advocates directed variation; organisms perceive felt needs, adapt their bodies accordingly, and pass these modifications directly to offspring.

Two additional postulates, generally considered part and parcel of the Darwinian world view, are intimately related to the claim for creativity, but are not absolute prerequisites or necessary deductive consequences: (i) *Gradualism*. If creativity resides in a step-by-step process of selection from a pool of random variants, then evolutionary change must be dominantly continuous and descendants must be linked to ancestors by a long chain of smoothly intermediate phenotypes. Darwin's own gradualism precedes his belief in natural selection and has deeper roots (8); it dominated his world view and provided a central focus for most other theories that he proposed, including the origin of coral atolls by subsidence of central islands, and formation of vegetable mold by earthworms (9, 10). (ii) *The adaptationist program*. If selection becomes creative by superintending, generation by generation, the continuous incorporation of favorable variation into altered forms, then evolutionary change must be fundamentally adaptive. If evolution were saltational, or driven by internally generated biases in the direction of variation, adaptation would not be a necessary attribute of evolutionary change.

The argument for creativity rests on relative frequency, not exclusivity. Other factors must regulate some cases of evolutionary change — randomness as a direct source of modification, not only of raw material, for example. The Darwinian strategy does not deny other factors, but attempts to circumscribe their domain to few and unimportant cases.

2) Selection operates through the differential reproductive success of individual organisms (the "struggle for existence" in Darwin's terminology). Selection is an interaction among individuals; there are no higher-order laws in nature, no statements about the "good" of species or ecosystems. If species survive longer, or if ecosystems appear to display harmony and balance, these features arise as a by-product of selection among individuals for reproductive success.

Although evolutionists, including many who call themselves Darwinians, have often muddled this point (*11*), it is a central feature of Darwin's logic (*12*). It underlies all his colorful visual imagery including the metaphor of the wedge (*13*, p. 67), or the true struggle that underlies an appearance of harmony: "we behold the face of nature bright with gladness," but . . . (*13*, p. 62). Darwin developed his theory of natural selection by transferring the basic argument of Adam Smith's economics into nature (14): an ordered economy can best be achieved by letting individuals struggle for personal profits, thereby permitting a natural sifting of the most competitive (laissez-faire); an ordered ecology is a transient balance established by successful competitors pursuing their own Darwinian edge.

As a primary consequence, this focus upon individual organisms leads to reductionism, not to ultimate atoms and molecules of course, but of higher-order, or macroevolutionary, processes to the accumulated struggles of individuals. Extrapolationism is the other side of the same coin — the claim that natural selection within local populations is the source of all important evolutionary change.

Darwinism and the Modern Synthesis

Although Darwin succeeded in his first goal, and lies in Westminster Abbey for his success in establishing the fact of evolution, his theory of natural selection did not triumph as an orthodoxy until long after his death. The Mendelian component to the modern, or Neo-Darwinian, theory only developed in our century. Moreover, and ironically, the first Mendelians emphasized macromutations and were non-Darwinians on the issue of creativity as discussed above.

The Darwinian resurgence began in earnest in the 1930's, but did not crystallize until the 1950's. At the last Darwinian centennial, in 1959 (both the 100th anniversary of the *Origin of Species* and the 150th of Darwin's birth), celebrations throughout the world lauded the "modern synthesis" as Darwinism finally triumphant (15).

Julian Huxley, who coined the term (16), defined the "modern synthesis" as an integration of the disparate parts of biology about a Darwinian core (17). Synthesis occurred at two levels: (i) The Mendelian research program merged with Darwinian traditions of natural history, as Mendelians recognized the importance of micromutations and their correspondence with Darwinian variation, and as population genetics supplied a quantitative mechanics for evolutionary change. (ii) The traditional disciplines of natural history, systematics, paleontology, morphology, and classical botany, for example (18), were integrated within the Darwinian core, or at least rendered consistent with it.

The initial works of the synthesis, particularly Dobzhansky's first (1937) edition of *Genetics and the Origin of Species*, were not firmly Darwinian (as defined above), and did not assert a dominant frequency for natural selection. They were more concerned with demonstrating that large-scale phenomena of evolution are consistent with the principles of genetics, whether Darwinian or not; and they therefore, for example, granted greater prominence to genetic drift than later editions of the same works would allow.

Throughout the late 1940's and 1950's, however, the synthesis hardened about its Darwinian core. Analysis of textbooks and, particularly, the comparison of first with later editions of the founding documents, demonstrates the emergence of natural selection and adaptation as preeminent factors of evolution. Thus, for example, G. G. Simpson redefined "quantum evolution" in 1953 as a limiting rate for adaptive phyletic transformation, not, as he had in 1944, as a higher-order analog of genetic drift, with a truly inadaptive phase between stabilized end points (19). Dobzhansky removed chapters and reduced emphasis upon rapid modification and random components to evolutionary change (20). David Lack reassessed his work on Darwin's finches and decided that minor differences among species are adaptive after all (21). His preface to the 1960 reissue of his monograph features the following statement (22):

> This text was completed in 1944 and . . . views on species-formation have advanced. In particular, it was generally believed when I wrote the book that, in animals, nearly all of the differences between subspecies of the same species, and between closely related species in the same genus, were without adaptive significance. . . . Sixteen years later, it is generally believed that all, or almost all, subspecific and specific differences are adaptive. . . . Hence it now seems probable that at least most of the seemingly nonadaptive differences in Darwin's finches would, if more were known, prove to be adaptive.

Mayr's definition of the synthesis, offered without rebuttal at a conference of historians and architects of the theory, reflects this crystallized version:

> The term "evolutionary synthesis" was introduced by Julian Huxley . . . to designate the general acceptance of two conclusions: gradual evolution can be explained in terms of small genetic changes ("mutations") and recombination, and the ordering of this genetic variation by natural selection; and the observed evolutionary phenomena, particularly macroevolutionary processes and speciation, can be explained in a manner that is consistent with the known genetic mechanisms (23).

This definition restates the two central claims of Darwinism discussed in the last section: Mayr's first conclusion, with its emphasis on gradualism, small genetic change, and natural selection, represents the argument for creativity; while the second embodies the claim for reduction. I have been challenged for erecting a straw man in citing this definition of the synthesis (24), but it was framed by a man who is both an architect and the leading historian of the theory, and it is surely an accurate statement of what I was taught as a graduate student in the mid-1960's. Moreover, these very words have been identified as the "broad version" of the synthesis (as opposed to a more partisan and restrictive stance) by White (25), a leading evolutionist and scholar who lived through it all.

The modern synthesis has sometimes been so broadly construed, usually by defenders who wish to see it as fully adequate to meet and encompass current critiques, that it loses all meaning by including everything. If, as Stebbins and Ayala claim, " 'selectionist' and 'neutralist' views of molecular evolution are competing hypotheses within the framework of the synthetic theory of evolution" (26), then what serious views are excluded? King and Jukes, authors of the neutralist theory, named it "non-Darwinian evolution" in the title of their famous paper (27). Stebbins and Ayala have tried to win an argument by redefinition. The essence of the modern synthesis must be its Darwinian core. If most evolutionary change is neutral, the synthesis is severely compromised.

What Is Happening to Darwinism

Current critics of Darwinism and the modern synthesis are proposing a good deal more than a comfortable extension of the theory, but much less than a revolution. In my partisan view, neither of Darwinism's two central themes will survive in their strict formulation; in that sense, "the modern synthesis, as an exclusive proposition, has broken down on both of its fundamental claims" (28). However, I believe that a restructured evolutionary theory will embody the essence of the Darwinian argument in a more abstract, and hierarchically extended form. The modern synthesis is incomplete, not incorrect.

Critique of Creativity: Gradualism

At issue is not the general idea that natural selection can act as a creative force; the basic argument, in principle, is a sound one. Primary doubts center on the subsidiary claims — gradualism and the adaptationist program. If most evolutionary changes, particularly large-scale trends, include major nonadaptive components as primary directing or channeling features, and if they proceed more in an episodic than a smoothly continuous fashion, then we inhabit a different world from the one Darwin envisaged.

Critiques of gradualist thought proceed on different levels and have different import, but none are fundamentally opposed to natural selection. They are therefore not directed against the heart of Darwinian theory, but against a fundamental subsidiary aspect of Darwin's own world view — one that he consistently conflated with natural selection, as in the following famous passage: "If it could be demonstrated that any complex organ existed, which could not possibly have been formed by numerous, successive, slight modifications, my theory would absolutely break down" (29).

At the levels of microevolution and speciation, the extreme saltationist claim that new species arise all at once, fully formed, by a fortunate macromutation would be anti-Darwinian, but no serious thinker now advances such a view, and neither did Richard Goldschmidt (30), the last major scholar to whom such an opinion is often attributed. Legitimate claims range from the saltational origin of key features by developmental shifts of dissociable segments of ontogeny (*31*) to the origin of reproductive isolation (speciation) by major and rapidly incorporated genetic changes that precede the acquisition of adaptive, phenotypic differences (*32*).

Are such styles of evolution anti-Darwinian? What can one say except "yes and no." They do not deny a creative role to natural selection, but neither do they embody the constant superintending of each event, or the step-by-step construction of each major feature, that traditional views about natural selection have advocated. If new *Baupläne* often arise in an adaptive cascade following the saltational origin of a key feature, then part of the process is sequential and adaptive, and therefore Darwinian; but the initial step is not, since selection does not play a creative role in building the key feature. If reproductive isolation often precedes adaptation, then a major aspect of speciation is Darwinian (for the new species will not prosper unless it builds distinctive adaptations in the sequential mode), but its initiation, including the defining feature of reproductive isolation, is not.

At the macroevolutionary level of trends, the theory of punctuated equilibrium (*33*) proposes that established species generally do not change substantially in phenotype over a lifetime that may encompass many million years (stasis), and that most evolutionary change is concentrated in geologically instantaneous events of branching speciation. These geological instants, resolvable (*34*) in favorable stratigraphic circumstances (so that the theory can be tested for its proposed punctuations as well as for its evident periods of stasis), represent amounts of microevolutionary time fully consistent with orthodox views about speciation. Indeed, Eldredge and I originally proposed punctuated equilibrium as the expected geological consequence of Mayr's theory of peripatric speciation. The non-Darwinian implications of punctuated equilibrium lie in its suggestions for the explanation of evolutionary trends (see below), not in the tempo of individual speciation events. Although punctuated equilibrium is a theory for a higher level of evolutionary change, and must therefore be agnostic with respect to the role of natural selection in speciation, the world that it proposes is quite different from that traditionally viewed by paleontologists (and by Darwin himself) as the proper geological extension of Darwinism.

The "gradualist-punctuationalist debate," the general label often applied to this disparate series of claims, may not be directed at the heart of natural selection, but it remains an important critique of the Darwinian tradition. The world is not inhabited exclusively by fools, and when a subject arouses intense interest and debate, as this one has, something other than semantics is usually at stake. In the largest sense, this debate is but one small aspect of a broader discussion about the nature of change: Is our world (to construct a ridiculously oversimplified dichotomy) primarily one of constant change (with structure as a mere incarnation of the moment), or is structure primary and constraining, with change as a "difficult" phenomenon, usually accomplished rapidly when a stable structure is stressed beyond its buffering capacity to resist and absorb. It would be hard to deny that the Darwinian tradition, including the modern synthesis, favored the first view while "punctuationalist" thought in general, including such aspects of classical morphology as D'Arcy Thompson's theory of form (35), prefers the second.

Critique of Creativity: Adaptation

The primary critiques of adaptation have arisen from molecular data, particularly from the approximately even ticking of the molecular clock, and the argument that natural populations generally maintain too much genetic variation to explain by natural selection, even when selection acts to preserve variation as in, for example, heterozygote advantage and frequency-dependent selection. To these phenomena, Darwinians have a response that is, in one sense, fully justified: Neutral genetic changes without phenotypic consequences are invisible to Darwinian processes of selection upon organisms and therefore represent a legitimate process separate from the subjects that Darwinism can treat. Still, since issues in natural history are generally resolved by appeals to relative frequency, the domain of Darwinism is restricted by these arguments.

But another general critique of the adaptationist program has been reasserted within the Darwinian domain of phenotypes (36). The theme is an old one, and not unfamiliar to Darwinians. Darwin himself took it seriously, as did the early, pluralistic accounts of the modern synthesis. The later, "hard" version of the synthesis relegated it to unimportance or lip service. The theme is two-pronged, both arguments asserting that the current utility of a structure permits no assumption that selection shaped it. First, the constraints of inherited form and developmental pathways may so channel any change that even though selection induces motion down permitted paths, the channel itself represents the primary determinant of evolutionary direction. Second, current utility permits no necessary conclusion about historical origin. Structures now indispensable for survival may have arisen for other reasons and been "coopted" by functional shift for their new role.

Both arguments have their Darwinian versions. First, if the channels are set by past adaptations, then selection remains preeminent, for all major structures are either expressions of immediate selection, or channeled by a phylogenetic heritage of previous selection. Darwin struggled mightily with this problem. Ultimately, in a neglected passage that I regard as one of the most crucial paragraphs in the *Origin of Species* (37), he resolved his doubts, and used this argument to uphold the great British tradition of adaptationism. Second, if coopted structures initially arose as adaptations for another function, then they too are products of selection, albeit in a regime not recorded by their current usage. We call this phenomenon preadaptation; as the primary solution to Mivart's taunt (38) about "the incipient stages of useful structures," it is a central theme of orthodox Darwinism.

But both arguments also have non-Darwinian versions, not widely appreciated but potentially fundamental. First, many features of organic architecture and developmental pathways have never been adaptations to anything, but arose as by-products or incidental consequences of changes with a basis in selection. Seilacher has suggested, for example, that the divaricate pattern of molluscan ornamentation may be nonadaptive in its essential design. In any case, it is certainly a channel for some fascinating subsidiary adaptations (39). Second, many structures available for cooptation did not arise as adaptations for something else (as the principle of preadaptation assumes) but were nonadaptive in their original construction. Evolutionary morphology now lacks a term for these coopted structures, and unnamed phenomena are not easily conceptualized. Vrba and I suggest that they be called exaptations (40), and present a range of potential examples from the genitalia of hyenas to redundant DNA.

Evolutionists admit, of course, that all selection yields by-products and incidental consequences, but we tend to think of these nonadaptations as a sort of evolutionary frill, a set of small and incidental modifications with no major consequences. I dispute this assessment and claim that the pool of nonadaptations must be far greater in extent than the direct adaptations that engender them. This pool must act as a higher-level analog of genetic variation, as a phenotypic source of raw material for further evolution. Nonadaptations are not just incidental allometric and pleiotropic effects on other parts of the body, but multifarious expressions potentially within any adapted structure. No one doubts, for example, that the human brain became large for a set of complex reasons related to selection. But, having reached its unprecedented bulk, it could, as a computer of some sophistication, perform in an unimagined range of ways bearing no relation to the selective reasons for initial enlargement. Most of human society may rest on these nonadaptive consequences. How many human institutions, for example, owe their shape to that most terrible datum that intelligence permitted us to grasp — the fact of our personal mortality.

I do not claim that a new force of evolutionary change has been discovered. Selection may supply all immediate direction, but if highly constraining channels are built of nonadaptations, and if evolutionary versatility resides primarily in the nature and extent of nonadaptive pools, then "internal" factors of organic design are an equal partner with selection. We say that mutation is the ultimate source of variation, yet we grant a fundamental role to recombination and the evolution of sexuality — often as a prerequisite to multicellularity, the Cambrian explosion and, ultimately, us. Likewise, selection may be the ultimate source of evolutionary change, but most actual events may owe more of their shape to its nonadaptive sequelae.

Is Evolution a Product of Selection among Individuals?

Although arguments for a multiplicity of units of selection have been advanced and widely discussed (41), evolutionists have generally held fast to the overwhelming predominance, if not exclusivity, of organisms as the objects sorted by selection—Dawkins' (42) attempt at further reduction to the gene itself notwithstanding. How else can we explain the vehement reaction of many evolutionists to Wynne-Edwards' theory of group selection for the maintenance of altruistic traits (43), or the delight felt by so many when the same phenomena were explained, under the theory of kin selection, as a result of individuals pursuing their traditional Darwinian edge. I am not a supporter of Wynne-Edwards' particular hypothesis, nor do I doubt the validity and importance of kin selection; I merely point out that the vehemence and delight convey deeper messages about general attitudes.

Nonetheless, I believe that the traditional Darwinian focus on individual bodies, and the attendant reductionist account of macroevolution, will be supplanted by a hierarchical approach recognizing legitimate Darwinian individuals at several levels of a structural hierarchy, including genes, bodies, demes, species, and clades.

The argument may begin with a claim that first appears to be merely semantic, yet contains great utility and richness in implication, namely the conclusion advanced by Ghiselin and later supported by Hull that species should be treated as individuals, not as classes (44). Most species function as entities in nature, with coherence and stability. And they display the primary characteristics of a Darwinian actor; they vary within their population (clade in this case), and they exhibit differential rates of birth (speciation) and death (extinction).

Our language and culture include a prejudice for applying the concept of individual only to bodies, but any coherent entity that has a unique origin, sufficient temporal stability, and a capacity for reproduction with change can serve as an evolutionary agent. The actual hierarchy of our world is a contingent fact of history, not a heuristic device or a logical necessity. One can easily imagine a world devoid of such hierarchy, and conferring the status of evolutionary individual upon bodies alone. If genes could not duplicate themselves and disperse among chromosomes, we might lack the legitimately independent level that the "selfish DNA" hypothesis establishes for some genes (45). If new species usually arose by the smooth transformation of an entire ancestral species, and then changed continuously toward a descendant form, they would lack the stability and coherence required for defining evolutionary individuals. The theory of punctuated equilibrium allows us to individuate species in both time and space; this property (rather than the debate about evolutionary tempo) may emerge as its primary contribution to evolutionary theory.

In itself, individuation does not guarantee the strong claim for evolutionary agency: that the higher-level individual acts as a unit of selection in its own right. Species might be individuals, but their differential evolutionary success might still arise entirely from natural selection acting upon their parts, that is, upon phenotypes of organisms. A trend toward increasing brain size, for example, might result from the greater longevity of big-brained species. But big-brained species might prosper only because the organisms within them tend to prevail in traditional competition.

But individuation of higher-level units is enough to invalidate the reductionism of traditional Darwinism — for pattern and style of evolution depend critically on the disposition of higher-level individuals, even when all selection occurs at the traditional level of organisms. Sewall Wright, for example, has often spoken of "interdemic selection" in his shifting balance theory (46), but he apparently uses this phrase in a descriptive sense and believes that the mechanism of change usually resides in selection among individual organisms, as when, for example, migrants from one deme swamp another. Still, the fact of deme structure itself — that is, the individuation of higher-level units within a species — is crucial to the operation of shifting balance. Without division into demes, and under panmixia, genetic drift could not operate as the major source of variation required by the theory.

We need not, however, confine ourselves to the simple fact of individuation as an argument against Darwinian reductionism. For the strong claim that higher-level individuals act as units of selection in their own right can often be made. Many evolutionary trends, for example, are driven by differential frequency of speciation (the analog of birth) rather than by differential extinction (the more usual style of selection by death). Features that enhance the frequency of speciation are often properties of populations, not of individual organisms, for example, dependence of dispersal (and resultant possibilities for isolation and speciation) on size and density of populations.

Unfortunately, the terminology of this area is plagued with a central confusion (some, I regret to say, abetted by my own previous writings). Terms like "interdemic selection" or "species selection" (47) have been used in the purely descriptive sense, when the sorting out among higher-level individuals may arise solely from natural selection operating upon organisms. Such cases are explained by Darwinian selection, although they are irreducible to organisms alone. The same terms have been restricted to cases of higher-level individuals acting as units of selection. Such situations are non-Darwinian, and irreducible on this strong criterion. Since issues involving the locus of selection are so crucial in evolutionary theory, I suggest that these terms only be used in the strong and restricted sense. Species selection, for example, should connote an irreducibility to individual organisms (because populations are acting as units of selection); it should not merely offer a convenient alternative description for the effects of traditional selection upon organisms.

The logic of species selection is sound, and few evolutionists would now doubt that it can occur in principle. The issue, again and as always in natural history, is one of relative frequency; how often does species selection occur, and how important is it in the panoply of evolutionary events. Fisher himself dismissed species selection because, relative to organisms, species are so few in number (within a clade) and so long in duration (*48*):

> The relative unimportance of this as an evolutionary factor would seem to follow decisively from the small number of closely related species which in fact do come into competition, as compared to the number of individuals in the same species; and from the vastly greater duration of the species compared to the individual.

But Fisher's argument rests on two hidden and questionable assumptions. (i) Mass selection can almost always be effective in transforming entire populations substantially in phenotype. The sheer number of organisms participating in this efficient process would then swamp any effect of selection among species. But if stasis be prevalent within established species, as the theory of punctuated equilibrium asserts and as paleontological experience affirms (overwhelmingly for marine invertebrates, at least), then the mere existence of billions of individuals and millions of generations guarantees no substantial role for directional selection upon organisms. (ii) Species selection depends on direct competition among species. Fisher argues for differential death (extinction) as the mechanism of species selection. I suspect, however, that differential frequency of speciation (selection by birth) is a far more common and effective mode of species selection. It may occur without direct competition between species, and can rapidly shift the average phenotype within a clade in regimes of random extinction.

J. Maynard Smith (*49*) has raised another objection against species selection: simply, that most features of organisms represent "things individual creatures do." How, he asks, could one attribute the secondary palate of mammals to species selection? But the origin of a feature is one thing (and I would not dispute traditional selection among organisms as the probable mechanism for evolving a secondary palate), and the spread of features through larger clades is another. Macroevolution is fundamentally about the combination of features and their differential spread. These phenomena lie comfortably within the domain of effective species selection. Many features must come to prominence primarily through their fortuitous phyletic link with high speciation rates. Mammals represent a lineage of therapsids that may have survived (while all others died) as a result of small body sizes and nocturnal habits. Was the secondary palate a key to their success, or did it piggyback on the high speciation rates often noted (for other reasons) in small-bodied forms. Did mammals survive the Cretaceous extinction, thereby inheriting the world from dinosaurs, as a result of their secondary palate, or did their small size again preserve them during an event that differentially wiped out large creatures.

Evolutionary Pattern by Interaction between Levels

The hierarchical model, with its assertion that selection works simultaneously and differently upon individuals at a variety of levels, suggests a revised interpretation for many phenomena that have puzzled people where they implicitly assumed causation by selection upon organisms. In particular, it suggests that negative interaction between levels might be an important principle in maintaining stability or holding rates of change within reasonable bounds.

The "selfish DNA" hypothesis, for example, proposes that much middle-repetitive DNA exists within genomes not because it provides Darwinian benefits to phenotypes, but because genes can (in certain circumstances) act as units of selection. Genes that can duplicate themselves and move among chromosomes will therefore accumulate copies of themselves for their own Darwinian reasons. But why does the process ever stop? The authors of the hypothesis (45) suggest that phenotypes will eventually "notice" the redundant copies when the energetic cost of producing them becomes high enough to entail negative selection at the level of organisms. Stability may represent a balance between positive selection at the gene level and the negative selection it eventually elicits at the organism level.

All evolutionary textbooks grant a paragraph or two to a phenomenon called "overspecialization," usually dismissing it as a peculiar and peripheral phenomenon. It records the irony that many creatures, by evolving highly complex and ecologically constraining features for their immediate Darwinian advantage, virtually guarantee the short duration of their species by restricting its capacity for subsequent adaptation. Will a peacock or an Irish elk survive when the environment alters radically? Yet fancy tails and big antlers do lead to more copulations in the short run of a lifetime. Overspecialization is, I believe, a central evolutionary phenomenon that has failed to gain the attention it deserves because we have lacked a vocabulary to express what is really happening: the negative interaction of species-level disadvantage and individual-level advantage. How else can morphological specialization be kept within bounds, leaving a place for drab and persistent creatures of the world. The general phenomenon must also regulate much of human society, with many higher-level institutions compromised or destroyed by the legitimate demands of individuals (high salaries of baseball stars, perhaps).

Some features may be enhanced by positive interaction between levels. Stenotopy in marine invertebrates, for example, seems to offer advantages at both the individual level (when environments are stable) and at the species level (boosting rates of speciation by brooding larvae and enhancing possibilities for isolation relative to eurytopic species with planktonic larvae). Why then do eurytopic species still inhabit our oceans? Suppression probably occurs at the still higher level of clades, by the differential removal of stenotopic branches in major environmental upheavals that accompany frequent mass extinctions in the geological record.

If no negative effect from a higher level suppressed an advantageous lower-level phenomenon, then it might sweep through life. Sex in eukaryotic organisms may owe its prominence to unsuppressed positive interaction between levels. The advantages of sex have inspired a major debate among evolutionists during the past decade. Most authors seek traditional explanation in terms of benefit to organisms (50), for example, better chance for survival of some offspring if all are not Xeroxed copies of an asexual parent, but the genetically variable products of two individuals. Some, however, propose a spread by species selection, for example, by vastly higher speciation rates in sexual creatures (51).

The debate has often proceeded by mutual dismissal, each side proclaiming its own answers correct. Perhaps both are right, and sex predominates because two levels interact positively and are not suppressed at any higher level. No statement is usually more dull and unenlightening than the mediator's claim, "you're both right." In this case, however, we must adopt a different view of biological organization itself to grasp the mediator's wisdom — and the old solution, for once, becomes interesting in its larger implication. We live in a world with reductionist traditions, and do not react comfortably to notions of hierarchy. Hierarchical theories permit us to retain the value of traditional ideas, while adding substantially to them. They traffic in accretion, not substitution. If we abandoned the "either – or" mentality that has characterized arguments about units of selection, we would not only reduce fruitless and often acrimonious debate, but we would also gain a deeper understanding of nature's complexity through the concept of hierarchy.

A Higher Darwinism?

What would a fully elaborated, hierarchically based evolutionary theory be called? It would neither be Darwinism, as usually understood, nor a smoothly continuous extension of Darwinism, for it violates directly the fundamental reductionist tradition embodied in Darwin's focus on organisms as units of selection.

Still, the hierarchical model does propose that selection operates on appropriate individuals at each level. Should the term "natural selection" be extended to all levels above and below organisms; there is certainly nothing unnatural about species selection. Some authors have extended the term (48), while others, Slatkin for example (52), restrict natural selection to its usual focus upon individual organisms: "Species selection is analogous to natural selection acting on an asexual population" (52).

Terminological issues aside, the hierarchically based theory would not be Darwinism as traditionally conceived; it would be both a richer and a different theory. But it would embody, in abstract form, the essence of Darwin's argument expanded to work at each level. Each level generates variation among its individuals; evolution occurs at each level by a sorting out among individuals, with differential success of some and their progeny. The hierarchical theory would therefore represent a kind of "higher Darwinism," with the substance of a claim for reduction to organisms lost, but the domain of the abstract "selectionist" style of argument extended.

Moreover, selection will work differently on the objects of diverse levels. The phenomena of one level have analogs on others, but not identical operation. For example, we usually deny the effectiveness of mutation pressure at the level of organisms. Populations contain so many individuals that small biases in mutation rate can rarely establish a feature if it is under selection at all. But the analog of mutation pressure at the species level, directed speciation (directional bias toward certain phenotypes in derived species), may be a powerful agent of evolutionary trends (as a macroevolutionary alternative to species selection). Directed speciation can be effective (where mutation pressure is not) for two reasons: first, because its effects are not so easily swamped (given the restricted number of species within a clade) by differential extinction; second, because such phenomena as ontogenetic channeling in phyletic size increase suggest that biases in the production of species may be more prevalent than biases in the genesis of mutations.

Each level must be approached on its own, and appreciated for the special emphasis it places upon common phenomena, but the selectionist style of argument regulates all levels and the Darwinian vision is extended and generalized, not defeated, even though Darwinism, strictly constructed, may be superseded. This expansion may impose a literal wisdom upon that famous last line of *Origin of Species*, "There is grandeur in this view of life."

Darwin, at the centenary of his death, is more alive than ever. Let us continue to praise famous men.

REFERENCES AND NOTES

1. I have argued [*Nat. Hist.* 91, 16 (April 1982)] that a third and larger theme captures the profound importance and intellectual power of Darwin's work in a more comprehensive way: his successful attempt to establish principles of reasoning for historical science. Each of his so-called "minor" works (treatises on orchids, worms, climbing plants, coral reefs, barnacles, for example) exhibits both an explicit and a covert theme — and the covert

theme is a principle of reasoning for the reconstruction of history. The principles can be arranged in order of decreasing availability of information, but each addresses the fundamental issue: how can history be scientific if we cannot directly observe a past process: (i) If we can observe present processes at work, then we should accumulate and extrapolate their results to render the past. Darwin's last book, on the formation of vegetable mold by earthworms (1881), is also a treatise on this aspect of uniformitarianism. (ii) If rates are too slow or scales too broad for direct observation, then try to render the range of present results as stages of a single historical process. Darwin's first book on a specific subject, the subsidence theory of coral atolls (1842), is (in its covert theme) a disquisition on this principle. (iii) When single objects must be analyzed, search for imperfections that record constraints of inheritance. Darwin's orchid book (1862), explicitly about fertilization by insects, argues that orchids are jury-rigged, rather than well built from scratch, because structures that attract insects and stick pollen to them had to be built from ordinary parts of ancestral flowers. Darwin used all three principles to establish evolution as well: (i) observed rates of change in artificial selection, (ii) stages in the process of speciation displayed by modern populations, and (iii) analysis of vestigial structures in various organisms. Thus, we should not claim that all Darwin's books are about evolution. Rather, they are all about the methodology of historical science. The establishment of evolution represents the greatest triumph of the method.

2. C. Darwin, *The Descent of Man* (Murray, London, ed. 2, 1889), p. 61.
3. G. J. Romanes, *Darwin, and After Darwin* (Longmans, Green, London, 1900), pp. 1–36.
4. Failure to recognize that all creationists accepted selection in this negative role led Eiseley to conclude falsely that Darwin had "borrowed" the principle of natural selection from his predecessor E. Blyth [L. Eiseley, *Darwin and the Mysterious Mr. X* (Dutton, New York, 1979)]. The Reverend William Paley's classic work *Natural Theology*, published in 1803, also contains many references to selective elimination.
5. By "random" in this context, evolutionists mean only that variation is not inherently directed towards adaptation, not that all mutational changes are equally likely. The word is unfortunate, but the historical tradition too deep to avoid.
6. E. D. Cope, *The Origin of the Fittest* (Appleton, New York, 1887).
7. L. G. Wilson, Ed., *Sir Charles Lyell's Scientific Journals on the Species Question* (Yale Univ. Press, New Haven, Conn., 1970), p. 369.
8. Darwin was convinced, for example, in part by reading a theological work arguing that extreme rapidity (as in the initial spread of Christianity) indicated a divine hand, that gradual and continuous change was the mark of a natural process [H. Gruber, *Darwin on Man* (Dutton, New York, 1974)].
9. C. Darwin, *The Structure and Distribution of Coral Reefs* (Smith, Elder, London, 1842).
10. ——, *The Formation of Vegetable Mould, Through the Action of Worms* (Murray, London, 1881).
11. The following works have done great service in identifying and correcting this confusion: G. C. Williams, *Adaptation and Natural Selection* (Princeton Univ. Press, Princeton, N.J., 1966); J. Maynard Smith, *The Evolution of Sex* (Cambridge Univ. Press, New York, 1978).
12. A persuasive case for Darwin's active interest in this subject and for his commitment to individual selection has been recently made by M. Ruse, *Ann. Sci.* 37, 615 (1980).

13. C. Darwin, *On the Origin of Species* (Murray, London, 1859).
14. S. S. Schweber, *J. Hist. Biol.* 10, 229 (1977).
15. S. Tax, Ed., *Evolution After Darwin* (Univ. of Chicago Press, Chicago, 1960), vols. 1–3.
16. J. Huxley, *Evolution, the Modern Synthesis* (Allen & Unwin, London, 1942).
17. For example: "The opposing factions became reconciled as the younger branches of biology achieved a synthesis with each other and with the classical disciplines: and the reconciliation converged upon a Darwinian center" (*16*, p. 25).
18. E. Mayr, *Systematics and the Origin of Species* (Columbia Univ. Press, New York, 1942); G. G. Simpson, *Tempo and Mode in Evolution* (Columbia Univ. Press, New York, 1944); B. Rensch, *Neuere Probleme der Abstammungslehre* (Enke, Stuttgart, 1947); G. L. Stebbins, *Variation and Evolution in Plants* (Columbia Univ. Press, New York, 1950).
19. S. J. Gould, in *The Evolutionary Synthesis*, E. Mayr and W. B. Provine, Eds. (Harvard Univ. Press, Cambridge, Mass., 1980), p. 153.
20. S. J. Gould, *Dobzhansky and the Modern Synthesis*, introduction to reprint of first (1937) edition of Th. Dobzhansky, *Genetics and the Origin of Species* (Columbia Univ. Press, New York, 1982).
21. D. Lack, *Darwin's Finches* (Harper Torchbook Edition, New York, 1960).
22. This statement appears as the first paragraph in the preface to (*21*).
23. E. Mayr, in *The Evolutionary Synthesis*, E. Mayr and W. B. Provine, Eds. (Harvard Univ. Press, Cambridge, Mass., 1980), p. 1.
24. S. Orzack, *Paleobiology* 7, 128 (1981).
25. M. J. D. White, *ibid.*, p. 287.
26. G. L. Stebbins and F. J. Ayala, *Science* 213, 967 (1981).
27. J. L. King and T. H. Jukes, *ibid.* 164, 788 (1969).
28. S. J. Gould, *Paleobiology* 6, 119 (1980).
29. C. Darwin (*13*, p. 189). On the day before publication of the *Origin of Species*, T. H. Huxley wrote to Darwin (letter of 23 November 1859): "You load yourself with an unnecessary difficulty in adopting *Natura non facit saltum* so unreservedly."
30. S. J. Gould, *The Uses of Heresy*, introduction to the republication of the 1940 edition of R. Goldschmidt, *The Material Basis of Evolution* (Yale Univ. Press, New Haven, Conn., 1982).
31. P. Alberch, *Am. Zool.* 20, 653 (1980).
32. M. J. D. White, *Modes of Speciation* (Freeman, San Francisco, 1978); G. L. Bush, S. M. Case, A. C. Wilson, J. L. Patton, *Proc. Natl. Acad. Sci. U.S.A.* 74, 3942 (1977).
33. N. Eldredge and S. J. Gould, in *Models in Paleobiology*, T. J. M. Schopf, Ed. (Freeman, Cooper, San Francisco, 1972), p. 82; S. J. Gould and N. Eldredge, *Paleobiology* 3, 115 (1977).
34. P. Williamson, *Nature (London)* 293, 437 (1981).
35. D'Arcy W. Thompson, *On Growth and Form* (Cambridge Univ. Press, New York, 1942).
36. S. J. Gould and R. C. Lewontin, *Proc. R. Soc. London Ser. B* 205, 581 (1979); G. V. Lauder, *Paleobiology* 7, 430 (1981).
37. It is the concluding comment of chapter 6, and reads, in part: "It is generally acknowledged that all organic beings have been formed on two great laws — Unity of Type, and the Conditions of Existence. . . . Natural selection acts by either now adapting the varying parts of each being to its organic and inorganic conditions of life; or by having adapted them during long-past periods of time. . . . Hence, in fact, the law of the Conditions of Existence is the higher law; as it includes, through the inheritance of former adaptations, that of Unity of Type."

38. St. G. Mivart, *On the Genesis of Species* (Macmillan, London, 1871).
39. A. Seilacher, *Lethaia* 5, 325 (1972).
40. S. J. Gould and E. S. Vrba, *Paleobiology*, in press.
41. R. C. Lewontin, *Annu. Rev. Ecol. Syst.* 1, 1 (1970).
42. R. Dawkins, *The Selfish Gene* (Oxford Univ. Press, New York, 1976).
43. V. C. Wynne-Edwards, *Animal Dispersion in Relation to Social Behavior* (Oliver & Boyd, Edinburgh, 1962).
44. M. Ghiselin, *Syst. Zool.* 23, 536 (1974); D. L. Hull, *Annu. Rev. Ecol. Syst.* 11, 311 (1980).
45. W. F. Doolittle and C. Sapienza, *Nature (London)* 284, 601 (1980); L. E. Orgel and F. H. C. Crick, *ibid.*, p. 604.
46. S. Wright, *Evolution and the Genetics of Populations* (Univ. of Chicago Press, Chicago, 1968–1978), vols. 1–4.
47. S. M. Stanley, *Macroevolution* (Freeman, San Francisco, 1979); *Proc. Natl. Acad. Sci. U.S.A.* 72, 646 (1975); also references in (33).
48. R. A. Fisher, *The Genetical Theory of Natural Selection* (Dover, ed. 2, New York, 1958), p. 50.
49. J. Maynard Smith, personal communication.
50. G. C. Williams, *Sex and Evolution* (Monographs in Population Biology, No. 8, Princeton Univ. Press, Princeton, N.J., 1975).
51. S. M. Stanley, *Science* 190, 382 (1975).
52. M. Slatkin, *Paleobiology* 7, 421 (1981).
53. I thank Ernst Mayr, Philip Kitcher, Montgomery Slatkin, and Steven Stanley for their most helpful comments. Malcolm Kottler kindly pointed out to me the passage from David Lack quoted in (*21, 22*).

EDITORIAL NOTE

Gould makes reference to "saltationism." This is a view of the evolutionary process which supposes that change goes instantaneously from one form (say *fox*) to another form (say *dog*). Although critics have argued that Gould's theory of punctuated equilibrium is saltationary because he sees change as continuous (albeit rapid and spasmodic), he has always denied the charge.

He refers also to "genetic drift," which is the claim that in small populations accidents of sampling can lead to random, nonadaptive change. Most who accept drift also accept selection, although some biologists have argued that at the molecular level much change lies beneath selection and thus drifts. This is the so-called "neutral theory" of evolution.

11

FRANCISCO J. AYALA*

Beyond Darwinism? The Challenge of Macroevolution to the Synthetic Theory of Evolution

1. Evolution: An Unfinished Theory

THE CURRENT THEORY of biological evolution (the "Synthetic Theory" or "Modern Synthesis") may be traced to Theodosius Dobzhansky's *Genetics and the Origin of Species*, published in 1937: a synthesis of genetic knowledge and Darwin's theory of evolution by natural selection. The excitement provoked by Dobzhansky's book soon became reflected in many important contributions which incorporated into the Modern Synthesis relevant fields of biological knowledge. Notable landmarks are Ernst Mayr's *Systematics and the Origin of Species* (1942), Julian S. Huxley's *Evolution: The Modern Synthesis* (1942), George Gaylord Simpson's *Tempo and Mode in Evolution* (1944), and G. Ledyard Stebbins' *Variation and Evolution in Plants* (1950). It seemed to many scientists that the theory of evolution was essentially complete and that all that was left was to fill in the details. This perception is reflected, for example, by Jacques Monod, in his widely known book, *Chance and Necessity* (1970, p. 139): "the elementary mechanisms of evolution have been not only understood in principle but identified with precision . . . the problem has been resolved and evolution now lies well to this side of the frontier of knowledge."

There can be little doubt that some components of the Synthetic

*From Peter D. Asquith and Thomas Nickles, eds., *PSA 1982*, Vol. 2 (East Lansing, MI: Philosophy of Science Association, 1983), pp. 275–291. Copyright © 1983 by the Philosophy of Science Association.

Theory are well established. In a nutshell, the theory proposes that mutation and sexual recombination furnish the raw materials for change; that natural selection fashions from these materials genotypes and gene pools; and that, in sexually reproducing organisms, the arrays of adaptive genotypes are protected from disintegration by reproductive isolating mechanisms (speciation). But Monod's unbridled optimism is unwarranted. Indeed,

> the causes of evolution, and the patterning of the processes that bring it about, are far from completely understood. We cannot predict the future course of evolution except in a few well-studied situations, and even then only short-range predictions are possible. Nor can we, again with a few isolated exceptions, explain why past evolutionary events had to happen as they did. A predictive theory of evolution is a goal for the future. Hardly any competent biologist doubts that natural selection is an important directing and controlling agency in evolution. Yet one current issue hotly debated is whether a majority or only a small minority of evolutionary changes are induced by selection. (Dobzhansky, Ayala, Stebbins, and Valentine, 1977, pp. 129–130.)

Perhaps no better evidence can be produced of the unfinished status of the theory of evolution than pointing out some of the remarkable discoveries and theoretical developments of recent years. Molecular biology has been one major source of progress. One needs only mention the recently acquired ability to obtain quantitative measures of genetic variation in populations and of genetic differentiation during speciation and phyletic evolution; the discontinuous nature of the coding sequences of eukaryotic genes with its implications concerning the evolutionary origin of the genes themselves; the dynamism of DNA increases or decreases in amount and of changes in position of certain sequences; etc. But notable advances have occurred as well, and continue to take place, in evolutionary ecology, theoretical and experimental population genetics, and in other branches of evolutionary knowledge.

Any active field of systematic knowledge is likely to be beset by many unconfirmed hypotheses, unsettled issues, and controversies. Evolutionary theory is no exception. During the 1970's perhaps no other issue was more actively debated than the "neutrality" theory of molecular evolution—whether the evolution of informational macromolecules (nucleic acids and proteins) is largely governed by random changes in the frequency of adaptively equivalent variants, rather than by natural selection. The 1980's have started with another actively, and at times acrimoniously, contested problem: whether in the geological scale evolution is a more or less gradual process, or whether instead it is "punctuated." The model of punctuated equilibrium proposes that morphological evolution happens in bursts, with most phenotypic change occurring during speciation events, so that new species are morphologically quite distinct from their ancestors, but do not thereafter change substantially in phenotype over a lifetime that may encompass many millions of years. The punctuational model is contrasted with the gradualistic model, which sees morphological change as a more or less gradual process, not strongly associated with speciation events (Figure 1).

Two important issues are raised by the model of punctuated equilibrium. The first is scientific: whether morphological change as observed in the paleontological record is essentially always associated with speciation events, i.e., with the splitting of a lineage into two or more lineages. The second issue is epistemological: whether owing to the punctuated character of paleontological evolution, macroevolution is an autonomous field of study, independent from population genetics and other disciplines that study microevolutionary processes. I shall here examine the two issues; the second one at greater length than the first.

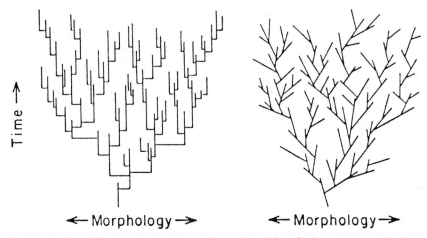

Figure 1. Simplified representation of two models of phenotypic evolution: punctuated equilibrium (left) and phyletic gradualism (right). According to the punctuated model, most morphological evolution in the history of life is associated with speciation events, which are geologically instantaneous. After their origin, established species generally do not change substantially in phenotype over a lifetime that may encompass many million years. According to the gradualist model, morphological evolution occurs during the lifetime of a species, with rapidly divergent speciation playing a lesser role. The figures are extreme versions of the models. Punctualism does not imply that phenotypic change never occurs between speciation events. Gradualism does not imply that phenotypic change is occurring continuously at a more or less constant rate throughout the life of a lineage, or that some acceleration does not take place during speciation, but rather that phenotypic change may occur at any time throughout the lifetime of a species.

2. Phenotypic Change and Speciation

I note, first, that whether phenotypic change in macroevolution occurs in bursts or is more or less gradual, is a question to be decided empirically. Examples of rapid phenotypic evolution followed by long periods of morphological stasis are known in the fossil record. But there are instances as well in which phenotypic evolution appears to occur gradually within a lineage. The question is the relative frequency of one or the other mode; and paleontologists disagree in their interpretation of the fossil record (Eldredge 1971, Eldredge and Gould 1972, Hallam 1978, Raup 1978, Stanley 1979, Gould 1980, and Vrba 1980, are among those who favor punctualism; whereas Kellogg 1975, Gingerich 1976, Levinton and Simon 1980, Schopf 1979 and 1981, Cronin, Boaz, Stringer and Rak 1981, and Douglas and Avise 1982, favor phyletic gradualism).

Whatever the paleontological record may show about the frequency of smooth, relative to jerky, evolutionary patterns, there is, however, one fundamental reservation that must be raised against the theory of punctuated equilibrium. This evolutionary model argues not only that most morphological change occurs in rapid bursts followed by long periods of phenotypic stability, but also that the bursts of change occur during the origin of new species. Stanley (1979, 1982), Gould (1982a, b), and other punctualists have made it clear that what is distinctive of the theory of punctuated equilibrium is this association between phenotypic change and speciation. One quotation should suffice: "Punctuated equilibrium is a specific claim about speciation and its deployment in geological time; *it should not be used as a synonym for any theory of rapid evolutionary change at any scale* . . . Punctuated equilibrium holds that accumulated speciation is the root of most major evolutionary change, and that what we have called anagenesis is usually no more than repeated cladogenesis (branching) filtered through the net of differential success at the species level." (Gould 1982a, pp. 84–85; italics added.)

Species are groups of interbreeding natural populations that are reproductively isolated from any other such groups (Mayr 1963, Dobzhansky *et al.* 1977). Speciation involves, by definition, the development of reproductive isolation between populations previously sharing in a common gene pool. But it is no way apparent how the fossil record could provide evidence of the development of reproductive isolation. Paleontologists recognize species by their different morphologies as preserved in the fossil record. New species that are morphologically indistinguishable from their ancestors (or from contemporary species) go totally unrecognized. Sibling species are common in many groups of insects, in rodents, and in other well studied organisms (Mayr 1963, Dobzhansky 1970, Nevo and Shaw 1972, Dobzhansky *et al.* 1977, White 1978, Benado, Aguilera, Reig and Ayala 1979). Moreover, morphological discontinuities in a time series of fossils are usually interpreted by paleontologists as speciation events, even though they may represent phyletic evolution within an established lineage, without splitting of lineages.

Thus, when paleontologists use evidence of rapid phenotypic change in favor of the punctuational model, they are committing a definitional fallacy. Speciation as seen by the paleontologist always involves substantial morphological change because paleontologists identify new species by the eventuation of substantial morphological change. Stanley (1979, p. 144) has argued that "rapid change is concentrated in small populations and . . . that such populations are likely to be associated with speciation and unlikely to be formed by constriction of an entire lineage." But the two points he makes are arguable. First, rapid (in the geological scale) change may occur in populations that are not small. Second, bottlenecks in population size are not necessarily rare (again, in the geological scale) within a given lineage.

One additional point needs to be made. Punctualists speak of evolutionary change "concentrated in geologically instantaneous events of branching speciation" (Gould 1982b, p. 383). But events that appear instantaneous in the geological time scale may involve thousands, even millions of generations. Gould (1982a, p. 84), for example, has made operational the fuzzy expression "geologically instantaneous" by suggesting that "it be defined as 1 percent or less of later existence in stasis. This permits up to 100,000 years for the origin of a species with a subsequent life span of 10 million years." But 100,000 years encompasses one million generations of any insect such as *Drosophila*, and tens or hundreds of thousands of generations of fish, birds, or mammals. Speciation events or morphological changes deployed during thousands of generations may occur by the slow processes of allelic substitution that are familiar to the population biologist. Hence, the problem faced by microevolutionary theory is not how to account for rapid paleontological change, because there is ample time for it, but why lineages persist for millions of years without apparent morphological change. Although other explanations have been proposed, it seems that stabilizing selection may be the process most often responsible for the morphological stasis of lineages (Stebbins and Ayala 1981, Charlesworth *et al.* 1982). Whether microevolutionary theory is sufficient to explain punctuated as well as gradual evolution is, however, a different question, to which I now turn.

3. From Punctuated Evolution to the Autonomy of Macroevolution

The second proposal made by the proponents of punctuated equilibrium is that macroevolution—the evolution of species, genera, and higher taxa—is an autonomous field of study, independent of microevolutionary theory (and the intellectual turf of paleontologists). This claim for autonomy has been expressed as a "decoupling" of macro-evolution from microevolution (e.g., Stanley 1979, pp. x, 187, 193) or as a rejection of the notion that microevolutionary mechanisms can be extrapolated to explain macroevolutionary processes (e.g., Gould 1982b, p. 383).

The argument for the autonomy of macroevolution is grounded precisely on the claim that large-scale evolution is punctuated, rather than gradual. Two quotations shall suffice. "If rapidly divergent speciation interposes discontinuities between rather stable entities (lineages), and if there is a strong random element in the origin of these discontinuities (in speciation), then phyletic trends are essentially *decoupled* from phyletic trends within lineages. Macroevolution is decoupled from microevolution." (Stanley 1979, p. 187, italics in the original.) "Punctuated equilibrium is crucial to the independence of macroevolution — for it embodies the claim that species are legitimate individuals, and therefore capable of displaying irreducible properties." (Gould 1982a, p. 94.)

According to the proponents of punctuated equilibrium, phyletic evolution proceeds at two levels. First, there is change within a population that is continuous through time. This consists largely of allelic substitutions prompted by natural selection, mutation, genetic drift, and the other processes familiar to the population geneticist, operating at the level of the individual organism. Thus, most evolution within established lineages rarely, if ever, yields any substantial morphological change. Second, there is the process of origination and extinction of species. Most morphological change is associated with the origin of new species. Evolutionary trends result from the patterns of origination and extinction of species, rather than from evolution within established lineages. Hence, the relevant unit of macroevolutionary study is the species rather than the individual organism. It follows from this argument that the study of microevolutionary processes provides little, if any, information about macroevolutionary patterns, the tempo and mode of large-scale evolution. Thus, macroevolution is autonomous relative to microevolution, much in the same way as biology is autonomous relative to physics. Gould (1982b, p. 384) has summarized the argument: "Individuation of higher-level units is enough to invalidate the reductionism of traditional Darwinism — for pattern and style of evolution depend critically on the disposition of higher-level individuals [i.e., species]."

The question raised is the general issue of reduction of one branch of science to another. But as so often happens with questions of reductionism, the issue of whether microevolutionary mechanisms can account for macroevolutionary processes is muddled by confusion of separate issues. Identification of the issues involved is necessary in order to resolve them and to avoid misunderstanding, exaggerated claims, or unwarranted fears.

The issue "whether the mechanisms underlying microevolution can be extrapolated" to macroevolution involves, at least, three separate questions: (1) whether microevolutionary processes *operate* (and have operated in the past) throughout the organisms which make up the taxa in which macroevolutionary phenomena are observed. (2) Whether the microevolutionary processes identified by population geneticists (mutation, random drift, natural selection) are sufficient to *account for* the morphological changes and other macroevolutionary phenomena observed in higher taxa, or whether additional microevolutionary processes need to be postulated. (3) Whether theories concerning evolutionary trends and other macroevolutionary patterns can be *derived* from knowledge of microevolutionary processes.

The distinctions that I have made may perhaps become clearer if I state them as they might be formulated by a biologist concerned with the question whether the laws of physics and chemistry can be extrapolated to biology. The first question would be whether the laws of physics and chemistry apply to the atoms and molecules present in living organisms. The second question would be whether interactions between atoms and molecules according to the laws known to physics and chemistry are sufficient to account for biological phenomena, or whether the workings of organisms require additional kinds of interactions between atoms and molecules. The third question would be whether biological theories can be derived from the laws and theories of physics and chemistry.

The first issue raised can easily be resolved. It is unlikely that any biologist would seriously argue that the laws of physics and chemistry do not apply to the atoms and molecules that make up living things. Similarly, it seems unlikely that any paleontologist or macroevolutionist would claim that mutation, drift, natural selection, and other microevolutionary processes do not apply to the organisms and populations which make up the higher taxa studied in macroevolution. There is, of course, an added snarl—macroevolution is largely concerned with phenomena of the past. Direct observation of microevolutionary processes in populations of long-extinct organisms is not possible. But there is no reason to doubt that the genetic structures of populations living in the past were in any fundamental way different from the genetic structures of living populations. Nor is there any reason to believe that the processes of mutation, random drift and natural selection, or the nature of the interactions between organisms and the environment would have been different in nature for, say, Paleozoic trilobites or Mesozoic ammonites than for modern molluscs or fishes. Extinct and living populations—like different living populations—may have experienced quantitative differences in the relative importance of one or another process, but the processes could have hardly been different in kind. Not only are there reasons to the contrary lacking, but the study of biochemical evolution reveals a remarkable continuity and gradual change of informational macromolecules (DNA and proteins) over the most diverse organisms, which advocates that the current processes of population change have persisted over evolutionary history (Dobzhansky *et al.* 1977).

4. Are Microevolutionary Processes Sufficient?

The second question raised above is considerably more interesting than the first: Can the microevolutionary processes studied by population geneticists account for macroevolutionary phenomena or do we need to postulate new kinds of genetic processes? The large morphological (phenotypic) changes observed in evolutionary history, and the rapidity with which they appear in the geological record, is one major matter of concern. Another issue is "stasis," the apparent persistence of species, with little or no morphological change, for hundreds of thousands or millions of years. The dilemma is that microevolutionary processes apparently yield small but continuous changes, while macroevolution as seen by punctualists occurs by large and rapid bursts of change followed by long periods without change.

Goldschmidt (1940, p. 183) argued long ago that the incompatibility is real: "The decisive step in evolution, the first step towards macroevolution, the step from one species to another, requires another evolutionary method than that of sheer accumulation of micromutations." Goldschmidt's solution was to postulate "systemic mutations," yielding "hopeful monsters" that, on occasion, would find a new niche or way of life for which they would be eminently preadapted. The progressive understanding of the nature and organization of the genetic material acquired during the last forty years excludes the "systemic mutations" postulated by Goldschmidt, which would involve transformations of the genome as a whole.

Single-gene or chromosome mutations are known that have large effects on the phenotype because they act early in the embryo and their effects become magnified through development. Examples of such "macromutations" carefully analyzed in *Drosophila* are "bithorax" and the homeotic mutants that transform one body structure, e.g., antennae, into another, e.g., legs. Whether the kinds of morphological differences that characterize different taxa are due to such "macromutations" or to the accumulation of several mutations with small-effect, has been examined particularly in plants where fertile interspecific, and even intergeneric, hybrids can be obtained. The results do not support the hypothesis that the establishment of macromutations is necessary for divergence at the macroevolutionary level (Stebbins 1950, Clausen 1951, Grant 1971, see Stebbins and Ayala 1981). Moreover, Lande (1981 and references therein; see also Charlesworth *et al.* 1982) has convincingly shown that major morphological changes, such as in the number of digits or limbs, can occur in a geologically rapid fashion through the accumulation of mutations each with a small effect. The analysis of progenies from crosses between races or species that differ greatly (by as much as 30 phenotypic standard deviations) in a quantitative trait indicates that these extreme differences can be caused by the cumulative effects of no more than 5 to 10 independently segregating genes.

The punctualists' claim that mutations with large phenotypic effects must have been largely responsible for macroevolutionary change is based on the rapidity with which morphological discontinuities appear in the fossil record (Stanley 1979, Gould 1980). But the alleged evidence does not necessarily support the claim. Microevolutionists and macroevolutionists use different time scales. As pointed out earlier, the "geological instants" during which speciation and morphological shifts occur may involve expands as long as 100,000 years. There is little doubt that the gradual accumulation of small mutations may yield sizable morphological changes during periods of that length.

Anderson's (1973) study of body size in *Drosophila pseudoobscura* provides an estimate of the rates of gradual morphological change produced by natural selection. Large populations, derived from a single set of parents, were set up at different temperatures and allowed to evolve on their own. A gradual, genetically determined, change in body size ensued, with flies kept at lower temperature becoming, as expected, larger than those kept at higher temperatures. After 12 years, the mean size of the flies from a population kept at 16°C had become, when tested under standard conditions, approximately 10 per cent greater than the size of flies kept at 27°C; the change of mean value being greater than the standard deviation in size at the time when the tests were made. Assuming 10 generations per year, the populations diverged at a rate of 8×10^{-4} of the mean value per generation.

Paleontologists have emphasized the "extraordinary high *net* rate of evolution that is the hallmark of human phylogeny" (Stanley 1979). Interpreted in terms of the punctualist hypothesis, human phylogeny would have occurred as a succession of jumps, or geologically instantaneous saltations, interspersed by long periods without morphological change. Could these bursts of phenotypic evolution be due to the gradual accumulation of small changes? Consider cranial capacity, the character undergoing the greatest relative amount of change. The fastest rate of net change occurred between 500,000 years ago, when our ancestors were represented by *Homo erectus* and 75,000 years ago, when Neanderthal man had acquired a cranial capacity similar to that of modern humans. In the intervening 425,000 years, cranial capacity evolved from about 900 cc in Peking man to about 1400 cc in Neanderthal people. Let us assume that the increase in brain size occurred in a single burst at the rate observed in *D. pseudoobscura* of 8×10^{-4} of the mean value per generation. The change from 900 cc to 1400 cc could have taken place in 540 generations or, assuming generously 25 years per generation, in 13,500 years. Thirteen thousand years are, of course, a geological instant. Yet, this evolutionary "burst" could have taken place by gradual accumulation of mutations with small effects at rates compatible with those observed in microevolutionary studies.

The known processes of microevolution can, then, account for macroevolutionary change, even when this occurs according to the punctualist model—i.e., at fast rates concentrated on geologically brief time intervals. But what about the problem of stasis? The theory of punctuated equilibrium argues that after the initial burst of morphological change associated with their origin, "species generally do not change substantially in phenotype over a lifetime that may encompass many million years" (Gould 1982b, p. 383). Is it necessary to postulate new processes, yet unknown to population genetics, in order to account for the long persistence of lineages with apparent phenotypic change? The answer is "no."

The geological persistence of lineages without morphological change was already known to Darwin, who wrote in the last edition of *The Origin of Species* (1872, p. 375). "Many species once formed never undergo any further change . . . ; and the periods, during which species have undergone modification, though long as measured by years, have probably been short in comparison with the periods during which they retain the same form." A successful morphology may remain unchanged for extremely long periods of time, even though successive speciation events—as manifested, e.g., by the existence of sibling species, which in many known instances have persisted for millions of years (Stebbins and Ayala 1981).

Evolutionists have long been aware of the problem of paleontological stasis and have explored a number of alternative hypotheses consistent with microevolutionary principles and sufficient to account for the phenomenon. Although the issue is far from definitely settled, the weight of the evidence favors stabilizing selection as the primary process responsible for morphological stasis of lineages through geological time (Stebbins and Ayala 1981, Charlesworth *et al.* 1982).

5. Emergence and Hierarchy Versus Reduction

Macroevolution and microevolution are *not* decoupled in the two senses so far expounded: identity at the level of events and compatibility of theories. First, the populations in which macroevolutionary patterns are observed are the same populations that evolve at the microevolutionary level. Second, macroevolutionary phenomena can be accounted for as the result of known microevolutionary processes. That is, the theory of punctuated equilibrium (as well as the theory of phyletic gradualism) is consistent with the theory of population genetics. Indeed, any theory of macroevolution that is correct must be compatible with the theory of population genetics, to the extent that this is a well established theory.

Now, I pose the third question raised earlier: can macroevolutionary theory be derived from microevolutionary knowledge? The answer can only be "no." If macroevolutionary theory were deducible from microevolutionary principles, it would be possible to decide between competing macroevolutionary models simply by examining the logical implications of microevolutionary theory. But the theory of population genetics is compatible with both, punctualism and gradualism; and, hence, logically it entails neither. Whether the tempo and mode of evolution occur predominantly according to the model of punctuated equilibria or according to the model of phyletic gradualism is an issue to be decided by studying macroevolutionary patterns, not by inference from microevolutionary processes. In other words, macroevolutionary theories are not reducible (at least at the present state of knowledge) to microevolution. Hence, macroevolution and microevolution are decoupled in the sense (which is epistemologically most important) that macroevolution is an autonomous field of study that must develop and test its own theories.

Punctualists have claimed autonomy for macroevolution because species—the units studied in macroevolution—are higher in the hierarchy of organization of the living world than individual organisms. Species, they argue, have therefore "emergent" properties, not exhibited by, nor predictable from, lower-level entities. In Gould's (1980, p. 121) words, the study of evolution embodies "a concept of hierarchy —a world constructed not as a smooth and seamless continuum, permitting simple extrapolation from the lowest level to the highest, but as a series of ascending levels, each bound to the one below it in some ways and independent in others . . . 'emergent' features not implicit in the operation of processes at lower levels, may control events at higher levels." Although I agree with the thesis that macroevolutionary theories are not reducible to microevolutionary principles, I shall argue that it is a mistake to ground this autonomy on the hierarchical organization of life, or on purported emergent properties exhibited by higher-level units.

The world of life is hierarchically structured. There is a hierarchy of levels that go from atoms, through molecules, organelles, cells, tissues, organs, multicellular individuals and populations, to communities. Time adds another dimension of the evolutionary hierarchy, with the interesting consequence that transitions from one level to another occur: as time proceeds the descendants of a single species may include separate species, genera, families, etc. But hierarchical differentiation of subject matter is neither necessary nor sufficient for the autonomy of scientific disciplines. It is not necessary, because entities of a given hierarchical level can be the subject of diversified disciplines: cells are appropriate subjects of study for cytology, genetics, immunology, and so on. Even a single event can be the subject matter of several disciplines. My writing of this paragraph can be studied by a physiologist interested in the workings of muscles and nerves, by a psychologist concerned with thought processes, by a philosopher interested in the epistemological question at issue, and so on. Nor is the hierarchical differentiation of subject matter a sufficient condition for the autonomy of scientific disciplines: relativity theory obtains all the way from sub-atomic particles to planetary motions and genetic laws apply to multi-cellular organisms as well as to cellular and even subcellular entities.

One alleged reason for the theoretical independence of levels within a hierarchy is the appearance of "emergent" properties, which are "not implicit in the operation of lesser levels, [but] may control events at higher levels." The question of emergence is an old one, particularly in discussions on the reducibility of biology to the physical sciences. The issue is, for example, whether the functional properties of the kidney are simply the properties of the chemical constituents of that organ. In the context of macroevolution, the question is: do species exhibit properties different from those of the individual organisms of which they consist? I have argued elsewhere (Dobzhansky *et al.* 1977, ch. 16), that questions about the emergence of properties are ill-formed, or at least unproductive, because they can only be solved by definition. The proper way of formulating questions about the relation-ship between complex systems and their component parts is by asking whether the properties of complex systems can be inferred from knowl-edge of the properties that their components have in isolation. The issue of emergence cannot be settled by discussions about the "nature" of things or their properties, but it is resolvable by reference to our *knowledge* of those objects.

Consider the following question: Are the properties of common salt, sodium chloride, simply the properties of sodium and chlorine when they are associated according to the formula NaCl? If among the prop-erties of sodium and chlorine I include their association into table salt and the properties of the latter, the answer is "yes"; otherwise, the answer is "no." But the solution, then, is simply a matter of definition; and resolving the issue by a definitional maneuver contributes little to understanding the relationships between complex systems and their parts.

Is there a rule by which one could decide whether the properties of complex systems should be listed among the properties of their compo-nent parts? Assume that by studying the components in isolation we can infer the properties they will have when combined with other compo-nent parts in certain ways. In such a case, it would seem reasonable to include the "emergent" properties of the whole among the properties of the component parts. (Notice that this solution to the problem implies that a feature that may seem emergent at a certain time, might not appear as emergent any longer at a more advanced state of knowl-edge.) Often, no matter how exhaustively an object is studied in isola-tion, there is no way to ascertain the properties it will have in association with other objects. We cannot infer the properties of ethyl alcohol, proteins, or human beings from the study of hydrogen, and thus it makes no good sense to list their properties among those of hydrogen. The important point, however, is that the issue of emergent properties is spurious and that it needs to be reformulated in terms of propositions expressing our knowledge. It is a legitimate question to ask whether the *statements* concerning the properties of organisms (but not the properties themselves) can be logically deduced from statements concerning the properties of their physical components.

The question of the autonomy of macroevolution, like other questions of reduction, can only be settled by empirical investigation of the logical consequences of propositions, and not by discussions about the "nature" of things or their properties. What is at issue is not whether the living world is hierarchically organized — it is; or whether higher level entities have emergent properties — which is a spurious question. The issue is whether in a particular case, a set of *propositions* formulated in a defined field of knowledge (e.g., macroevolution) can be derived from another set of propositions (e.g., microevolutionary theory). Scientific theories consist, indeed, of propositions about the natural world. Only the investigation of the logical relations between propositions can establish whether or not one theory or branch of science is reducible to some other theory or branch of science. This implies that a discipline which is autonomous at a given stage of knowledge may become reducible to another discipline at a later time. The reduction of thermodynamics to statistical mechanics became possible only after it was discovered that the temperature of a gas bears a simple relationship to the mean kinetic energy of its molecules. The reduction of genetics to chemistry could not take place before the discovery of the chemical nature of the hereditary material (I am not, of course, intimating that genetics can now be fully reduced to chemistry, but only that a partial reduction may be possible now, whereas it was not before the discovery of the structure and mode of replication of DNA).

Nagel (1961, see also Ayala 1968) has formulated the two conditions that are necessary and, jointly, sufficient to effect the reduction of one theory or branch of science to another. These are the condition of derivability and the condition of connectability.

The *condition of derivability* requires that the laws and theories of the branch of science to be reduced be derived as logical consequences from the laws and theories of some other branch of science. The *condition of connectability* requires that the distinctive terms of the branch of science to be reduced be redefined in the language of the branch of science to which it is reduced — this redefinition of terms is, of course, necessary in order to analyze the logical connections between the theories of the two branches of science.

Microevolutionary processes, as presently known, are compatible with the two models of macroevolution — punctualism and gradualism.

From microevolutionary knowledge, we cannot infer which one of those two macroevolutionary patterns prevails, nor can we deduce answers for many other distinctive macroevolutionary issues, such as rates of morphological evolution, patterns of species extinctions, and historical factors regulating taxonomic diversity. The condition of derivability is not satisfied: the theories, models, and laws of macroevolution cannot be logically derived, at least at the present state of knowledge, from the theories and laws of population biology.

In conclusion, then, macroevolutionary processes are underlain by microevolutionary phenomena and are compatible with microevolutionary theories, but macroevolutionary studies require the formulation of autonomous hypotheses and models (which must be tested using macroevolutionary evidence). In this (epistemologically) very important sense, macroevolution is decoupled from microevolution: macroevolution is an autonomous field of evolutionary study.

6. Whence Macroevolution and Paleontology?

I have argued that distinctive macroevolutionary issues, such as rates of morphological evolution or patterns of species origination and extinction, cannot presently be derived from microevolutionary theories and principles. It deserves notice, however, that the discipline of (morphological) macroevolution is notoriously lacking in theoretical constructs of great deductive import and generality. There is little by way of hypotheses, theories, and models that would allow one to make inferences about what should be the case in one or other group of organisms and that would lead to the corroboration or falsification of a given general hypothesis, theory, or model. There are, of course, models such as punctuated equilibrium, species selection (Stanley 1979), and the Red Queen Hypothesis (Van Valen 1973). But this amounts to considerably less than a substantial body of deductive theory, such as it exists in population genetics, population ecology, and other microevolutionary disciplines. Awareness of the autonomy of macroevolutionary studies may perhaps help paleontologists in developing their science into a mature theoretical discipline.

There is one sector of macroevolutionary problems where considerable advances have occurred in the last decade yielding propositions of great generality and import: the field of molecular evolution. General theoretical constructs include the hypothesis of the molecular clock and related concepts concerning rates of amino acid and nucleotide substitution; the constraints imposed upon fundamental biochemical processes (transcription, translation, the genetic code, . . .) and structures (ribosomes, membranes, mitotic apparatus . . .); and so on. These have lead to notable advances, for example, in the reconstruction of the topology and timing of phylogenetic history, including the identification of a new phylum (the archeobacteria).

It may seem invidious to call attention to the successes of molecular biology in solving macroevolutionary issues. But it does corroborate my claim that hierarchical distinction of the objects studied is no grounds for the autonomy of scientific theories. Molecular macroevolution is not an autonomous field of study, because it can be reduced to microevolutionary theories, even though it concerns macroevolutionary entities. The hypothesis of the molecular clock of evolution, for example, is founded on concepts — such as mutation rate, population size, and genetic drift — distinctive of the field of population genetics.

REFERENCES

Anderson, W. W. (1973). "Genetic Divergence in Body Size Among Experimental Populations of *Drosophila Pseudoobscura* Kept at Different Temperatures." *Evolution* 27: 278–284.

Ayala, F. J. (1968). "Biology as an Autonomous Science." *American Scientist* 56: 207–221.

Benado, M.; Aguilera, M.; Reig, D. A.; and Ayala, F. J. (1979). "Biochemical Genetics of Venezuelan Spiny Rats of the *Proechimys Guainae* and *Proechimys Trinitatis* Superspecies." *Genetica* 50: 89–97.

Charlesworth, B.; Lande, R.; and Slatkin, M. (1982). "A Neo-Darwinian Commentary on Macroevolution." *Evolution* 36: 474–498.

Clausen, J. (1951). *Stages in the Evolution of Plant Species*. Ithaca, New York: Cornell University Press.

Cronin, J. E.; Boaz, N. T.; Stringer, C. B.; and Rak, Y. (1981). "Tempo and Mode in Hominid Evolution." *Nature* 292: 113–122.

Darwin, C. R. (1872). *The Origin of Species*. 6th ed. London, England: J. Murray. (As reprinted New York: Random House, 1936.)

Dobzhansky, Th. (1937). *Genetics and the Origin of Species*. (*Columbia Biological Series, Number XI.*) 2nd ed. 1941, 3rd ed. 1951. New York: Columbia University Press.

———. (1970). *Genetics of the Evolutionary Process*. New York: Columbia University Press.

———; Ayala, F. J.; Stebbins, G. L.; and Valentine, J. W. (1977). *Evolution*. San Francisco, California: W. H. Freeman & Co.

Douglas, M. E. and Avise, J. C. (1982). "Speciation Rates and Morphological Divergence in Fishes. Tests of Gradual Versus Rectangular Modes of Evolutionary Change." *Evolution* 36: 224–232.

Eldredge, N. (1971). "The Allopatric Model and Phylogeny in Paleozoic Invertebrates." *Evolution* 25: 156–167.

——— and Gould, S. J. (1972). "Punctuated Equilibria: An Alternative to Phyletic Gradualism." In *Models in Paleobiology*. Edited by T. J. M. Schopf. San Francisco: Freeman, Cooper Co. Pages 82–115.

Gingerich, P. D. (1976). "Paleontology and Phylogeny: Patterns of Evolution at the Species Level in Early Tertiary Mammals." *American Journal of Science* 276: 1–28.

Goldschmidt, R. B. (1940). *The Material Basis of Evolution*. New Haven, Connecticut: Yale University Press.

Gould, S. J. (1980). "Is a New and General Theory of Evolution Emerging?" *Paleobiology* 6: 119–130.

———. (1982a). "The Meaning of Punctuated Equilibrium and its Role in Validating a Hierarchical Approach to Macroevolution." In *Perspectives on*

Evolution. Edited by R. Milkman. Sunderland, Massachusetts: Sinauer Press. Pages 83–104.

———. (1982b). "Darwinism and the Expansion of Evolutionary Theory." *Science* 216: 380–387.

Grant, V. (1971). *Plant Speciation.* New York: Columbia University Press.

Hallam, A. (1978). "How Rare is Phyletic Gradualism and What is its Evolutionary Significance? Evidence from Jurassic Bivalves." *Paleobiology* 4: 16–25.

Huxley, J. S. (1942). *Evolution: The Modern Synthesis.* New York: Harper Publishing Company.

Kellogg, D. E. (1975). "The Role of Phyletic Change in the Evolution of *Pseudocubus Vema* (Radiolaria)." *Paleobiology* 1: 359–370.

Lande, R. (1981). "The Minimum Number of Genes Contributing to Quantitative Variation Between and Within Populations." *Genetics* 99: 541–553.

Levinton, J. S. and Simon, C. M. (1980). "A Critique of the Punctuated Equilibria Model and Implications for the Detection of Speciation in the Fossil Record." *Systematic Zoology* 29: 130–142.

Lewin, R. (1980). "Evolution Theory Under Fire." *Science* 210: 883–887.

Mayr, E. (1942). *Systematics and the Origin of Species.* New York: Columbia University Press.

———. (1963). *Animal Species and Evolution.* Cambridge, Massachusetts: Harvard University Press.

Monod, J. (1970.) *Le hasard et la nécessité.* Paris: Éditions du Seuil. (As reprinted in *Chance and Necessity.* (trans.) A. Wainhouse. New York: Vintage Books, 1972.)

Nagel, E. (1961). *The Structure of Science.* New York: Harcourt, Brace and World, Inc.

Nevo, E. and Shaw, C. R. (1972). "Genetic Variation in a Subterranean Mammal, *Spalax Ehrenbergi.*" *Biochemical Genetics* 7: 235–241.

Raup, D. M. (1978). "Cohort Analysis of Generic Survivorship." *Paleobiology* 4: 1–15.

Schopf, T.J.M. (1979). "Evolving Paleontological Views on Deterministic and Stochastic Approaches." *Paleobiology* 5: 337–352.

———. (1981). "Punctuated Equilibrium and Evolutionary Stasis." *Paleobiology* 7: 156–166.

Simpson, G. G. (1944). *Tempo and Mode in Evolution.* New York: Columbia University Press.

Stanley, S. M. (1979). *Macroevolution: Pattern and Process.* San Francisco, California: W. H. Freeman, Co.

———. (1982). "Macroevolution and the Fossil Record." *Evolution* 36: 460–473.

Stebbins, G. L. (1950). *Variation and Evolution in Plants.* New York: Columbia University Press.

——— and Ayala, F. J. (1981). "Is a New Evolutionary Synthesis Necessary?" *Science* 213: 967–971.

Van Valen, L. (1973). "A New Evolutionary Law." *Evolutionary Theory* 1: 1–30.

Vrba, E. S. (1980). "Evolution, Species, and Fossils: How Does Life Evolve?" *South African Journal of Science* 76: 61–84.

White, M. J. D. (1978). *Modes of Speciation.* San Francisco, California: W. H. Freeman, Co.

PROBLEMS OF CLASSIFICATION

12

ERNST MAYR*

Species Concepts and Their Application

DARWIN'S CHOICE OF title for his great evolutionary classic, *On the Origin of Species*, was no accident. The origin of new "varieties" within species had been taken for granted since the time of the Greeks. Likewise the occurrence of gradations, of "scales of perfection" among "higher" and "lower" organisms, was a familiar concept, though usually interpreted in a strictly static manner. The species remained the great fortress of stability, and this stability was the crux of the anti-evolutionist argument. "Descent with modification," true biological evolution, could be proved only by demonstrating that one species could originate from another. It is a familiar and often-told story how Darwin succeeded in convincing the world of the occurrence of evolution and how — in natural selection — he found the mechanism that is responsible for evolutionary change and adaptation. It is not nearly so widely recognized that Darwin failed to solve the problem indicated by the title of his work. Although he demonstrated the modification of species in the time dimension, he never seriously attempted a rigorous analysis of the problem of the multiplication of species, of the splitting of one species into two. I have examined the reasons for this failure (Mayr, 1959a) and found that foremost among them was Darwin's uncertainty about the nature of species. The same can be said of those authors who

*Ernst Mayr, *Populations, Species, and Evolution*, Chapter 2, pp. 10–20. Reprinted by permission of the publishers from *Populations, Species, and Evolution* by Ernst Mayr, Cambridge, Massachusetts: Harvard University Press, Copyright © 1963, 1970 by the President and Fellows of Harvard College.

attempted to solve the problem of speciation by saltation or other heterodox hypotheses. They all failed to find solutions that are workable in the light of the modern appreciation of the population structure of species. An understanding of the nature of species, then, is an indispensable prerequisite for the understanding of the evolutionary process.

Species Concepts

The term *species* is frequently used to designate a class of similar things to which a name has been attached. Most often this term is applied to living organisms, such as birds, fishes, flowers, or trees, but it has also been used for inanimate objects and even for human artifacts. Mineralogists speak of species of minerals, physicists of nuclear species; interior decorators consider tables and chairs species of furniture. The application of the same term both to organisms and to inanimate objects has led to much confusion and an almost endless number of species definitions (Mayr, 1963, 1969); these, however, can be reduced to three basic species concepts. The first two, mainly applicable to inanimate objects, have considerable historical significance, because their advocacy was the cause of much past confusion. The third is the species concept now prevailing in biology.

1. The Typological Species Concept

The typological species concept, going back to the philosophies of Plato and Aristotle (and thus sometimes called the essentialist concept), was the species concept of Linnaeus and his followers (Cain, 1958). According to this concept, the observed diversity of the universe reflects the existence of a limited number of underlying "universals" or types (*eidos* of Plato). Individuals do not stand in any special relation to one another, being merely expressions of the same type. Variation is the result of imperfect manifestations of the idea implicit in each species. The presence of the same underlying essence is inferred from similarity, and morphological similarity is, therefore, the species criterion for the essentialist. This is the so-called morphological species concept. Morphological characteristics do provide valuable clues for the determination of species status. However, using degree of morphological difference as the primary criterion for species status is completely different from utilizing morphological evidence together with various other kinds of evidence in order to determine whether or not a population deserves species rank under the biological species concept. Degree of morphological difference is not the decisive criterion in the ranking of taxa as species. This is quite apparent from the difficulties into which a morphological–typological species concept leads in taxonomic practice. Indeed, its own adherents abandon the typological species concept whenever they discover that they have named as a separate species something that is merely an individual variant.

2. The Nominalistic Species Concept

The nominalists (Occam and his followers) deny the existence of "real" universals. For them only individuals exist; species are man-made abstractions. (When they have to deal with a species, they treat it as an individual on a higher plane.) The nominalistic species concept was popular in France in the eighteenth century and still has adherents today. Bessey (1908) expressed this viewpoint particularly well: "Nature produces individuals and nothing more . . . species have no actual existence in nature. They are mental concepts and nothing more . . . species have been invented in order that we may refer to great numbers of individuals collectively."

Any naturalist, whether a primitive native or a trained population geneticist, knows that this is simply not true. Species of animals are not human constructs, nor are they types in the sense of Plato and Aristotle; but they are something for which there is no equivalent in the realm of inanimate objects.

From the middle of the eighteenth century on, the inapplicability of these two medieval species concepts (1 and 2 above) to biological species became increasingly apparent. An entirely new concept, applicable only to species of organisms, began to emerge in the later writings of Buffon and of many other naturalists and taxonomists of the nineteenth century (Mayr, 1968).

3. The Biological Species Concept

This concept stresses the fact that species consist of populations and that species have reality and an internal genetic cohesion owing to the historically evolved genetic program that is shared by all members of the species. According to this concept, then, the members of a species constitute (1) *a reproductive community.* The individuals of a species of animals respond to one another as potential mates and seek one another for the purpose of reproduction. A multitude of devices ensures intraspecific reproduction in all organisms. The species is also (2) *an ecological unit* that, regardless of the individuals composing it, interacts as a unit with other species with which it shares the environment. The species, finally, is (3) *a genetic unit* consisting of a large intercommunicating gene pool, whereas an individual is merely a temporary vessel holding a small portion of the contents of the gene pool for a short period of time. These three properties raise the species above the typological interpretation of a "class of objects" (Mayr, 1963, p. 21). The species definition that results from this theoretical species concept is: *Species are groups of interbreeding natural populations that are reproductively isolated from other such groups.*

The development of the biological concept of the species is one of the earliest manifestations of the emancipation of biology from an inappropriate philosophy based on the phenomena of inanimate nature. The species concept is called biological not because it deals with biological taxa, but because the definition is biological. It utilizes criteria that are meaningless as far as the inanimate world is concerned.

When difficulties are encountered, it is important to focus on the basic biological meaning of the species: A species is a protected gene pool. It is a Mendelian population that has its own devices (called isolating mechanisms) to protect it from harmful gene flow from other gene pools. Genes of the same gene pool form harmonious combinations because they have become coadapted by natural selection. Mixing the genes of two different species leads to a high frequency of disharmonious gene combinations; mechanisms that prevent this are therefore favored by selection. Thus it is quite clear that the word "species" in biology is a relational term. A is a species in relation to B or C because it is reproductively isolated from them. The biological species concept has its primary significance with respect to sympatric and synchronic populations (existing at a single locality and at the same time), and these — the "nondimensional species" — are precisely the ones where the application of the concept faces the fewest difficulties. The more distant two populations are in space and time, the more difficult it becomes to test their species status in relation to each other, but also the more irrelevant biologically this becomes.

The biological species concept also solves the paradox caused by the conflict between the fixity of the species of the naturalist and the fluidity of the species of the evolutionist. It was this conflict that made Linnaeus deny evolution and Darwin the reality of species (Mayr, 1957). The biological species combines the discreteness of the local species at a given time with an evolutionary potential for continuing change.

The Species Category and Species Taxa

The advocacy of three different species concepts has been one of the two major reasons for the "species problem." The second is that many authors have failed to make a distinction between the definition of the species category and the delimitation of species taxa (for fuller discussion see Mayr, 1969).

A *category* designates a given rank or level in a hierarchic classification. Such terms as "species," "genus," "family," and "order" designate categories. A category, thus, is an abstract term, a class name, while the organisms placed in these categories are concrete zoological objects.

Organisms, in turn, are classified not as individuals, but as groups of organisms. Words like "bluebirds," "thrushes," "songbirds," or "vertebrates" refer to such groups. These are the concrete objects of classification. Any such group of populations is called a *taxon* if it is considered sufficiently distinct to be worthy of being formally assigned to a definite category in the hierarchic classification. *A taxon is a taxonomic group of any rank that is sufficiently distinct to be worthy of being assigned to a definite category.*

Two aspects of the taxon must be stressed. A taxon always refers to specified organisms. Thus *the* species is not a taxon, but any given species, such as the Robin (*Turdus migratorius*) is. Second, the taxon must be formally recognized as such, by being described under a designated name.

Categories, which designate a rank in a hierarchy, and taxa, which designate named groupings of organisms, are thus two very different kinds of phenomena. A somewhat analogous situation exists in our human affairs. Fred Smith is a concrete person, but "captain" or "professor" is his rank in a hierarchy of levels.

The Assignment of Taxa to the Species Category

Much of the task of the taxonomist consists of assigning taxa to the appropriate categorical rank. In this procedure there is a drastic difference between the species taxon and the higher taxa. Higher taxa are defined by intrinsic characteristics. Birds is the class of feathered vertebrates. Any and all species that satisfy the definition of "feathered vertebrates" belong to the class of birds. An essentialist (typological) definition is satisfactory and sufficient at the level of the higher taxa. It is, however, irrelevant and misleading to define species in an essentialistic way because the species is not defined by intrinsic, but by *relational* properties.

Let me explain this. There are certain words that indicate a relational property, like the word "brother." Being a brother is not an inherent property of an individual, as hardness is a property of a stone. An individual is a brother only with respect to someone else. The word "species" likewise designates such a relational property. A population is a species with respect to all other populations with which it exhibits the relationship of reproductive isolation — noninterbreeding. If only a single population existed in the entire world, it would be meaningless to call it a species.

Noninterbreeding between populations is manifested by a gap. It is this gap between populations that coexist (are sympatric) at a single locality at a given time which delimits the species recognized by the local naturalist. Whether one studies birds, mammals, butterflies, or snails near one's home town, one finds each species clearly delimited and sharply separated from all other species. This demarcation is sometimes referred to as the species delimitation *in a nondimensional system* (a system without the dimensions of space and time).

Anyone can test the reality of these discontinuities for himself, even where the morphological differences are slight. In eastern North America, for instance, there are four similar species of the thrush genus *Catharus* (Table 1), the Veery (*C. fuscescens*), the Hermit Thrush (*C. guttatus*), the Olive-Backed or Swainson's Thrush (*C. ustulatus*), and the Gray-Cheeked Thrush (*C. minimus*). These four species are sufficiently similar visually to confuse not only the human observer, but also silent

Table 1. *Characteristics of Four Eastern North American Species of* Catharus
(from Dilger 1956)

Characteristics compared	C. fuscescens	C. guttatus	C. ustulatus	C. minimus
Breeding range	Southernmost	More northerly	Boreal	Arctic
Wintering area	No. South America	So. United States	C. America to Argentina	No. South America
Breeding habitat	Bottomland woods with lush undergrowth	Coniferous woods mixed with deciduous	Mixed or pure tall coniferous forests	Stunted northern fir and spruce forests
Foraging	Ground and arboreal (forest interior)	Ground (inner forest edges)	Largely arboreal (forest interior)	Ground (forest interior)
Nest	Ground	Ground	Trees	Trees
Spotting on eggs	Rare	Rare	Always	Always
Relative wing length	Medium	Short	Very long	Medium
Hostile call	*veer* *pheu*	*chuck* *seeeep*	*peep* *chuck-burr*	*beer*
Song	Very distinct	Very distinct	Very distinct	Very distinct
Flight song	Absent	Absent	Absent	Present

males of the other species. The species-specific songs and call notes, however, permit easy species discrimination, as observationally substantiated by Dilger (1956). Rarely do more than two species breed in the same area, and the overlapping species, $f + g$, $g + u$, and $u + m$, usually differ considerably in their foraging habits and niche preference, so that competition is minimized with each other and with two other thrushes, the Robin (*Turdus migratorius*) and the Wood Thrush (*Hylocichla mustelina*), with which they share their geographic range and many ecological requirements. In connection with their different foraging and migratory habits the four species differ from one another (and from other thrushes) in the relative length of wing and leg elements and in the shape of the bill. There are thus many small differences between these at first sight very similar species. Most important, no hybrids or intermediates among these four species have ever been found. Each is a separate genetic, behavioral, and ecological system, separated from the others by a complete biological discontinuity, a gap.

Difficulties in the Application of the Biological Species Concept

The practicing taxonomist often has difficulties when he endeavors to assign populations to the correct rank. Sometimes the difficulty is caused by a lack of information concerning the degree of variability of the species with which he is dealing. Helpful hints on the solution of such practical difficulties are given in the technical taxonomic literature (Mayr, 1969).

More interesting to the evolutionist are the difficulties that are introduced when the dimensions of time and space are added. Most species taxa do not consist merely of a single local population but are an aggregate of numerous local populations that exchange genes with each other to a greater or lesser degree. The more distant that two populations are from each other, the more likely they are to differ in a number of characteristics. I show elsewhere (Mayr, 1963, ch. 10 and 11) that some of these populations are incipient species, having acquired some but not all characteristics of species. One or another of the three most characteristic properties of species taxa—reproductive isolation, ecological difference, and morphological distinguishability—is in such cases only incompletely developed. The application of the species concept to such incompletely speciated populations raises considerable difficulties. There are six wholly different situations that may cause difficulties.

(1) *Evolutionary continuity in space and time.* Widespread species may have terminal populations that behave toward each other as distinct species even though they are connected by a chain of interbreeding populations. Cases of reproductive isolation among geographically distant populations of a single species are discussed in Mayr, 1963, ch. 16.

(2) *Acquisition of reproductive isolation without corresponding morphological change.* When the reconstruction of the genotype in an isolated population has resulted in the acquisition of reproductive isolation, such a population must be considered a biological species. If the correlated morphological change is very slight or unnoticeable, such a species is called a sibling species (Mayr, 1963, ch. 3).

(3) *Morphological differentiation without acquisition of reproductive isolation.* Isolated populations sometimes acquire a degree of morphological divergence one would ordinarily expect only in a different species. Yet some such populations, although as different morphologically as good species, interbreed indiscriminately where they come in contact. The West Indian snail genus *Cerion* illustrates this situation particularly well (Fig. 1).

(4) *Reproductive isolation dependent on habitat isolation.* Numerous cases have been described in the literature in which natural populations acted toward each other like good species (in areas of contact) as long as their habitats were undisturbed. Yet the reproductive isolation broke down as soon as the characteristics of these habitats were changed, usually by the interference of man. Such cases of secondary breakdown of isolation are discussed in Mayr, 1963, ch. 6.

(5) *Incompleteness of isolating mechanisms.* Very few isolating mechanisms are all-or-none devices (see Mayr, 1963, ch. 5). They are built up step by step, and most isolating mechanisms of an incipient species are imperfect and incomplete. Species level is reached when the pro-

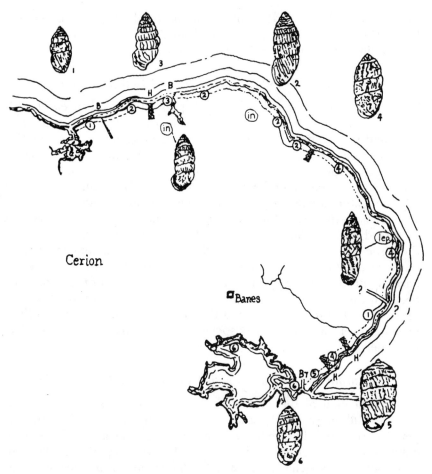

Figure 1. The distribution pattern of populations of the halophilous land snail *Cerion* on the Banes Peninsula in eastern Cuba. Numbers refer to distinctive races or "species." Where two populations come in contact (with one exception) they hybridize (*H*), regardless of degree of difference. In other cases contact is prevented by a barrier (*B*). *In* = isolated inland population.

cess of speciation has become irreversible, even if some of the secondary isolating mechanisms have not yet reached perfection (see Mayr, 1963, ch. 17).

(6) *Attainment of different levels of speciation in different local populations*. The perfecting of isolating mechanisms may proceed in different populations of a polytypic species (one having several subspecies) at different rates. Two widely overlapping species may, as a consequence, be completely distinct at certain localities but may freely hybridize at others. Many cases of sympatric hybridization discussed in Mayr, 1963, ch. 6, fit this characterization (see Mayr, 1969, for advice on handling such situations).

These six types of phenomena are consequences of the gradual nature of the ordinary process of speciation (excluding polyploidy; see Mayr, 1963, p. 254). Determination of species status of a given population is difficult or arbitrary in many of these cases.

Difficulties Posed by Uniparental Reproduction

The task of assembling individuals into populations and species taxa is very difficult in most cases involving uniparental (asexual) reproduction. Self-fertilization, parthenogenesis, pseudogamy, and vegetative reproduction are forms of uniparental reproduction. The biological species concept, which is based on the presence or absence of interbreeding between natural populations, cannot be applied to groups with obligatory asexual reproduction because interbreeding of populations is nonexistent in these groups. The nature of this dilemma is discussed in more detail elsewhere (Mayr, 1963, 1969). Fortunately, there seem to be rather well-defined discontinuities among most kinds of uniparentally reproducing organisms. These discontinuities are apparently produced by natural selection from the various mutations that occur in the asexual lines (clones). It is customary to utilize the existence of such discontinuities and the amount of morphological difference between them to delimit species among uniparentally reproducing types.

The Importance of a Nonarbitrary Definition of Species

The clarification of the species concept has led to a clarification of many evolutionary problems as well as, often, to a simplification of practical problems in taxonomy. The correct classification of the many different kinds of varieties (phena), of polymorphism (Mayr, 1963, ch. 7), of polytypic species (ibid. ch. 12), and of biological races (ibid. ch. 15) would be impossible without the arranging of natural populations and phenotypes into biological species. It was impossible to solve, indeed even to state precisely, the problem of the multiplication of species until the biological species concept had been developed. The genetics of speciation, the role of species in large-scale evolutionary trends, and other major evolutionary problems could not be discussed profitably until the species problem was settled. It is evident then that the species problem is of great importance in evolutionary biology and that the growing agreement on the concept of the biological species has resulted in a uniformity of standards and a precision that is beneficial for practical as well as theoretical reasons.

The Biological Meaning of Species

The fact that the organic world is organized into species seems so fundamental that one usually forgets to ask why there are species, what their meaning is in the scheme of things. There is no better way of answering these questions than to try to conceive of a world without species. Let us think, for instance, of a world in which there are only individuals, all belonging to a single interbreeding community. Each individual is in varying degrees different from every other one, and each individual is capable of mating with those others that are most similar to it. In such a world, each individual would be, so to speak, the center of a series of concentric rings of increasingly more different individuals. Any two mates would be on the average rather different from each other and would produce a vast array of genetically different types among their offspring. Now let us assume that one of these recombinations is particularly well adapted for one of the available niches. It is prosperous in this niche, but when the time for mating comes, this superior genotype will inevitably be broken up. There is no mechanism that would prevent such a destruction of superior gene combinations, and there is, therefore, no possibility of the gradual improvement of gene combinations. The significance of the species now becomes evident. The reproductive isolation of a species is a protective device that guards against the breaking up of its well-integrated, coadapted gene system. Organizing organic diversity into species creates a system that permits genetic diversification and the accumulation of favorable genes and gene combinations without the danger of destruction of the basic gene complex. There are definite limits to the amount of genetic variability that can be accommodated in a single gene pool without producing too high a proportion of inviable recombinants. Organizing genetic diversity into protected gene pools — that is, species — guarantees that these limits are not overstepped. This is the biological meaning of species.

REFERENCES

Bessey, C. E., 1908, The taxonomic aspect of the species. *American Naturalist*, 42: 218–224.

Cain, A. J., 1958, Logic and memory in Linnaeus's system of taxonomy. *Proc. Linn. Soc.*, London, 169: 144–163.

Dilger, W. C., 1956, Hostile behavior and reproductive isolating mechanisms in the avian genera *Catharus* and *Hylocichla*. *Auk*, 73: 313–353.

Mayr, E., 1957, Species concepts and definitions. *Amer. Assoc. Adv. Sci.*, Publ. No. 50: 1–22, Washington, D.C.

———, 1959a, Darwin and the evolutionary theory in biology. In *Evolution and Anthropology: A Centennial Approach*, Anthropological Society of America, Washington, D.C.

———, 1963, *Animal Species and Evolution*, Cambridge, Harvard University Press.

———, 1968, Illiger and the biological species concept. *J. Hist. Biol.*, 1: 163–178.

———, 1969, *Principles of Systematic Zoology*, New York, McGraw-Hill.

13

DAVID L. HULL*

The Ontological Status of Species as Evolutionary Units

1. The Nature of the Species Problem

REFERENCE TO THE species problem today might sound quaint and vaguely anachronistic. Perhaps the species problem was of some importance ages ago in the philosophical dispute between nominalists and essentialists or a century ago in biology when Darwin introduced his theory of organic evolution, but it certainly is of no contemporary interest. But "species," like the terms "gene," "electron," "non-local simultaneity," and "element," is a theoretical term embedded in a significant scientific theory. At one time the nature of the physical elements was an important issue in physics. The transition from the elements being defined in terms of gross traits, to specific density, to molecular weight, to atomic number was important in the development of atomic theory. The transition in biology from genes being defined in terms of unit characters, to production of enzymes, to coding for specific polypeptides, to structurally defined segments of nucleic acid was equally important in the growth of modern genetics. A comparable transition is taking place with respect to the species concept and is equally important.

It is easy enough to say that species evolve; it is not so easy to explain in detail the exact nature of these evolutionary units. Evolutionary theory is currently undergoing a period of rapid and fundamental re-formulation, and our conception of biological species as evolutionary units is being modified accordingly. There is nothing unusual about a theoretical term changing its meaning as the theory in which it is embedded changes, but sometimes such development has an added dimension. Not only is the meaning of the theoretical term altered, but the ontological status of the entities to which it refers is also modified. For centuries, philosophers and scientists alike have treated species as secondary substances, universals, classes, etc. Nothing has seemed clearer than the relation between particular organisms and their species. In contemporary terminology, this relation is termed **class membership**. Species are classes defined by means of the covariation of the traits which their members possess. On this same interpretation, organisms are individuals and the species category itself is a class of classes.

However, species have proved to be very peculiar classes. Their membership is constantly undergoing change as new organisms are born and old ones die. At any one time, one can rarely discover a set of traits which is possessed by all the members of a species and by no members of some other species. In addition, the members of successive generations of the same species are usually characterized by slightly different sets of traits. These facts of life have forced philosophers to view the names of particular biological species (as well as all taxa) as cluster concepts. However, in this paper, I would like to argue for a radically different solution to the species problem. Just as relativity theory necessitated shifting such notions as space and time from one ontological category to another, evolutionary theory necessitates a similar shift in the ontological status of biological species. If species are units of evolution, then they cannot be interpreted as classes; they are individuals. The purpose of this paper is to show why evolutionary theory requires such a change in the ontological status of species.

There are three basic premises to the argument which I will present. Two are quite familiar; one is a variation on a familiar theme; none is totally uncontroversial. In this paper, however, I will not attempt to defend these premises. Instead I will show the consequences which follow for the nature of biological species *if* they are accepted.

2. Three Basic Premises

My first premise is that the ontological status of theoretical entities is theory-dependent.[1] Does time really flow like a river independent of the existence and distribution of material bodies? Are species really individuals? I don't think that such questions can be answered without reference to a particular scientific theory. A particular atom of gold is an individual and not a property, universal, process, relation, or what have you, because atomic theory requires atoms to be viewed in this way. The ontological status of electrons is equivocal for similar reasons, the theory in this case being quantum mechanics. A segment of DNA bounded by initiation and termination codons is an individual because molecular genetics requires genes to be interpreted in this way. If species are to be viewed as individuals, it will be because current evolutionary theory necessitates such a conceptualization.

The concept of an individual in philosophy is very broad, including such entities as sense data and bare particulars. In this paper, I will limit myself to a narrower use of the term to refer just to those entities which are characterized by unity and continuity, specifically spatiotemporal unity and continuity (see Hull, 1975). This is the sense in which organisms, planets, houses, and atoms are individuals. Such individuals have unique beginnings and endings in time, reasonably discrete boundaries, internal coherence at any one time, and continuity through time. There are individuals for which such spatiotemporal unity and continuity are not required; e.g. nations, political parties, ideas. The United States did not become any the less an individual when Alaska and Hawaii became states. In addition, there are situations in which a nation ceases to exist for a time and then the "same" nation comes into being again later. However, species as units of evolution fulfill the stricter requirements. They are "individuals" in the same sense that organisms are individuals. Both are localized in space and time.

On the surface, the distinction between a class and an individual could not seem sharper. A class, on the one hand, is the sort of thing that can have members. The name of a class is a class term defined intensionally by means of the properties which its members possess. The relation between a member and its class is the intransitive class–membership relation. An individual, on the other hand, is the sort of thing that can have parts. The name of an individual is a proper name possessing ideally no intension at all. The relation between a part and the individual of which it is part is the transitive part–whole relation. However, classes can be construed in ways which make them all but indistinguishable from individuals. Classes are often defined in terms of simple one-place predicates, like the class of atoms with atomic number 79. Classes can also be defined in terms of relations, sometimes a relation between two classes, sometimes a relation which the members of a class have to each other, sometimes a relation which each member of the class has to a specified focus. For example, "planet" can be defined as any relatively large non-luminous body revolving around a star. Although the relation mentioned in this definition is spatiotemporal, the class itself is spatiotemporally unrestricted because planets can revolve around any star whatsoever.

The problematic cases are those "classes" defined in terms of a spatiotemporal relation which the "members" have to each other or to a specified individual. For example, "forest" could be defined in terms of a sufficiently large number of trees no further apart than a certain distance from at least one other tree in the complex. On this definition, the term "forest" would be a spatiotemporally unrestricted class, but each particular forest would not be. Similarly, "tributary system" could be defined in terms of those rivers which flow into one main river. On this definition, "tributary system" would be a spatiotemporally unrestricted class, but each particular tributary system would not be. Such complexes as the Black Forest and the Mississippi River tributary system can be treated as classes only at the expense of collapsing the distinction between classes and individuals, an excellent reason for not doing so.[2] The extension of this analysis to "species" defined in terms of descent and the names of particular species should be obvious.

The final consideration central to my argument concerns the nature of scientific laws. According to the traditional conception, natural laws must be spatiotemporally unrestricted. To the extent that a scientific law is true, it must be true for all entities falling within its domain anywhere and at any time. The regularities in nature which scientific laws are designed to capture cannot vary from place to place or from time to time. I must hasten to add, however, that my acceptance of this traditional notion of a scientific law does not commit me to a raft of other beliefs commonly associated with it. Laws are not all there is to science. Descriptions, for example, are also important. Nor do I think that recourse to scientific laws is the only way in which an event can be explained. The issue of the nature of scientific laws must be raised, however, because evolutionary theory is often cited as evidence against the view that laws must be spatiotemporally unrestricted. But properly construed, the laws which go to make up evolutionary theory are as spatiotemporally unrestricted as any other laws of nature. Species evolve, languages evolve, our understanding of the empirical world evolves, but the laws of nature are eternal and immutable.[3]

3. Evolutionary Theory—Is It Different?

The vague feeling which philosophers have had that evolutionary theory is somehow different from other process theories stems from three sources, two of them clearly mistaken. One source of the belief that the laws of evolutionary theory are spatiotemporally restricted stems from the process–product confusion, from confusing evolutionary processes such as mutation, selection, and evolution with their product — phylogeny. The statement that mammals arose from several species of reptile is clearly historical in the sense that it describes a temporal succession of events, but such descriptions or phylogenetic sequences are no more part of evolutionary theory than a description of the successive stages in an eclipse is part of celestial mechanics.

A second source of the vague feeling which philosophers have had that evolutionary theory is peculiar is their conceiving of species as classes. In the 19th century, philosophers such as John Stuart Mill distinguished between laws of succession and laws of co-existence. They viewed the apparent universal distribution of traits among the organisms which make up biological species as the best example of such laws of co-existence. Thus, for 19th century philosophers, the evolution of species meant the evolution of laws. This line of reasoning is still common today, especially among social scientists.[4] But if species are interpreted as individuals, the evolution of species poses no problem for the traditional conception of a scientific law. A description of an individual as it develops through time is hardly a candidate for the status of a scientific law. If individuals are localized in space and time and scientific laws must be spatiotemporally unrestricted, then no law of nature can contain essential reference to a particular individual. If species (as well as all taxa) are interpreted as individuals developing through time, then any statement which contains essential reference to such individuals can no more count as a scientific law than Kepler's laws — if Kepler's laws had been true only for the sun and its planets.[5] In point of fact, evolutionary theory contains no reference to particular taxa, just what one would expect if taxa are actually individuals and not classes. On this view, "All swans are white" could not count as a scientific law even if it were true.

However, there is a sense in which evolutionary theory might turn out to be peculiar. Biological evolution is a selection process. If selection processes are different in kind from other processes, then selection theories such as evolutionary theory might turn out to be different in form from ordinary process theories. To the extent that there is a difference, it seems to be this: selection processes require the existence of at least two partially independent processes operating at different levels of organization on different time scales. Hence, one finds biologists contrasting "ecological time" with "geological time," a strange manner of speaking to say the least. In a selection process, variation and selective retention result in the evolution of some unit of much larger scope. The entities which vary and which are selected come into being, reproduce themselves, and pass away in such a fashion that these larger units gradually change. However, the resolution of this particular problem has no special relevance to the thesis of this paper. Whether or not selection processes turn out to be reducible to ordinary processes, the consequences for the ontological status of biological species remains the same: they must be interpreted as individuals.

4. Species as Individuals

According to the traditional formula, three levels of organization are involved in organic evolution. Genes mutate, organisms compete and are selected, and species evolve. We now know that this formula is too simple. Mutation can be as slight as the change of a single base pair or as major as the gain or loss of entire chromosomes. Competition and selection can take place at a variety of levels from macromolecules (including genes) and cells (including gametes) to organisms and kin groups. In addition, an important level of organization exists between kin groups and entire species — the population. Populations are the effective units of evolution. There is no question that genes, cells, organisms and kin groups are related by the part–whole relation. Genes are part of cells, cells are part of organisms, and organisms are part of kin groups. The point of the dispute is whether the relation changes abruptly from part–whole to class–membership above the level of organisms and possibly kin groups.

Several issues are involved in this dispute, including the existence of group selection. Organisms and kin groups form units of selection. Can populations and possibly even entire species form units of selection? Populations and species evolve. Can entities at lower levels of organization like colonies also form units of evolution? What kind of organization is required for something to function as a unit of selection? A unit of evolution? How do units of selection differ from units of evolution? Must evolution always occur at levels of organization higher than that at which selection is taking place? What does it mean to say that something is a "unit" of selection or a "unit" of evolution? Can such units be classes as well as individuals?

Considerable disagreement exists among biologists over the answers to the preceding questions, increased to some extent by certain unfortunate terminological conventions like using the terms "organism" and "individual" interchangeably. For example, E. O. Wilson (1974, p. 184) argues that the preceding "are not trivial questions. They address a theoretical issue seldom made explicit in biology":

> In zoology the very word colony implies that the members of the society are physically united, or differentiated into reproductive and sterile castes, or both. When both conditions exist to an advanced degree, as they do in many of these animals, the society can equally well be viewed as a superorganism or even an organism. The dilemma can therefore be expressed as follows: At what point does a society become so well integrated that it is no longer a society? On what basis do we distinguish the extremely modified members of an invertebrate colony from the organs of a metazoan animal?

Theodosius Dobzhansky (1970, p. 23) has long argued that:

> A species, like a race or a genus or a family, is a group concept and a category of classification. A species is, however, also something else: a superindividual biological system, the perpetuation of which from generation to generation depends on the reproductive bonds between its members.

Michael Ghiselin (1974) objects to terming species "organisms" lest the term imply that species can function as units of selection the way that organisms do. He prefers the generic term "individual." However, the message for our purposes is the same. There is something about evolutionary processes that requires that the units of mutation, selection, and evolution be treated as individuals integrated by the part–whole relation.

All versions of evolutionary theory from Darwin to the present have included a strong principle of heredity. It is not enough for a gene to be able to mutate; it must be able to replicate and pass this change on. It is not enough for an organism to cope more successfully in its environment than its competitors; it must also be able to reproduce itself and pass on these variations to its progeny. But exactly the same observations can be made about populations and species, possibly not as units of selection but certainly as units of evolution.

The role of spatiotemporal unity and continuity in the evolutionary process is easily overlooked, especially when it is being described in terms of populations. The term "population" is systematically ambiguous in the biological literature. In its broadest sense, a population is merely a collection of individuals of any sort characterized by the distribution of one or more traits of these individuals. Although the members of populations in this broad sense could be chosen at random, usually they are selected on the basis of some criterion; e.g. people on welfare, herbivorous quadrupeds, stars increasing in brightness. As biologists such as Ernst Mayr (1963) have repeatedly emphasized, the populations which function in the evolutionary process are "populations" in a much more restricted sense of the term. Descent is required. But descent presupposes replication and reproduction, and these processes in turn presuppose spatiotemporal proximity and continuity. When a single gene undergoes replication to produce two new genes or a single cell undergoes mitotic division to produce two new cells, the end products are spatiotemporally continuous with the parent entity. In sexual reproduction, the propagules, if not the parent organisms themselves, must come into contact. The end result is the successive modification of the "same" population.

Populations are made up of successive generations of organisms. These generations may be temporally disjoint or largely overlapping, but a certain degree of genetic continuity is required for a population to function as a population in the evolutionary process. To be sure, new organisms can migrate into a population and others leave, changing the genetic composition of the population. New genes can be introduced by means of mutation. But such changes cannot be too massive or too sudden without disrupting the evolutionary process. Given the differences in time spans required for mutation, selection and evolution, sufficient continuity is required to allow for the cumulation of the adaptive changes necessary for evolution. Channeling is required, sufficient channeling to warrant conceptualizing such lineages as individuals.

Identity in populations is determined by the same considerations which determine identity in organisms. Constancy of neither substance nor essence is required in either case; spatiotemporal continuity is. Just as all the cells which comprise an organism can be changed while that organism remains the "same" organism, all the organisms which comprise a population can be changed while that population remains the "same" population, just so long as such changes are gradual. Just as all the traits which characterize an organism at one stage of its development can change without that organism ceasing to be the "same" organism, all the distributions of traits which characterize a population at one stage in its evolution can change without that population ceasing to be the "same" population, just so long as such changes are gradual.[6]

The factors responsible for spatiotemporal continuity in evolution are fairly straightforward; the factors which promote evolutionary unity are not. In order for selection to result in evolutionary change, it must act on successive generations of the "same" population. But the units of evolution must also be sufficiently cohesive at any one time to evolve as units. Two issues are at stake: the mechanisms which tend to promote evolutionary unity and the degree of unity necessary for an entity to count as an individual. Certain biologists have argued that in order for organisms to form populations or species, they must reproduce sexually. Gene exchange is the only mechanism capable of producing evolutionary unity (see Mayr, 1963; Dobzhansky, 1970; Ghiselin, 1974). Others have argued that gene exchange is not all that powerful a force in promoting evolutionary unity. The same selection pressures which produce unity in asexual species produce it in sexual species as well. Gene exchange is only of minor significance (see Meglitsch, 1954; Ehrlich and Raven, 1969).

From the discussion so far, it might seem that populations and species are far from paradigm individuals. In most cases, there is no one instant at which a species comes into existence or becomes extinct. Both speciation and phyletic evolution usually take hundreds, if not thousands, of generations. One mechanism which has been suggested for narrowing the borderline between a newly emerging species and its parent species is Mayr's Founder Principle. According to this principle, speciation in sexual species occurs always or usually by means of the isolation of one or a few organisms. Such isolates rarely succeed in forming a population, but when they do, the resulting population tends to be quite different from its parent species and can undergo additional rapid change. The end result is that the transition between two species is reduced both in time and with respect to the number of organisms involved.

Nor are the organisms which go to make up a species always in close proximity, let alone spatially contiguous. However, the exigencies of reproduction require at least periodic proximity of the relevant entities, whether organisms or merely their propagules. But before the notion that species are spatiotemporal individuals is dismissed, it should be noted that exactly the same observations can be made with respect to organisms as individuals, albeit on a reduced scale; and organisms are supposed to be paradigm individuals. Anyone who thinks that the spatiotemporal boundaries of organisms are that much sharper than those of species should read up on slime molds and grasses. Any problem which can be found in the spatiotemporal unity and continuity of species also exists for particular organisms. If organisms can count as individuals in the face of such difficulties, so can species.[7]

5. Conclusion

I think that the preceding considerations give ample support for the conclusion that species are not spatiotemporally unrestricted classes. However, doubt may remain whether or not they should be interpreted as individuals. Perhaps they belong in some hybrid category like "individualistic classes," as Leigh Van Valen has suggested, or "complex particulars," as Fred Suppe (1974) has proposed, but the evaluation of such suggestions must wait for further elucidation of these notions. For now, species fit as naturally into the idealized category of a spatiotemporally localized individual as do particular organisms. Once again, if organisms are individuals, so are species.

NOTES

1. The claim that ontological status of theoretical entities is theory-dependent is not the same thing as the well-known claim that they are theory-laden, though the rationale behind the two claims is much the same.
2. Another reason for not considering complexes defined in terms of a spatiotemporal relation between spatiotemporally localized individuals as genuine classes is the role which class terms play in science. The chief use of class terms in science is to function in scientific laws. A scientific classification is important to the extent that it produces theoretically significant classes. In fact, I think that the class–individual distinction is best made in terms of the contrasting roles played by classes and individuals in scientific laws.
3. If it was discovered that the laws of nature, as we currently conceive them, actually changed through time, our conception of a scientific law would not have to be modified if these changes were themselves regular. Hence, statements characterizing these regular changes would become the new basic laws of nature.
4. The highly touted book by M. D. Sahlins and E. R. Service (1960) relies throughout on confusing evolutionary processes with the products of such processes. They consistently argue that political laws evolve because political systems evolve. See also T. L. Thorson (1970) for the same crude mistake.
5. The distinction assumed in this discussion is between laws of nature and true accidental generalizations. It is only fair to acknowledge, however, that this distinction is currently one of the most problematic in philosophy of science. At the risk of some circularity, I suspect that the only way in which this distinction can be made is by reference to the actual or eventual inclusion in a scientific theory. Any generalization, true though it may be, cannot count as a scientific law if it remains isolated from all other such generalizations.

6. The subject matter of this paragraph deserves a paper of its own. Neither organisms nor species require constancy of substance or essence for individual identity, but as individuals, organisms and species differ in some important respects. Organisms possess a program which directs and circumscribes their development to some extent. The development of species is much more open-ended. One might wish to argue that an organism's genome is its essence. If so, then this essence is an individual essence and an essence of a very peculiar sort, since genomes are neither eternal nor immutable and two organisms can have the same individual essence without becoming the same individual. Finally, rapid change can occur in a single generation in a population without that population ceasing to be the same population under special circumstances; e.g. if gene frequencies differ markedly in males and females.

7. Comparable difficulties exist for the philosophical notion of the "self." Unlike organisms and species, most of the problem cases for the "self" as an individual are hypothetical; see Derek Parfit (1971).

BIBLIOGRAPHY

Dobzhansky, T.: 1970, *Genetics of the Evolutionary Process*, Columbia Univ. Press, New York.

Ehrlich, P. R. and Raven, P. H.: 1969, "Differentiation of Populations," *Science* 165, 1228–1231.

Ghiselin, M. T.: 1974, "A Radical Solution to the Species Problem," *Systematic Zoology*, 536–544.

Hull, D. L.: 1975, "Central Subjects and Historical Narratives," *History and Theory* 14, 253–274.

Lewontin, R. C.: 1970, "The Units of Selection," *Annual Review of Ecology and Systematics* 1, 1–18.

Mayr, E.: 1963, *Animal Species and Evolution*, Harvard Univ. Press, Cambridge, Mass.

Meglitsch, P. A.: 1954, "On the Nature of Species," *Systematic Zoology* 3, 49–65.

Parfit, D.: 1971, "Personal Identity," *The Philosophical Review* 80, 3–27.

Sahlins, M. D. and Service, E. R.: 1960, *Evolution and Culture*, Univ. of Michigan Press, Ann Arbor, Mich.

Suppe, F.: 1974, "Some Philosophical Problems in Biological Speciation and Taxonomy," in J. A. Wojciechowski (ed.), *Conceptual Basis of the Classification of Knowledge*, Verlag Dokumentation, Pullach/Muchen.

Thorson, T. L.: 1970, *Biopolitics*, Holt, Rinehart & Winston, Inc., New York.

Wilson, E. O.: 1974, "The Perfect Societies," *Science* 184, 54.

14

ARTHUR L. CAPLAN*

Have Species Become Déclassé?[1]

1. Two Views of the Ontology of Species

THERE IS NO more popular pastime in the literature of the philosophy of biology than analyzing the concept of species. This is partly due to the fact that the concept is such a prominent one in presentations of evolutionary theory. It is also a result of the fact that philosophers and biologists have struggled without a great deal of success to formulate a definition of species that would be acceptable for all the diverse purposes of the biological sciences.

For some time, philosophers of biology assumed that, whatever problems of definition and explication exist regarding the concept of a species, the ontological status of the concept was certain. Species have long been viewed as classes. Indeed, they have been viewed as paradigmatic examples of classes of a special variety. Since the traits of organisms vary from creature to creature as well as from generation to generation, a special set of properties must be used to group or aggregate organisms into species. Unlike most classes of objects, there often is no single property or trait which is present in all the members of a species. Rather, there are sets of properties some number of which are instantiated in any given organism (Beckner 1968, pp. 60–66; Ruse 1973, pp. 122–139).

*From Arthur L. Caplan, "Have Species Become Déclassé?," P. Asquith and R. Ciere (eds.), *PSA 1980*, Vol. 1 (East Lansing, MI: Philosophy of Science Association, 1980), pp. 71–82. Copyright © 1980 by the Philosophy of Science Association.

Thus, species in biology have been viewed as good examples of cluster concepts (Bambrough 1960–61). The properties used to group individuals into classes are such that, while no single organism possesses all of them, each organism possesses some of the set or cluster. Thus, for example, if the set of properties used to define the class of organisms we know as *Cygnus olor* are white color, long necks, downy feathers, large wings, black eye and beak markings, short legs, and good swimming ability, traditionally it has not been viewed as necessary that every bird possess all of these traits to be included in this species. It is entirely possible that a swan may be found which lacks one or another of these various properties. Nevertheless, the bird might still be classified as a member of the species *Cygnus olor*. As long as a given organism possessed a high proportion of the relevant diagnostic properties, the species membership of the organism was thought to be secure (Beckner 1968, pp. 60–72).

Recently, however, there has been a shift in the thinking of a number of philosophers and biologists concerning the ontological status of species. They have argued that if we are to justifiably group creatures with diverse traits into species, we must begin to rethink the ontological status of the species concept.

The proposal which has captured the allegiance of a number of scholars (Ghiselin 1975; Hull 1976, 1978; Mayr 1976; Reed 1979; Van Valen 1976) in the past few years is that, rather than viewing species as classes, they be viewed as spatiotemporally localized individuals. By individuals proponents of this ontological shift mean "spatiotemporally localized, cohesive and continuous entities" (Hull 1978, p. 336). On this view organisms would not be members of the classes of species to which they belong. Instead, they would have the relationship of a part to a whole in the way that cells, tissues, and organs are parts of and not members of individual human beings.

The cohesion and continuity used to individuate species as individuals resides in the fact that organisms can be grouped according to their lineage or phylogeny. Since evolution occurs partly as a consequence of the (imperfect) transmission of genes from one organism to another, it is possible to trace continuous lines of descent among many different organisms. As David Hull has observed ". . . organisms form lineages. The relevant organismal units in evolution are not sets of organisms defined in terms of structural similarity but lineages formed by the imperfect copying processes of reproduction" (1978, p. 341). It is the continuity and coherence of organic lineages that provide the grounds for thinking of species as individuals.

Those who have advocated this change in the ontological status of species have not done so lightly; they are well aware of the fact that viewing species as individuals represents a drastic break from previous thinking about the species concept (Ghiselin 1975; Hull 1976, 1978). However, they feel that such an ontological shift is necessitated both by

current theoretical work in evolutionary biology, and, by the conceptual benefits to be garnered from this maneuver. In the remainder of this paper I shall try to show that (a) the reasons advanced for viewing species as individuals rather than classes are not persuasive, (b) a reasonable explication can be given of species as classes that is consistent with the tenets of the modern synthetic theory of evolution, and, (c) that a cost/benefit computation of the two ontological views of the species concept actually favors the classic rather than the individualistic interpretation.

2. The Rationale for Viewing Species as Individuals

One of the most important reasons underlying the claim that species are best viewed as individuals and not classes or sets of organisms is that current evolutionary theory has as one of its main implications the conclusion that species are best treated as individuals. This implication is drawn from two disparate sources: (a) the specific claims about species and speciation made by evolutionists and (b) the role played by species in evolutionary theorizing.

While there is a good deal of uncertainty as to the exact nature of the tenets of the modern synthetic theory of evolution, there is no dispute concerning the fact that evolutionary change is an expected outcome of this theory. This is a result of the fact that in a world of scarcity containing a number of promiscuous creatures, there are enough sources of variation and selection present to guarantee continuous change in the phenotypic constitution of these creatures over long periods of time. Continuing phenotypic change is to be expected as soon as the mechanisms and makeup of the biological world are even roughly discerned (Caplan 1979).

If this is so then the class view of species is doomed. Essentialism is, quite simply, not compatible with the continuous change model of population evolution dictated by current evolutionary theory. There are, in the long run, simply not going to be any essences which persist long enough to permit systematists to aggregate organisms into groups (Griffiths 1974). Even if one gets clever and tries to use clusters of properties to construct polythetic definitions, the probability is very great that *all* of the observable traits of creatures will eventually evolve to the point where any proposed cluster of properties would be useless as the criteria for classification.[2]

Not only is continuous change an implication of evolutionary theory, it is also an empirical fact. Few contemporary ornithologists would use the cluster of traits now thought diagnostic of *Cygnus olor* to identify the distant avian ancestors of this species. There are few essentialists in the trenches of paleontologists.

Not only does evolutionary theory dictate the occurrence of continuous change, it also mandates the need for continuity. As Hull observes, "Evolution is a selection process, and selection processes require continuity" (1976, p. 190). If selection is to act upon variation to produce new phenotypic alterations, it must do so by acting upon variants produced by genotypes which are transmitted from one generation to the next. The reproduction and replication of organisms require precisely the sort of continuity and spatiotemporal localization appropriate to individuals. Since species ultimately result from selection among genetic lineages, and, since genetic lineages are individuals, species must, to be selected, be individuals as well.

If it is true that species are the basic units of evolutionary change and if species can be seen as equivalent to segments of evolutionary lineages or phylogenies, then the view that species are best understood as individuals appears quite credible. Devotees of this line of reasoning note that if species are really treated as individuals by evolutionary theorists, this fact should be reflected in their language, both in describing and in individuating species. And, indeed, numerous examples can be found of evolutionary theorists referring to species as "super-organisms" or "homeostatic systems." Also biologists do often refer to new species as "splitting off," "budding," "fusing," and, even, as "unique" — all terminology more appropriate to the individuation of individuals than it is to the individuation of classes (Ghiselin 1975, pp. 542–543; Hull 1978, pp. 344–350).

3. Difficulties with These Arguments

It is certainly true that evolutionary theory, when supplemented with the appropriate empirical information for describing the contingencies of the past and present biotic world, provides much support for the view that change is an expected outcome of evolutionary processes. However, the notion that species are classes is quite capable of accommodating this fact.

It is not the case that the ontological interpretation of species as classes implies any commitment to essentialism. No property or properties need be deemed eternal or immutable in trying to generate traits by which organisms can be grouped or aggregated. The idea of family resemblance and the related notion of a cluster concept were specifically introduced into the analysis of the species concept to provide a means of grouping organisms into classes despite the fact of continuous evolutionary change in species membership. Merely noting the implications of evolutionary theory or the facts of phylogeny relative to change does not per se invalidate the utility of these modes of classification.

Nor is it the case that grouping individuals into classes is incompatible with complete and continuous change in the properties used to define a given class. When the vast majority of class defining properties are no longer instantiated by any organisms it is reasonable to assume that the class is either extinct or has evolved into a different class. Thus, at some point Archaeopteryx crossed the class line dividing reptiles and birds. Reptilian properties gradually disappeared while avian characteristics emerged.

Surprisingly, if evolutionary theory mandates continuous and irreversible change in the traits of organisms, the concept of a class is more appropriate to describing this state of affairs than is the concept of individuals. Individuals have clear, datable, unambiguous origins and terminations. Their parts are usually contiguous in space and time. Species, however, have fuzzy beginnings and ambiguous demises. Their members are rarely spatially contiguous. If one views them over time they are likely to either disappear with a whimper or slowly transmute into a new variety. Neither process is clear, datable, or localizable as would be required in individuating individuals. The individual Charles Darwin has a datable beginning and end. The evolution of reptiles into birds is not amenable to similar spatiotemporal demarcations.

It is true that biologists talk about species as budding, fusing, and splitting. However, the criteria they look to more than any other are the capacity for interbreeding and descent from a common ancestor. Biological language can be opaque in such uses because it is not always clear whether in talking about species biologists have reference to the taxonomic category "species" or to the actual groups of organisms (taxa) to be seen, directly, in nature, and, by inference, in the fossil record.

The taxonomic category "species" refers to a particular level of biological organization in a classification. It makes no reference to any actual organism or organisms. Indeed, the category "species" is clearly treated as a class concept in biology. Its defining properties are (a) the ability to produce fertile offspring between members of a group and (b) descent from a common ancestor (Mayr 1970; Simpson 1961). Most of the references to species made by biologists are to the category of species and not the taxa of the actual biological world. Statements to the effect that the "species" is the central unit of evolution, or, about the budding or splitting of species are claims about species as a general category. Certainly no evolutionist believes that claims such as "the species is the primary locus of selection" refer in an oblique way to a finite and denumerable list of extant and extinct species taxa. The dual function of the term species as a category and as a description of actual taxa in nature is misleading. The role played by species in evolutionary theory is, thus, ambiguous as to whether it refers to a particular level of biological organization characterized by descent and gene exchange, or, to the properties of a particular group of creatures which possess this type of biological organization.

4. Are Species Taxa Individuals or Classes?

Traditionally species taxa have been grouped on the basis of phenotypic properties. Observable traits of morphology, physiology, or behavior have been used to lump or group individual organisms into species. When a systematist attempts to decide whether a given fly is a member of *Drosophilia melanogaster* or *Drosophila persimilis* the decision is based upon a comparison of the traits of the unidentified fly with the traits of known individuals from these two species. The search for similarities and differences in phenotypic traits is at the heart of most classifications of taxa in biology (Bock 1977; Sokal 1973).

However, the story is more complex than this. In many cases biologists are concerned with establishing the cause of a particular similarity between organisms in classifying them. Biologists are particularly concerned to know whether the similarities they observe among creatures are a consequence of common ancestry at some earlier point in phylogeny. The issue of whether a given trait is homologous, analogous, or simply an artifact of measurement or allometry is one that occupies a good deal of attention in the classification and analysis of taxa (Bock 1977; Mayr 1969).

The focus on similarities and their proper classification as homologous or analogous is illustrative of the fact that the grounds for grouping organisms into species is less a matter of establishing coherence, continuity, or, functional role than it is a matter of similarity among the traits of organisms. Similarity of traits and establishing the causes of these observed similarities seem decisive criteria for grouping organisms into species.

When biologists address themselves to the question of why certain organisms manifest similarities among their traits they tend to offer two major explanations – similar genotypes as a result of common descent (homology), or, similar life histories and environmental circumstances (analogy). The fact that organisms exist with similar and relatively stable properties is thus held to be a result, not of the ontological nature of species, but to similarities and stabilities in the causes and circumstances affecting organisms (Stebbins 1977). Similar causes produce similar effects. So the similarities in the traits of organisms can be attributed to the stability and permanence of certain genotypes and environments. While it is true that genes and environments change, this fact is in no way incompatible with the observation that similar genes in similar environments will produce organisms endowed with similar traits.

It is unnecessary to posit any new ontological classification to explain the unity and coherence of species taxa. The unity and coherence of species is a direct consequence of evolutionary theory which pinpoints the relevant causes of species unity and coherence. The theory also reveals the fact that phenotypic traits are best understood as criteria for establishing commonalities at a more basic level. Commonality of genotype and environment are pivotal for understanding overt similarities among organisms. Species taxa are perhaps best understood as groups of individuals which share similar genotypes and similar environments. The hidden "essences" of species taxa are the genotypes and environments which produce the similarities of traits we observe among organisms. The problem is that the resolving power of phenotypic properties for discerning similarities of genotype or environment is relatively poor (Lorenz 1970; Mayr 1969). Nevertheless, the characterization of a species taxon as a class aggregated on the basis of environmental and genotypic similarities seems quite adequate for explaining the unity and coherence of species and the kind of references biologists make in describing species taxa.

5. Ontological Costs and Benefits

One of the most important arguments cited in favor of the designation of species as individuals concerns the role played by species in the laws and generalizations of current evolutionary theory, or rather the absence of a role for species in the generalizations of evolutionary theory.

Most philosophical analyses of laws in natural science depict them as describing timeless regularities in nature. Reference to particulars is seen as simply incompatible with the universality required of scientific laws (Hempel 1966, pp. 33–40; Ruse 1970, 1977). Thus, if species are treated as individuals, they cannot appear in the generalizations of evolutionary theory. But since, this argument continues, phrases such as "swans are white" rarely grace the pages of modern textbooks of evolutionary theory, the validity of the species as individuals view is confirmed (Hull 1978, pp. 353–355). Treating species as individuals will free evolutionary biology from the charge that it can have no true laws because lawlike claims about albinotic mice, black crows, and other assorted creatures of zoonomic concern will no longer count as evidence for the provincial nature of evolutionary biology.

There is something quite peculiar about this line of argument. The argument explains away a problem that does not really exist and claims it as a benefit. Few biologists seriously believe that generalizations about the colors of swans or crows are central tenets of evolutionary theory, or, mainstays in evolutionary explanations.

Insofar as there is a need to explain the absence of references to specific taxa in the laws of evolutionary biology, perfectly sound reasons for this absence can be given that require no radical ontological shifts. First, descriptive generalizations about observable empirical phenomena are rarely viewed as plausible candidates for the status of laws in any science. In physics descriptions of the behavior of balls rolling down inclined planes are viewed as *explananda* — things to be explained and not the explanations of physical science. The fact that no evolutionary biologist posits laws about bird colors or the breeding patterns of mice can be understood as a consequence of the view that observable empirical regularities *require* explanation.

Second, the distinction between species as a category and species as a means of denoting a particular taxon is relevant to understanding the absence of certain types of generalizations in evolutionary theory. Biologists study taxa they suspect possess enough properties (genotypes and environments) to count as species. They then generalize, on the basis of the observations that they make concerning these taxa, to hypotheses about any group of organisms that might be organized and related in this fashion — the species category. When the term species does in fact appear, in such evolutionary generalizations as "two species can rarely occupy the same niche," "speciation is only the multiplication of species," or, "populations of a species may start to diverge before the appearance of an external barrier," the concept "species" is meant to refer to the *category* species and not to a finite number of spatiotemporally localized groups of taxa.

Those who object that biology is not a science or is, at best, a provincial science (Munson 1975), conflate the categorical and taxonomic senses of the species concept. The properties used to group taxa into species may be limited in their proper predication to entities which occupy a small segment of space and time. However, the criteria used to characterize the species category — commonality of descent and the ability to exchange genetic information — can, and have (Williams 1970) been described without reference to the particularities of individuals. And it is the latter sense which dominates biological usage in evolutionary generalizations.

There are further twists and turns to the argument about the benefits to be obtained from allowing species to be treated as individuals in scientific theorizing in biology. A key benefit that is held to accrue to this view of species is a broadening of the concept of scientific explanation. Hull argues that:

> Because many scientists, especially those working in historical disciplines . . . can rarely derive the sequences of events which they investigate from any laws of nature, philosophers tend to dismiss the efforts of these scientists as not being genuinely explanatory . . . . Historical narratives describing the evolution of mammals, the splitting of Pangea into the various continents, and the rise and fall of the Third Reich certainly seem explanatory. . . . Historical narratives can be just as explanatory as derivations from scientific laws even though they concern unique sequences of events. (1976, p. 188).

If species are individuals then their continuity and coherence can provide the "glue" necessary for constructing explanatory narratives in historical sciences.

The problem here is that the supposed benefit of viewing species as individuals is not contingent upon ontology. Debates about the nature and structure of historical explanations are legion. But the fact is that there is sufficient unity and continuity in the common properties of the members of a class to allow class terms to serve as the subjects of narrative statements. Think about the sagas that have been spun about the glorious achievements of the Boston Celtics, the Green Bay Packers, and the graduates of Harvard College. All of these entities are classes, yet they all act as unifying reference points for historical narratives of great complexity and, even, explanatory power.

Not only can classes serve the role of the subjects of historical accounts, but, the use of species in this role is quite compatible with the modes of explanation current in evolutionary theory. Evolutionists often do describe the changes that have taken place in a particular phylogeny or lineage segment. But it would be inaccurate to say that such narratives are the end of explanatory matters. For modern evolutionists (Simpson 1961; Williams 1966; Wilson 1975) believe that all of the trends and patterns exhibited in phylogeny can be explained by means of atemporal laws governing the processes of genetic duplication, gene transmission, development, and selection. The key properties of evolving species are genotype and environment — class categorizing properties that allow for the analysis of historical sequences of events by atemporal causal laws. The explanation of historical sequences of events by subsuming each event under a set of various atemporal nomological generalizations is the most distinctive of all the explanatory patterns found in evolutionary biology. Tracing phylogenies and spinning historical narratives may be legitimate explanatory techniques, but it is only because such techniques are place-holders for the actual application of the atemporal mechanistic laws of population biology, genetics, ecology and demography to events in the history of the characteristic genotypic and environmental properties of the members of species.

Perhaps the strangest of all the benefits that has been claimed in the name of rugged species individualism is the fact that such a view results in the elimination of any science from the roster of science that purports to study only one species. Surely it would be ludicrous to have a discipline devoted to the study of a single individual — no possible laws or generalizations could emerge from such a sample. Thus, much of social science, ethology and various sub-fields in biology such as primatology, parasitology, and ornithology will either be consigned to the scientific dustbin, or, to the unglamorous status of explanatory narration.

The proposition that theories in the social and biological sciences which explain events in single species ought be dismissed as unscientific is likely to receive a rather cool reception from practitioners in these areas. Many scientists appear to be quite content to construct nomic generalizations about small numbers of species. The touting of the recategorization of the social sciences as a welcome benefit of a species as individuals view appears especially dubious at a time when many scientists are renewing their efforts to construct systematic theories of human social behavior.

Perhaps the difficulties associated with generalizations in social science are better understood as problems with predicates and not ontology. Attempts to link the phenotypic properties of a class of entities into laws may fail simply because such efforts overlook the causally efficacious role of genotypic and environmental properties in producing such traits. Social scientists and others concerned with single species may need to construct new nomic generalizations that attempt to link the truly significant attributes of such classes. Indeed, many biologists and anthropologists (Alexander 1975; Chagnon and Irons 1979; Wilson 1975) have recently advanced this claim in some of the sociobiological literature. Scope, range, and particular reference may be false leads in understanding the difficulties confronting those who study mice and men.

The costs of treating species as individuals do not appear to be neutralized by corresponding gains in conceptual clarity. Nor do current biological theories or usage seem to demand or support any ontological shift. Without demonstrable need or benefit there would seem to be no reason for declassifying species.

NOTES

1. I would like to thank Walter Bock, David Hull, Janet Caplan, and Carola Mone for their helpful comments on an earlier draft of this paper.
2. This view overlooks the fact that the properties of cluster concepts can themselves be viewed as classes. One solution to accommodating continuous change with the view of species as classes is to view cluster concepts as classes and attempt to locate "characteristic" properties for particular clusters (See Suppe 1973).

REFERENCES

Alexander, R. D. (1975). "The Search For a General Theory of Behavior." *Behavioral Science* 10: 77–100.

Bambrough, R. (1960–61). "Universals and Family Resemblances." *Proceedings of the Aristotelian Society* 61: 207–222. (As reprinted in *Universals and Particulars*. Edited by Michael Loux. Garden City, NY: Doubleday & Co., 1970. Pages 109–127.)

Beckner, M. (1968). *The Biological Way of Thought*. Berkeley: University of California.

Bock, W. (1977). "Foundations and Methods of Evolutionary Classification." In *Major Patterns in Vertebrate Evolution*. Edited by M. Hecht. New York: Plenum. Pages 851–895.

Caplan, A. L. (1979). "Darwinism and Deductivist Models of Theory Structure." *Studies in the History and Philosophy of Science* 10: 341–353.

Chagnon, N. A. and Irons, W. (eds.). (1979). *Evolutionary Biology and Human Social Behavior: An Anthropological Perspective*. North Scituate, Mass.: Duxbury Press.

Ghiselin, M. T. (1975). "A Radical Solution to the Species Problem." *Systematic Zoology* 23: 536–544.

Griffiths, G. C. D. (1974). "On the Foundations of Biological Systematics." *Acta Biotheoretica* 23: 85–131.

Hempel, C. G. (1966). *Philosophy of Natural Science*. New York: Prentice-Hall.
Hull, D. L. (1976). "Are Species Really Individuals?" *Systematic Zoology* 25: 536–544.
———. (1978). "A Matter of Individuality." *Philosophy of Science* 45: 335–360.
Lorenz, K. (1970). *Studies in Animal and Human Behavior*, Vol. I. Trans. Robert Martin. Cambridge: Harvard University Press. (Originally published as *Über Tierisches und Menschliches Verhalten: Gesammelte Abhandlungen*, Band I. Munich: R. Piper Verlag, 1970.)
Mayr, E. (1969). *Principles of Systematic Zoology*. New York: McGraw-Hill.
———. (1970). *Populations, Species, and Evolution*. Cambridge: Harvard.
———. (1976). "Is the Species a Class or an Individual?" *Systematic Zoology* 25: 192.
Munson, R. (1975). "Is Biology a Provincial Science?" *Philosophy of Science* 42: 428–447.
Reed, E. S. (1979). "The Role of Symmetry in Ghiselin's Radical Solution to the Species Problem." *Systematic Zoology* 28: 71–78.
Ruse, M. (1970). "Are There Laws in Biology?" *Australasian Journal of Philosophy* 48: 234–246.
———. (1973). *The Philosophy of Biology*. London: Hutchinson.
———. (1977). "Is Biology Different From Physics?" In *Logic, Laws and Life*. (*University of Pittsburgh Series in the Philosophy of Science*, Vol. 6.) Edited by R. G. Colodny. Pittsburgh: University of Pittsburgh Press. Pages 89–128.
Simpson, G. G. (1950). "Evolutionary Determinism." In *This View of Life*. New York: Harcourt, Brace & World. Pages 176–189. (As reprinted in *Man and Nature*. Edited by R. Munson. New York: Dell, 1971. Pages 200–212.)
———. (1961). *Principles of Animal Taxonomy*. New York: Columbia University Press.
Sokal, R. R. (1973). "The Species Problem Reconsidered." *Systematic Zoology* 22: 360–374.
Stebbins, G. L. (1977). *Processes of Organic Evolution*. Englewood Cliffs: Prentice-Hall.
Suppe, F. (1973). "Facts and Empirical Truth." *Canadian Journal of Philosophy* 3: 197–212.
Williams, G. C. (1966). *Adaptation and Natural Selection*. Princeton: Princeton University Press.
Williams, M. B. (1970). "Deducing the Consequences of Evolution: A Mathematical Model." *Journal of Theoretical Biology* 29: 343–385.
Wilson, E. O. (1975). *Sociobiology: The New Synthesis*. Cambridge, Mass.: Harvard University Press.
Van Valen, L. (1976). "Individualistic Classes." *Philosophy of Science* 43: 539–541.

15

MARK RIDLEY*

Principles of Classification

WHAT IS THE proper relation of the theory of evolution and the classification of living things? The strongest possible relation would be one of practical necessity, if classification were practically impossible without the theory of evolution. The facts of history alone show that this relation does not hold. People had successfully classified animals and plants for two millennia before evolution was ever accepted. The simplest act of classification, indeed, requires no theory at all, let alone the theory of evolution; it merely requires that groups be recognized, defined, and named. A group, in this simple sense, is a collection of organisms that share a particular defining trait; the group Chordata for instance contains all animals that possess a notochord, a hollow dorsal nerve chord, and segmented muscles. Classification, as the definition and naming of groups, is in principle easy, but it is also important. It is even essential. Biologists could not communicate or check their discoveries if their specimens had not been classified into publicly recognized groups.

If communication were the only purpose of classification it would not matter what groups were defined, provided that the definitions were agreed upon. Chordata happens to be a group that is generally recognized, but by the same method we could define other groups that are not normally recognized. We might, for instance, define the Ocellata as the group of all living things that possess eyes. It would contain most vertebrates, many insects and crustaceans, some molluscs and worms, and some other odd invertebrates. The Ocellata has not, so far as I know, ever been considered as a taxonomic group; but, if classification is only a matter of defining and naming groups, we might ask why it is not.

That question brings us to the fundamental problem of classification. Different traits define different groups. We could just accept some groups and not others. We could agree that Ocellata is intrinsically as good a group as Chordata, but declare that it just so happens that we have decided to recognize Chordata but not Ocellata. Classification would then be subjective. And if that satisfied us, evolution would not only be practically unnecessary but completely unnecessary, for there would be no more to classification than its practice.

Most biologists, however, are not so easily satisfied. They would prefer the choice of groups to be principled rather than subjective. Then, if we do recognize Chordata but not Ocellata, it must be because some principle shows that the Chordata are an acceptable group, but the Ocellata are not. A perfect principle would unambiguously show whether any group was acceptable, and admit no conflict between acceptable groups. No groups would then be chosen subjectively: they would all be chosen by reference to the principle. Even in the absence of a perfect principle, a principle might still be useful if it narrowed down the number of groups that were acceptable. Once the need for a principle of the choice of traits is recognized, a third relation of the theory of evolution and classification is opened up between practical necessity and complete dispensability. If evolution supplied the only valid principle of choosing traits it would be philosophically necessary for classification, or if merely one among several principles, philosophically desirable. We have seen that evolution is practically unnecessary: from now on we shall be concerned with whether it is philosophically necessary, philosophically desirable, or completely unnecessary.

We can assume that the classification of life will at all events be hierarchical. A hierarchical classification is one whose groups are contained completely within more inclusive groups, with no overlap: humans (for example) are contained within the genus *Homo*, which is contained within the order of primates, which is contained within the class Mammalia, which is contained within the sub-phylum Vertebrata, which is within the phylum Chordata, which is within the kingdom Animalia. In principle biological classifications might not be hierarchical, but in practice they nearly all are. We are not ignoring a contentious practical issue.

What, then, could supply a principle for the hierarchical classification of life? Two kinds of answers have been offered: a *phenetic* hierarchy, or a *phylogenetic* one. A phenetic hierarchy is one of the similarity of form of the groups being classified; it is defined by any traits, such as leg length, skin colour, number of spines on back, or some collection of them. A phylogenetic hierarchy is one of the pattern of evolutionary descent; **groups are** formed according to recency of common ancestry.

The phenetic and phylogenetic principles may agree or disagree, **according** to the species considered. Figure 1 shows the phenetic and **phylogenetic** relations of three sets of three species. The phenetic and phylogenetic classifications are the same if the rate of evolution is approximately constant and its direction is divergent, as is probably true of a human, a chimp, and a rabbit (Figure 1a): the human and the

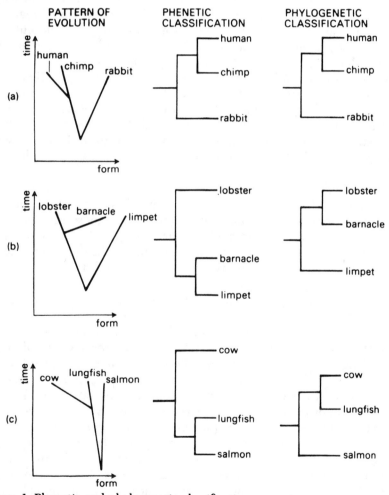

Figure 1. Phenetic and phylogenetic classification.
The pattern of evolution (left), phenetic classification (centre), and phylogenetic classification (right) for three cases. Phenetic and phylogenetic classification agree in the case of (a) human, chimp, and rabbit; but disagree when there is convergence, as in the case of (b) barnacle, limpet, and lobster, or when there is differential divergence, as in the case of (c) salmon, lungfish, and cow.

chimp share a more recent common ancestor and resemble each other more closely than does either with the rabbit. The two principles disagree when there is convergence or differential rates of divergent evolution. A barnacle, a limpet, and a lobster illustrate the case of convergence (Figure 1b): the barnacle and limpet are phenetically closer, but the barnacle has a more recent common ancestor with the lobster. The barnacle has converged, during evolution, on to the molluscan form. The salmon, lungfish, and cow illustrate the other source of disagreement (Figure 1c): a lungfish is phenetically more like a salmon than a cow; but it shares a more recent common ancestor with a cow than with a salmon. The evolutionary line leading from lungfish to cows has changed so rapidly that cows now look utterly different from their piscine ancestors. Lungfish indeed have hardly changed at all in 400 million years; they are often called living fossils.

Clearly, evolution is a necessary assumption of phylogenetic classification: if organisms did not have evolutionary relations, we could not classify according to them. But evolution is not an assumption of the phenetic system: we could classify organisms by their similarity of appearance whether they shared a common ancestor or had been separately created. In the phenetic system, classification is by similarity of appearance, not evolution. If the phylogenetic principle is invalid, evolution will be completely unnecessary in classification; if both principles are valid, evolution will be philosophically desirable; if only the phylogenetic principle is valid, evolution will be philosophically necessary. If neither principle is valid, we shall have to fall back on the kind of subjectivity that we aimed to escape from. Such is the significance of the question of whether the two principles are valid.

Phenetic and phylogenetic classification have each grown into a whole school, complete with its own philosophical self-justifications, techniques, and advocates. Phenetic classification is advocated by the school of numerical taxonomy; phylogenetic by the school called cladism. There is another important school of classification as well. It is a school often called evolutionary taxonomy, whose practitioners are more numerous than either numerical or cladistic taxonomy. But despite its importance, we need not concern ourselves with it here. Its hierarchical classifications mix phenetic and phylogenetic components; the "evolutionary" classification of the barnacle, limpet, and lobster is phylogenetic, but of lungfish, salmon, and cow is phenetic. In covering the pure extremes, we shall cover the arguments of the mixed school too. What we want to know are the justifications of numerical taxonomy, which should justify phenetic classification, and of cladism, which should justify phylogenetic classification. Let us take phenetic classification first.

We have already met the difficulty of classification by an arbitrarily chosen trait. It is that some traits define some groups, other traits other groups; eyes defined the (customarily unrecognized) Ocellata, noto-chords the (customarily recognized) Chordata. If our only principle is to pick traits and define groups by them, we are left with a subjective choice among conflicting groups like Ocellata and Chordata. The numerical taxonomic school, which flourished from the late fifties into the sixties, believed that it had an answer to this problem. It would classify not by single traits, but by as many traits as possible. It would study dozens, even hundreds, of traits, which it would then average in order to define its groups. It came with a repertory of statistical procedures designed to realize that end. The general kind of statistic is what is called a multivariate cluster statistic. Given many measurements of many traits in the units to be classified, the cluster statistic averages all the measurements, to form groups (or "clusters") of units according to their similarity in all the traits. The groups in the classification are said to be defined by their "overall morphological similarity." It was believed that if many traits were used, the groups discerned by the cluster statistic would be less arbitrary. Whereas groups defined by a few different traits may contain very different members, as did Chordata and Ocellata, groups defined by a large number of traits (it was thought) would have more consistent memberships.

We must consider a cluster statistic in more detail. The clusters are formed according to what is called the "distance" between the units being classified. A distance in this sense is not the distance from one place to another, such as five miles, but is the difference between the values of a trait in two groups. Suppose that we are classifying two species. If the legs of one species are on average six inches long and those of the other four inches, the distance between the two species with respect to leg length is two inches. Numerical taxonomy, however, does not operate with only one trait. It uses dozens. The distance between the groups is therefore measured as the average for all the traits. If we had also measured skin colour in the two species and the distance between the colours had been 0.3 units, then the average distance for the leg length and skin colour combined is $(2 + 0.3)/2 = 1.15$ units. This figure is called the "mean trait distance"; it would also be possible to use the Euclidean distance, which is measured, in two dimensions, by Pythagoras's theorem. The method can be applied for an indefinitely large number of traits and species (or whatever unit). The cluster statistic can then set to work.

The cluster statistic forms groups (or "clusters") by successively aggregating the units with the shortest distances to each other. It forms a hierarchy of clusters as more and more distant units are added in. The numerical taxonomy, or phenetic classification, of the species will then be exactly defined. The classification *is* the hierarchic output of the statistic.

The advantages which numerical taxonomy claims for itself are its objectivity and repeatability. Any taxonomist could take the same group of animals or plants, measure many traits quantitatively, feed the measurements into the cluster statistic, and the same classification would always emerge. Numerical taxonomists claimed that by contrast the methods used to reconstruct phylogenetic trees were hopelessly vague and woolly. We have not yet come to those techniques; but we can look now at how well the method of numerical taxonomy stands up by its own criterion. Is numerical taxonomy, and its resulting phenetic classification, really objective and repeatable?

The answer is that it is not. The reason is abstract, but too important to ignore. It was pointed out most powerfully by an Australian entomologist, L. A. S. Johnson, in 1968. When I wrote above that the cluster statistic simply forms a hierarchy by adding in turn the next least distant group, I ignored a problem. There is more than one cluster statistic, because there is more than one way of recognizing the "nearest" group. As we shall see, these different cluster statistics define different groups. If numerical taxonomy is truly objective, its own principle must dictate which cluster statistic should be used and which classificatory groups recognized. If it does not its own claim to repeatability will be exploded. It will be hoist on its own petard.

We can illustrate the point by two different statistics, called a nearest neighbour statistic and an average neighbour statistic. These are just two among many, which makes the real problem even worse than what follows; but we can use only two statistics to illustrate the nature, if not the extent, of the problem. Nearest neighbour statistics form successively more inclusive groups by combining the sub-groups with the nearest neighbour to each other. We can see it in Figure 2, along with the average neighbour statistic, which forms more inclusive groups not from those subgroups with the nearest *nearest* neighbour, but from those with the nearest *average* neighbour.

In Figure 2, the nearest neighbour and average neighbour cluster statistics produce different hierarchies. In many cases the two statistics will produce hierarchies of the same shape, even if they do differ quantitatively. But sometimes they will not. The two statistics then give different classifications. (The Figure only has two dimensions, which might be, say, leg length and skin colour; as we have seen, numerical taxonomists rely on many more than two traits. But that simplification is only to fit the printed page; it does not matter for the general point. Indeed the problem grows worse as more dimensions are introduced).

Figure 2. Two cluster statistics in disagreement. Seven species (nos. 1–7) have been measured for two traits, leg length and skin colour, and plotted in a two-dimensional space, with their leg length on the x-axis and their skin colour on the y-axis. Two cluster statistics, a nearest neighbour statistic and an average neighbour statistic, have been used to classify the seven species. The resulting classifications are shown below: (a) by nearest neighbour (b) by average neighbour. Both statistics first recognize the two groups, one of species 1, 2, 3, and the other of species 5, 6, 7. They disagree about the classification of species 4.

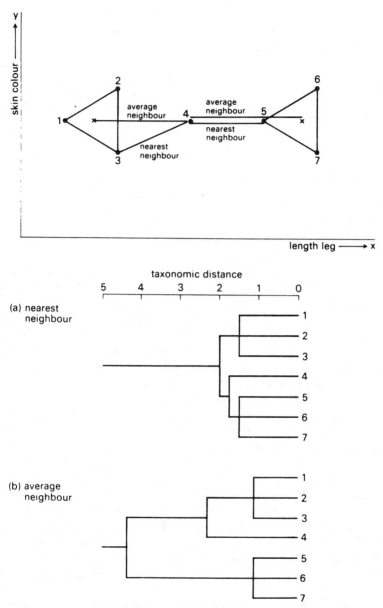

The nearest neighbour statistic joins it to the group with the nearest *nearest* neighbour: it compares the two distances labelled nearest neighbour: the one to species 5 is shorter: species 4 is classified with species 5, 6, 7 (classification (a)). The average neighbour statistic joins it to the group with the nearest *average* neighbour. The two points marked x are the average distances from species 4 to each group. The average neighbour statistic compares the two distances labelled average neighbour: the distance to the group of species 1, 2, and 3 is shorter, and species 4 is therefore classified with them (classification (b)).

The principle of numerical taxonomy provides no guidance among the different cluster statistics. It implies no criterion by which to choose among the different hierarchies produced by different statistics. The principle of numerical taxonomy is to classify according to "overall morphological similarity," but overall morphological similarity can only be measured by a cluster statistic. There is no higher measure of overall morphological similarity against which the different cluster statistics can be compared. When different statistics conflict, the practical numerical taxonomist has to decide which one he prefers. He can make a choice, of course; but it will have to be subjective.

The principle of phenetic classification, therefore, is a failure. Numerical taxonomy successfully removed subjectivity from the choice of traits, but only to see it pop up again (in a less obvious but equally destructive form) in the choice of cluster statistic. If phenetic relations cannot provide a valid principle for the hierarchical classification of living things, what about phylogeny?

Here there is more hope. Unlike the hierarchy of phenetic resemblance (or overall morphological similarity), the phylogenetic hierarchy does exist independently of our techniques to measure it. The phylogenetic tree is a unique hierarchy. It really is true, of any two species, that they either do or do not share a more recent common ancestor with each other than with another species. In phylogenetic classification, there is no problem of subjective choice among different possible hierarchies. There is only one correct phylogeny. If the evidence suggests that one classification is more like it than another, that is the classification to choose. The Chordata are allowed, but the Ocellata are not, because the group of chordates share a common ancestor but the group of species that possess eyes do not. Sometimes there will not be enough evidence to say whether one classification is more phylogenetic than another: then we should either have to make a provisional subjective choice, or refuse to choose until more evidence becomes available. The advantage of the phylogenetic principle is that it does possess a higher criterion to compare the evidence with: the phylogenetic hierarchy. This advantage had been vaguely understood by many evolutionary taxonomists, but it was first thoroughly thought through by the German entomologist Hennig. Cladism is sometimes called Hennigian classification.

Now that we have solved the philosophical problem we are left with the practical one. How can phylogenetic relations be discovered? Evolution took place in the past. Unlike phenetic relations, phylogenetic relations cannot be directly observed. They have to be inferred. But how?

For each species, we need to know which other species it shares its most recent common ancestor with, for that is the species with which it should be classified. How can we discover it? The method proposed by Hennig (and previously applied by many others) is to look for traits that are evolutionary innovations. During evolution, traits change from time to time. According to whether a particular trait is an earlier or a later evolutionary stage, it can be called a primitive or a derived trait. Most traits pass through several stages in evolution, and whether a particular stage is primitive or derived depends on which other stage it is being compared with; it is primitive with respect to later stages, but derived with respect to earlier ones. Consider as an example the evolution of the vertebrate limb. The most primitive stage is its absence; then, in fish, it appears as a fin; in amphibians the fin evolves into the tetrapodan pentadactyl (five-digit) limb; it has stayed like that in most vertebrates, but in some lizards and independently in some ungulates the number of digits on the limb has been reduced from five to four, three, two, or even (in horses) one. If we compare amphibians with fish, the pentadactyl limb is the derived state and fins the primitive state of this trait; but if we compare an amphibian with a horse, the five-toed state becomes primitive and the one-toed state in the horse is now the derived. Similarly if we compare the hand of a human with the front foot of a horse, the pentadactyl human hand is in the primitive state relative to the single-toed equine foot.

Such is the meaning of primitive and derived states. The distinction is necessary, in Hennig's system, in order that the derived traits can be selected for use in classification. The derived traits are selected because shared derived traits indicate common ancestry, whereas shared primitive traits do not. Let us stay with the same example. Suppose that we wish to classify a five-toed lizard, a horse, and an ape in relation to each other. The ape and the lizard share the trait "five-toed," but this does not indicate that they share a more recent common ancestor than either does with the horse: the trait is primitive and does not indicate common ancestry within the group of horse, ape, lizard. Whenever there is an evolutionary innovation, it is retained (until the next evolutionary change) by the species descended from the innovatory species: shared derived traits do indicate common ancestry. That is why they are used to discover phylogenetic relations.

We can now move a further step in the search for a method. The problem has now become to distinguish primitive from derived traits. In the case of the vertebrate limb we assumed that the course of evolution was known. But how could it be discovered to begin with? There are several techniques. We need not consider them all. Let us look at one in detail to demonstrate that the distinction can be made. Let us consider outgroup comparison. The simplest case has one trait, with two states, in two species; lactation and its absence, for instance, in a horse and a toad. The problem is whether lactation in horses is derived with respect to its absence in toads, or its absence in toads is derived with respect to its presence in horses. The solution, by outgroup comparison, is obtained by examining the state of some related species, called the outgroup. The outgroup should be a species which is not more closely related to one of the two species than the other, that is why it is an *out*group: it is separated from the species under consideration. In this case any fish or invertebrate would do but a cow would not, because it is more closely related to the horse than it is to the toad. By the method of outgroup comparison, that trait is taken to be primitive which is found in the outgroup. Whether we took a species of fish or an invertebrate, the answer would be the same. The outgroup lacks lactation: lactation is the derived state.

The result of outgroup comparison is uncertain. It will be wrong whenever there is unrecognized convergence. Thus if we compared the trait "body shape" in dolphins and dogs with some such outgroup as a fish we should determine that the dog had the derived shape. Actually the dolphin does. No one would be mistaken in the case of dolphins and fish; but other more subtle cases of convergence surely exist which are not as easy to recognize, and in them outgroup comparison will be misleading. But although it can go wrong, it is probably better than nothing. Shared traits are probably more often due to common ancestry than to convergence: but only more often, not always.

Advocates of phenetic classification often remark that phylogenetic classification is impossible because its techniques are circular. In the case of outgroup comparison, for instance, in order to apply the technique we needed to know that the outgroup (the fish) was less related to the toad and the horse than either to each other. It appears that we need to know the classification before we can apply the techniques; which would be quite a problem since the technique is supposed to be used to discover the classification. The problem, however, is not as destructive as it appears.

The argument of outgroup comparison is not circular. It works by what is often called successive approximation. It is the method by which theories are developed in all sciences. As new facts are collected and considered, they are examined in the light of the present theory. If they fit it, confidence in the theory is increased. If they do not, they may suggest a new theory, which can be used in considering yet further evidence. There is a continual reexamination of the theory in the light of new evidence, and when the theory is changed, our interpretation of all previous evidence should change too. This is not circular reasoning: it is testing a theory. In outgroup comparison, we can start with some crude idea of which species is an outgroup; if further evidence fits the crude idea, the hypothesis is (tentatively) confirmed, and it can then be used in interpreting further facts. Let us consider a hypothetical example.

Let us suppose that we have six species, and we suspect that one of them is less closely related than the other five. That one can be used as an outgroup. Comparison with it can be used to classify the other five, whose relations are not yet known. We first examine a trait in all six species. We take the state in the outgroup to be primitive for the group of five. Figure 3 shows the procedure. There are two points to notice. One is that the procedure can start from a very vague starting point; we do not need a firm classification to apply outgroup comparison. The other is that, if further evidence demands it, we can modify our initial ideas. If one trait after another suggests that species 6 is *not* separate from 1–5 then we can modify the classification, and put 6 in its appropriate place. All the previous steps would then have to be reconsidered. As the analysis proceeds any error at the beginning has a decreasing effect; the initial errors are gradually discovered, and their damaging consequences removed. Such is the method of successive approximation. It is the common method of scientific theory-building: only sciences that completely lack theories can do without the feedback between the interpretation of facts and the testing of theories.

Outgroup comparison is not the only cladistic method. Another method supposes that, as the organism develops, the evolutionarily derived stages appear after the primitive stages. The backbone of vertebrates is a derived state relative to its absence in invertebrates; and the backbone develops after its absence in a vertebrate embryo. Like outgroup comparison, the embryological criterion is imperfect but better than nothing. Another method is to look at the order in which the traits appear in the fossil record. The most powerful technique is to take all these methods together, and use all the evidence available. I do

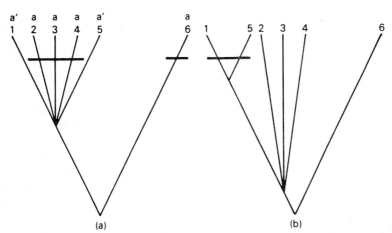

Figure 3. Outgroup comparison by successive approximation. (a) Five species (nos. 1–5) are to be classified, and it is thought that another species (6) is less related to them than any of the five to each other. Trait A is compared in all six species: species 2, 3, and 4 have it in state *a*, as does species 6, but species 1 and 5 have another state *a'*; by outgroup comparison *a'* is reasoned to be a derived state. (b) Species 1 and 5 are classified together, for they share a derived trait. The relations of species 2, 3, and 4 remain unknown; but the procedure can be repeated when evidence for new traits becomes available.

not wish to give the impression that the techniques of reconstructing phylogeny are perfect. They are far from that. Many problems remain, especially that of how to reconcile conflicting information from different traits. But although the system has difficulties, it is probably not altogether impractical. The cladistic evidence suggests that humans share a more recent common ancestor with chimps than with butterflies, and few biologists would deny that the evidence is correct in this case.

Derived traits, therefore, can be distinguished from primitive ones. Groups can be defined by shared derived traits. The cladistic system of phylogenetic classification is workable. But although cladistic groups can be recognized and defined, and although the classification (so far as it is phylogenetic) will be philosophically sound, it will only be valid for any one time. Evolution is continually going on; the traits of lineages continually change. What will define a group at one time may not define it at another. The traits defining groups are temporarily contingent, not essential. Biological groups do not have Aristotelian "essences." The phylogenetic group Vertebrata may happen to be defined, at present, by the possession of bones (or cartilage), but that may no longer be true in a million years' time. A descendant of a vertebrate species may lack bones, and the trait will cease to define the evolutionary group. The traits that define groups are not eternal and inevitable: they just happen to be useful sometimes.

For the same reason, difficulties arise when species from different times (particularly fossils and present-day species) are put in the same classification. It can be done, but it is often awkward. Because traits are continually changing, groups with members from more than one geological period cannot be defined by constant traits. Furthermore, because (as we shall see in the next chapter) species may vary in space, the traits used in a classification strictly only apply to one place. But spatial variation, being more limited than temporal variation, is less of a practical problem. Spatial and temporal changes in traits are both only difficulties in the practice, not in the philosophy, or phylogenetic classification. The phylogenetic relations of a species remain real. The difficulty is in their discovery; phylogenetic relations have to be inferred from shared derived traits. But because traits do not remain constant in time and space, they are not perfectly reliable guides to the true phylogenetic relations of a species.

Because phylogenetic classifications are defined by shared traits, it is tempting to think that phylogenetic classification is really a form of phenetic classification. There is a measure of truth in this idea. Phylogenetic classification is defined by shared traits, and to that extent is phenetic. But it could not be otherwise. All classifications are phenetic in this sense. The proper description of a classificatory system as phylogenetic or phenetic should not be according to the techniques that it uses, but the hierarchy that it seeks to represent. There is an utterly different philosophy behind the two systems. One tries to represent the branching pattern of evolution; the other tries to represent the pattern of morphological similarity.

Moreover, different *kinds* of traits define phylogenetic and phenetic classifications. Phylogenetic classification supposes that some traits — shared derived traits in particular — are better indicators of phylogeny than are others. Phylogenetic classification, at least of the cladistic variety, uses only shared derived traits. Phenetic classification, by contrast, indiscriminately uses both primitive and derived traits. Both for reasons of philosophy and technique therefore it is misleading to call phylogenetic classification a form of phenetic classification.

Phylogenetic classification is philosophically preferable to its only competitor, phenetic classification. It is also a practical possibility. Its main problems are in its techniques, which are (as yet) far from perfect. But the techniques of cladism have been improved, even within their short history, and further work, particularly on molecules, should improve them further. The difficulty in phenetic classification is more fundamental. Its claim to preference, which is its objectivity, is a false claim. It is left with little to recommend itself by. Phenetic classification should, I think, be avoided whenever possible. Classifications, if they are to be objective, must represent phylogeny.

If this conclusion is correct, evolution and classification are closely related. The relation is one of philosophical necessity. Evolution is required to justify the kind of classification that is practiced. Evolution is not merely desirable, but necessary, because phylogeny is the only known principle of classification. If we were content with merely subjective classification, evolution would be unnecessary. But if we are not — if we seek a principled classification — evolution becomes essential. It underwrites the entire philosophy of phylogenetic classification. Without evolution, phylogenetic classification, and its method of searching for derived traits, would be as subjective as any other technique. Only because we can assume that evolution is true, can we even begin to think about phylogenetic classification. Then we need techniques to detect phylogeny. The source of those techniques is our understanding of how traits change in evolution. The theory of evolution, therefore, not only guarantees the philosophy of classification: it is also the breeding ground of taxonomic techniques.

TELEOLOGY: HELP OR HINDRANCE?

16

GEORGE C. WILLIAMS*

Adaptation and Natural Selection

ANY BIOLOGICAL MECHANISM produces at least one effect that can properly be called its goal: vision for the eye or reproduction and dispersal for the apple. There may also be other effects, such as the apple's contribution to man's economy. In many published discussions it is not at all clear whether an author regards a particular effect as the specific function of the causal mechanism or merely as an incidental consequence. In some cases it would appear that he has not appreciated the importance of the distinction. In this book I will adhere to a terminological convention that may help to reduce this difficulty. Whenever I believe that an effect is produced as the function of an adaptation perfected by natural selection to serve that function, I will use terms appropriate to human artifice and conscious design. The designation of something as the *means* or *mechanism* for a certain *goal* or *function* or *purpose* will imply that the machinery involved was fashioned by selection for the goal attributed to it. When I do not believe that such a relationship exists I will avoid such terms and use words appropriate to fortuitous relationships such as *cause* and *effect*. This is a convention in general use already, perhaps unconsciously, and its appropriateness is supported in discussions by Muller (1948), Pittendrigh (1958), Simpson (1962), and others.

Thus I would say that reproduction and dispersal are the goals or functions or purposes of apples and that the apple is a means or mechanism by which such goals are realized by apple trees. By contrast, the apple's contributions to Newtonian inspiration and the economy of Kalamazoo County are merely fortuitous effects and of no biological interest.

It is often easy, in practice, to perceive functional design intuitively, but unfortunately disputes sometimes arise as to whether certain effects are produced by design or merely as by-products of some other function. The formulation of practical definitions and sets of objective criteria will not be easy, but it is a problem of great importance and will have to be faced. An excellent beginning was made by Sommerhoff (1950), but apparently no one has built upon the foundation he provided. In this book I will rely on informal arguments as to whether a presumed function is served with sufficient precision, economy, efficiency, etc. to rule out pure chance as an adequate explanation.

A frequently helpful but not infallible rule is to recognize adaptation in organic systems that show a clear analogy with human implements. There are convincing analogies between bird wings and airship wings, between bridge suspensions and skeletal suspensions, between the vascularization of a leaf and the water supply of a city. In all such examples, conscious human goals have an analogy in the biological goal of survival, and similar problems are often resolved by similar mechanisms. Such analogies may forcefully occur to a physiologist at the beginning of an investigation of a structure or process and provide a continuing source of fruitful hypotheses. At other times the purpose of a mechanism may not be apparent initially, and the search for the goal becomes a motivation for further study. Adaptation is assumed in such cases, not on the basis of a demonstrable appropriateness of the means to the end but on the indirect evidence of complexity and constancy. Examples are (or were) the rectal glands of sharks, cypress "knees," the lateral lines of fishes, the anting of birds, the vocalization of porpoises.

The lateral line is a good illustration. This organ is a conspicuous morphological feature of the great majority of fishes. It shows a structural constancy within taxa and a high degree of histological complexity. In all these features it is analogous to clearly adaptive and demonstrably important structures. The only missing feature, to those who first concerned themselves with this organ, was a convincing story as to how it might make an efficient contribution to survival. Eventually painstaking morphological and physiological studies by many workers demonstrated that the lateral line is a sense organ related in basic mechanism to audition (Dijkgraaf, 1952, 1963). The fact that man does not have this sense organ himself, and had not perfected artificial receptors in any way analogous, was a handicap in the attempt to understand the organ. Its constancy and complexity, however, and the consequent conviction that it must be useful in some way, were incentives and guides in the studies that eventually elucidated the workings of an important sensory mechanism.

I have stressed the importance of the use of such concepts as biological means and ends because I want it clearly understood that I think that such a conceptual framework is the essence of the science of biology. Much of this book, however, will constitute an attack on what I consider unwarranted uses of the concept of adaptation. This biological principle should be used only as a last resort. It should not be invoked when less onerous principles, such as those of physics and chemistry or that of unspecific cause and effect, are sufficient for a complete explanation.

For an example that I assume will not be controversial, consider a flying fish that has just left the water to undertake an aerial flight. It is clear that there is a physiological necessity for it to return to the water very soon; it cannot long survive in air. It is, moreover, a matter of common observation that an aerial glide normally terminates with a return to the sea. Is this the result of a mechanism for getting the fish back into water? Certainly not; we need not invoke the principle of adaptation here. The purely physical principle of gravitation adequately explains why the fish, having gone up, eventually comes down. The real problem is not how it manages to come down, but why it takes it so long to do so. To explain the delay in returning we would be forced to recognize a gliding mechanism of an aerodynamic perfection that must be attributed to natural selection for efficiency in gliding. Here we would be dealing with adaptation.

LITERATURE CITED

DIJKGRAAF, V.S., 1952, Bau and Funktionen der Seitenorgane und des Ohrlabyrinthes bei Fischen, *Experientia* 8:205–216.
——, 1963, The functioning and significance of the lateral-line organs, *Biol. Rev. Cambridge Phil. Soc.* 38:51–105.
MULLER, H. J., 1948, Evidence of the precision of genetic adaptation, *Harvey Lectures* 43:165–229.
PITTENDRIGH, COLIN S., 1958, Adaptation, natural selection, and behavior, Chap. 18 (pp. 390–416) in: *Behavior and Evolution,* A. Roe, G. G. Simpson, eds., Yale University Press, viii, 557 pp.
SIMPSON, G. G. 1962, Biology and the nature of life, *Science* 139:81–88.
SOMMERHOFF, G., 1950, *Analytical Biology,* Oxford University Press, viii, 207 pp.

17

PAUL J. KRAMER*

Misuse of the Term Strategy

DURING THE PAST decade biologists have become increasingly teleological in the terminology used to describe the activities of living organisms. This tendency is especially noticeable in discussions of evolution, life cycles, and the reactions of organisms to environmental stresses. It is manifested in such titles as "The Reproductive Strategy of Higher Plants," "The Strategy of the Red Algal Life History," and "Plant Strategies and Vegetation Processes." Too many papers have an anthropomorphic approach to biological problems in terms of a search for adaptations that will enable organisms to attain "goals" such as increased drought or cold tolerance. One scientist wrote, "We may usefully view a plant as an intricate control system in which responses to stress are strategies directed toward achieving certain goals."

How can plants or animals have goals, or strategies to attain these goals, which really exist only in the minds of the writers? Dictionaries define strategy as the development of plans of action to attain desired ends, usually social, economic, military, or political. How can a plant or animal, either as an individual or as a species, predict changes in the environment or know what it should do when confronted by an environmental crisis? Plants subjected to water stress often develop smaller, thicker leaves, with thicker cuticle, and have a smaller leaf area and a larger root–shoot ratio. However, these changes did not

*From Paul J. Kramer, "Misuse of the Term Strategy," *Bioscience*, Vol. 34, no. 7 (Washington: American Institute of Biological Sciences, 1984). Copyright © 1984 American Institute of Biological Sciences.

come about for the purpose of increasing drought tolerance, but because water stress causes physiological and biochemical changes in plants that affect their structure. These reactions to water stress result from screening by natural selection among the numerous mutations and recombinations that have occurred in the evolution of a particular kind of plant. As A. D. Bradshaw wrote in 1965, successful organisms represent a combination of characters that minimizes deleterious effects and maximizes advantageous effects of structure and function. In other words, plant structure and function are compromises to conflicting pressures such as the necessity of absorbing carbon dioxide while regulating water loss. They are not the product of a carefully planned strategy, but the result of the screening of random variations by natural selection.

Some writers seem to regard the use of teleological terminology such as, "striving to attain goals," as a way of catching the reader's attention. Others apparently use it metaphorically as a convenient method of examining problems. However, it is dangerous because it results in careless thinking and writing, and it misleads readers not trained in science who often mistake the metaphor for the truth. A humanist acquaintance, the late Harry Levy, suggested a warning that, "The attribution of purpose to plants is not intended literally, and if so taken is dangerous to your mental health." Scientists can have goals and can develop research strategies to attain them, but plants cannot, unless we are willing to grant that they have intelligence and can make decisions. Terms such as "strategy" and "tactics" are philosophically objectionable when applied to plants and lower animals, and are best left to politicians, the military, and athletic coaches.

18

FRANCISCO J. AYALA*

Teleological Explanations

TELEOLOGY (FROM THE Greek *telos* = end) is "the use of design, purpose, or utility as an explanation of any natural phenomenon" (*Webster's Third New International Dictionary*, 1966). An object or a behavior is said to be teleological or telic when it gives evidence of design or appears to be directed toward certain ends. The behavior of human beings is often teleological. A person who buys an airplane ticket, reads a book, or cultivates the earth is trying to achieve a certain end: getting to a given city, acquiring knowledge, or getting food. Objects and machines made by people also are usually teleological: a knife is made for cutting, a clock is made for telling time, a thermostat is made to regulate temperature. Features of organisms are teleological as well: a bird's wings are *for* flying, eyes are for seeing, kidneys are constituted for regulating the composition of the blood. The features of organisms that may be said to be teleological are those that can be identified as adaptations, whether they are structures like a wing or a hand, or organs like a kidney, or behaviors like the courtship displays of a peacock. Adaptations are features of organisms that have come about by natural selection because they serve certain functions and thus increase the reproductive success of their carriers.

Inanimate objects and processes (other than those created by men) are not teleological because they are not directed toward specific ends,

*From *Evolution* by Theodosius Dobzhansky et al., pp. 497–504. Copyright © 1977 W. H. Freeman and Company. Reprinted with permission.

they do not exist to serve certain purposes. The configuration of a sodium chloride molecule depends on the structure of sodium and chlorine, but it makes no sense to say that that structure is made up so as to serve a certain end. The shape of a mountain is the result of certain geological processes, but it did not come about so as to serve a certain end. The motion of the earth around the sun results from the laws of gravity, but it does not exist in order to satisfy certain ends or goals. We may use sodium chloride as food, a mountain for skiing, and take advantage of the seasons, but the use that we make of these objects or phenomena is not the reason why they came into existence or why they have certain configurations. On the other hand, a knife and a car exist and have particular configurations precisely in order to serve the ends of cutting and transportation. Similarly, the wings of birds came about precisely because they permitted flying, which was reproductively advantageous. The mating display of peacocks came about because it increased the chances of mating and thus of leaving progeny.

The previous comments point out the essential characteristics of telic phenomena, i.e., phenomena whose existence and configuration can be explained teleologically. We may now propose the following definition. *Teleological explanations account for the existence of a certain feature in a system by demonstrating the feature's contribution to a specific property or state of the system.* Teleological explanations require that the feature or behavior contribute to the existence or maintenance of a certain state or property of the system. Moreover, and this is the essential component of the concept, the contribution *must be the reason why the feature or behavior exists at all.*

The configuration of a molecule of sodium chloride contributes to its property of tasting salty and therefore to its use as food, not vice versa; the potential use of sodium chloride as food is not the reason why it has a particular molecular configuration or tastes salty. The motion of the earth around the sun is the reason why seasons exist; the existence of the seasons is not the reason why the earth moves about the sun. On the other hand, the sharpness of a knife can be explained teleologically because the knife has been created precisely to serve the purpose of cutting. Motorcars and their particular configurations exist because they serve transportation, and thus can be explained teleologically. (Not all features of a car contribute to efficient transportation — some features are added for aesthetic or other reasons. But as long as a feature is added because it exhibits certain properties — like appeal to the aesthetic preferences of potential customers — it may be explained teleologically. Nevertheless, there may be features in a car, a knife, or any other man-made object that need not be explained teleologically. That knives have handles may be explained teleologically, but the fact that a particular handle is made of pine rather than oak might simply be due to the availability of material. Similarly, not all features of organisms have teleological explanations.)

Many features and behaviors of organisms meet the requirements of teleological explanation. The hand of man, the wings of birds, the structure and behavior of kidneys, the mating displays of peacocks are examples already given. In general, as pointed out above, those features and behaviors that are considered adaptations are explained teleologically. This is simply because adaptations are features that come about by natural selection. Among alternative genetic variants that may arise by mutation or recombination, the ones that become established in a population are those that contribute more to the reproductive success of their carriers. The effects on reproductive success are usually mediated by some function or property. Wings and hands acquired their present configuration through long-term accumulation of genetic variants adaptive to their carriers. An alternative feature may be due to a single gene mutation, e.g., the presence of normal hemoglobin rather than hemoglobin S in humans. One amino acid substitution in the beta chain in humans results in hemoglobin molecules less efficient for oxygen transport. The general occurrence in human populations of normal rather than S hemoglobins is explained teleologically by the contribution of hemoglobin to effective oxygen transport and thus to reproductive success. The difference between peppered-gray and melanic moths is due to one or only a few genes. The replacement of gray moths by melanics in polluted regions is explained teleologically by the fact that melanism decreases the probability of predation in such regions. The predominance of peppered forms in nonpolluted regions is similarly explained.

Not all features of organisms need to be explained teleologically, since not all come about as a direct result of natural selection. Some features may become established by genetic drift, by chance association with adaptive traits, or in general by processes other than natural selection. Proponents of the neutrality theory of protein evolution argue that many alternative protein variants are adaptively equivalent. Most evolutionists would admit that at least in certain cases the selective differences between alternative protein variants must be virtually nil, particularly when population size is very small. The presence in a population of one amino acid sequence rather than another adaptively equivalent to the first would not then be explained teleologically. Needless to say, in such cases there would be amino acid sequences that would not be adaptive. The presence of an adaptive protein rather than a nonadaptive one would be explained teleologically; but the presence of one protein rather than another among those adaptively equivalent would not require a teleological explanation.

Natural and Artificial Teleology

In the previous section some man-made objects and adaptive traits of organisms served as examples of teleological phenomena. We may now distinguish several kinds of teleological phenomena (Ayala, 1968b, 1970). Actions or objects are *purposeful* when the end-state or goal is consciously intended by an agent. Thus, a man mowing his lawn is acting teleologically in the purposeful sense; a lion hunting deer and a bird building a nest have at least the appearance of purposeful behavior. Objects resulting from purposeful behavior exhibit *artificial* (or *external*) teleology. A knife, a table, a car, and a thermostat are examples of systems exhibiting artificial teleology: their teleological features were consciously intended by some agent.

Systems with teleological features that are not due to the purposeful action of an agent but result from some natural process exhibit *natural* (or *internal*) teleology. The wings of birds have a natural teleology; they serve an end, flying, but their configuration is not due to the conscious design of any agent. We may distinguish two kinds of natural teleology: *determinate* or necessary, and *indeterminate* or nonspecific. Determinate natural teleology exists when a specific end-state is reached in spite of environmental fluctuations. The development of an egg into a chicken, or of a human zygote into a human being, are examples of determinate natural teleological processes. The regulation of body temperature in a mammal is another example. In general, the homeostatic processes of organisms are instances of determinate natural teleology. Two types of homeostasis are usually distinguished — physiological and developmental — although intermediate conditions exist. Physiological homeostatic reactions enable organisms to maintain certain physiological steady states in spite of environmental shocks. The regulation of the concentration of salt in blood by the kidneys, or the hypertrophy of muscle owing to strenuous use, are examples of physiological homeostasis. Developmental homeostasis refers to the regulation of the different paths that an organism may follow in the progression from fertilized egg to adult. The process can be influenced by the environment in various ways, but the characteristics of the adult individual, at least within a certain range, are largely predetermined in the zygote.

Indeterminate or nonspecific teleology occurs when the end-state served is not specifically predetermined, but rather is the result of selection of one from among several available alternatives. For teleology to exist, the selection of one alternative over another must be deterministic and not purely stochastic. But what alternatives happen to be present may depend on environmental and/or historical circumstances and thus the specific end-state is not generally predictable. Indeterminate teleology results from a mixture of stochastic (at least from the point of view of the teleological system) and deterministic events. The adaptations of organisms are teleological in this indeterminate sense. The wings of birds require teleological explanations: the genetic constitutions responsible for their configuration came about because wings serve to fly and flying contributes to the reproductive success of birds. But there was nothing in the constitution of the remote ancestors of birds that would necessitate the appearance of wings in their descendants. Wings came about as the consequence of a long sequence of events, where at each stage the most advantageous alternative was selected among those that happened to be available; but what alternatives were available at any one time depended at least in part on chance events. In spite of the role played by stochastic events in the phylogenetic history of birds, it would be mistaken to say that wings are not teleological features. Again, there are differences between the teleology of an organism's adaptations and the nonteleological potential uses of natural inanimate objects. A mountain may have features appropriate for skiing, but those features did not come about so as to provide skiing slopes. On the other hand, the wings of birds came about precisely because they serve flying. The explanatory reason for the existence of wings and their configuration is the end they serve — flying — which in turn contributes to the reproductive success of birds. If wings did not serve an adaptive function they would have never come about or would gradually disappear over the generations.

The indeterminate character of the outcome of natural selection over time is due to a variety of nondeterministic factors. The outcome of natural selection depends, first, on what alternative genetic variants happen to be available at any one time. This in turn depends on the stochastic processes of mutation and recombination, and also on the past history of any given population. (What new genes may arise by mutation and what new genetic constitutions may arise by recombination depend on what genes happen to be present — which depends on previous history.) The outcome of natural selection depends also on the conditions of the physical and biotic environment. Which alternatives among available genetic variants may be favored by selection depends on the particular set of environmental conditions to which a population is exposed.

Some evolutionists have rejected teleological explanations because they have failed to recognize the various meanings that the term "teleology" may have (Pittendrigh, 1958; Mayr, 1965, 1974b; Williams, 1966; Ghiselin, 1974). These biologists are correct in excluding certain forms of teleology from evolutionary explanations, but they err when they claim that teleological explanations should be excluded altogether from evolutionary theory. In fact, they themselves often use teleological explanations in their works, but fail to recognize them as such, or prefer to call them by some other name, such as "teleonomic." Teleological explanations, as explained above, are appropriate in evolutionary theory, and are recognized by most evolutionary biologists and philosophers of science who have thoughtfully considered the question (Beckner, 1959; Nagel, 1961; Simpson, 1964; Dobzhansky, 1970; Ayala, 1968b, 1970; Wimsatt, 1972; Hull, 1974). Which kinds of teleological explanations are appropriate and which ones are inappropriate with respect to various biological questions may be briefly specified.

Mayr (1965) has pointed out that teleological explanations have been applied to two different sets of biological phenomena. "On the one hand is the production and perfection throughout the history of the

animal and plant kingdoms of ever-new and ever-improved DNA programs of information. On the other hand is the testing of these programs and their decoding throughout the lifetime of each individual.
There is a fundamental difference between end-directed behavioral
activities or developmental processes of an individual or system, which
are controlled by a program, and the steady improvement of the genetically coded programs. This genetic improvement is evolutionary adaptation controlled by natural selection." The "decoding" and "testing"
of genetic programs of information are the issues considered, respectively, by developmental biology and functional biology. The historical
and causal processes by which genetic programs of information come
about are the concern of evolutionary biology. Grene (1974) uses the
term "instrumental" for the teleology of organs that act in a functional
way, such as the hand for grasping; "developmental" for the teleology
of such processes as the maturation of a limb; and "historical" for the
process (natural selection) producing teleologically organized systems.

Organs and features such as the eye and the hand have determinate
(and internal) natural teleology. These organs serve determinate ends
(seeing or grasping) but have come about by natural processes that did
not involve the conscious design of any agent. Physiological homeostatic reactions and embryological development are processes that also
have determinate natural teleology. These processes lead to end-states
(from egg to chicken) or maintain properties (body temperature in a
mammal) that are on the whole determinate. Thus, Mayr's "decoding"
of DNA programs of information and Grene's "instrumental" and "developmental" teleology, when applied to organisms, are cases of determinate natural teleology (Mayr prefers to use the term "teleonomy"
for this type of teleology). Human tools (such as a knife), machines
(such as a car), and servomechanisms (such as a thermostat) also have
determinate teleology, but of the artificial kind, since they have been
consciously designed.

The process of natural selection is teleological but only in the sense
of indeterminate natural teleology. It is not consciously intended by
any agent, nor is it directed towards specific or predetermined end-
states. Yet the process is far from random or completely indeterminate.
Among the genetic alternatives available at any one time, natural selection favors those that increase the reproductive success of their carriers
in the particular environmental circumstances in which the organisms
live. Reproductive success is, of course, mediated by some adaptive
function, say flying, that is determined by the genetic variants that are
favored by natural selection.

Some authors exclude teleological explanations from evolutionary
biology because they believe that teleology exists only when a specific
goal is purposefully sought. This is not so. Terms other than "teleology" could be used for natural (or internal) teleology, but this might in
the end add more confusion than clarity. Philosophers as well as scien-

tists use the term "teleological" in the broader sense, to include explanations that account for the existence of an object in terms of the end-state or goal that they serve.

The process of evolution by natural selection is not teleological in the purposeful sense. Thomas Aquinas and the natural theologians of the nineteenth century erroneously claimed that the directive organization of living beings evinces the existence of a Designer. The adaptations of organisms can be explained as the result of natural processes without recourse to consciously intended end-products. There is purposeful activity in the world, at least in man; but the existence and particular structures of organisms, including man, need not be explained as the result of purposeful behavior.

Lamarck (1809) erroneously thought that evolutionary change necessarily proceeded along determined paths from simpler to more complex organisms. Similarly, the evolutionary philosophies of Bergson (1907), Teilhard de Chardin (1959), and such theories as *nomogenesis* (Berg, 1926), *aristogenesis* (Osborn, 1934), *orthogenesis*, and the like are erroneous because they all claim that evolutionary change necessarily proceeds along determined paths. These theories mistakenly take embryological development as the model of evolutionary change, regarding the teleology of evolution as determinate. Although there are teleologically determinate processes in the living world, like embryological development and physiological homeostasis, the evolutionary origin of living beings is teleological only in the indeterminate sense. Natural selection does not in any way direct evolution toward any particular organisms or toward any particular properties.

Teleological explanations are fully compatible with causal explanations (Nagel, 1961; Ayala, 1970). It is possible, at least in principle, to give a causal account of the various physical and chemical processes in the development of an egg into a chicken, or of the physicochemical, neural, and muscular interactions involved in the functioning of the eye. It is also possible in principle to describe the causal processes by which one genetic variant becomes eventually established in a population. But these causal explanations do not make it unnecessary to provide teleological explanations where appropriate. Both teleological and causal explanations are called for in such cases.

One question biologists ask about features of organisms is "What for?" That is, "What is the function or role of a particular structure or process?" The answer to this question must be formulated teleologically. A causal account of the operation of the eye is satisfactory as far as it goes, but it does not tell all that is relevant about the eye, namely that it serves to see. Moreover, evolutionary biologists are interested in the question why one particular genetic alternative rather than others came to be established in a population. This question also calls for teleological explanations of the type: "Eyes came into existence because they serve to see, and seeing increases reproductive success of certain organisms in particular circumstances." In fact, eyes came about in several independent evolutionary lineages: cephalopods, arthropods, vertebrates.

There are two questions that must be addressed by a teleological account of evolutionary events. First, there is the question of how a genetic variant contributes to reproductive success; a teleological account states that an existing genetic constitution (say, the allele coding for a normal hemoglobin beta chain) enhances reproductive fitness better than alternative constitutions. Then there is the question of how the specific genetic constitution of an organism enhances its reproductive success; a teleological explanation states that a certain genetic constitution serves a particular function (for example, the molecular composition of hemoglobin has a role in oxygen transport).

Both questions call for teleological hypotheses that can be empirically tested. It sometimes happens, however, that information is available on one or the other question but not for both. In population genetics the fitness effects of alternative genetic constitutions can often be measured while the mediating adaptive function responsible for the fitness differences may be difficult to identify. We know, for example, that in *Drosophila pseudoobscura* different inversion polymorphisms are favored by natural selection at different times of the year (Chapter 4) but we are largely ignorant of the physiological processes involved. In a historical account of evolutionary sequences the problem is occasionally reversed: the function served by an organ or structure may be easily identified, but it may be difficult to ascertain why the development of that feature enhanced reproductive success and thus was favored by natural selection. One example is the large brain of man, which makes possible culture and other important human attributes. We may advance hypotheses about the reproductive advantages of increased brain size in the evolution of man, but these hypotheses are notoriously difficult to test empirically.

Teleological explanations in evolutionary biology have great heuristic value. They are also occasionally very facile, precisely because they may be difficult to test empirically. Every effort should be made to formulate teleological explanations in a fashion that makes them readily subject to empirical testing. When appropriate empirical tests cannot be formulated, evolutionary biologists should use teleological explanations only with the greatest restraint (see Williams, 1966).

LITERATURE CITED

Ayala, F. J. 1968b. Biology as an autonomous science. *Amer. Sci.*, 56:207–221.
———. 1970. Teleological explanations in evolutionary biology. *Phil. of Sci.*, 37:1–15.
Beckner, M. 1959. *The Biological Way of Thought.* Columbia University Press.
Berg, E. S. 1926. *Nomogenesis or Evolution Determined by Law.* London; reissued 1969, M.I.T. Press.

Bergson, H. 1907. *L'Évolution Créatrice.* [1911] Creative Evolution, New York.

Dobzhansky, Th. 1970. *Genetics of the Evolutionary Process.* Columbia University Press.

Ghiselin, M. 1974. *The Economy of Nature and the Evolution of Sex.* University of California Press.

Grene, M. 1974. *The Understanding of Nature. Essays in the Philosophy of Biology.* Reidel, Boston.

Hull, D. 1974. *Philosophy of Biological Science.* Prentice-Hall, Englewood Cliffs, New Jersey.

Lamarck, J. B. 1809. *Zoological Philosophy,* Translated by H. Elliot. Reprinted 1963, Hafner, New York.

Mayr, E. 1965. Cause and effect in biology. In *Cause and Effect,* D. Lerner, ed., Free Press, New York, pp. 33–50.

———. 1974b. Teleological and teleonomic, a new analysis. In *Boston Studies in the Philosophy of Science,* XIV, R. S. Cohen and M. W. Wartofsky, eds., pp. 91–117. Reidel, Boston.

Nagel, E. 1961. *The Structure of Science.* Harcourt, Brace, and World, New York.

Osborn, H. F. 1934. Aristogenesis, the creative principle in the origin of species. *Amer. Nat.,* 68:193–235.

Pittendrigh, C. S. 1958. Adaptation, natural selection, and behavior. In *Behavior and Evolution,* A. Roe and G. G. Simpson, eds., pp. 390–416. Yale University Press.

Simpson, G. G. 1964. *This View of Life.* Harcourt, Brace, and World, New York.

Teilhard de Chardin, P. 1959. *The Phenomenon of Man.* Harper, New York.

Williams, G. C. 1966. *Adaptation and Natural Selection.* Princeton University Press.

Wimsatt, W. C. 1972. Teleology and the logical structure of function statements. *Stud. Hist. Phil. Sci.,* 3:1–80.

MOLECULAR BIOLOGY

19

KENNETH F. SCHAFFNER*

Chemical Systems and Chemical Evolution: The Philosophy of Molecular Biology

CONTEMPORARY BIOLOGY COMPRISES many different branches of inquiry, ranging from a descriptive anatomy through historical and taxonomical evolutionary studies, to the physical and chemical accounts of molecular genetics. As the number of significant accomplishments in molecular biology continues to mount, attention is likely to be diverted from more "classical" areas of biology. Partly in response to this growing interest in physical and chemical analyses of biological phenomena, and partly because of certain philosophical (including metaphysical) points of view, a number of biologists and some philosophers have argued that biology is autonomous. They claim that in a number of respects and for a number of reasons biology is not reducible to physics and chemistry.[1-8] Though they do not deny the utility of physical and chemical analyses, it is the contention of the nonreductionists or the antireductionists that biology is not simply a complicated physics and chemistry.

In this essay I should like to criticize such arguments — not by citing those varied arguments and examining their premises in the context of contemporary biological knowledge[9] — but rather by developing a general account of explanation in molecular biology that introduces a number of considerations that are often overlooked in disputes about the autonomy and reducibility of biology. I also hope in developing this

*From Kenneth F. Schaffner, "Chemical Systems and Chemical Evolution: The Philosophy of Molecular Biology," *American Scientist,* Vol. 57, no. 4 (New Haven, CT: Scientific Research Society, 1969), pp. 410–420. By permission of the publisher.

general account to show how it implicitly figures in contemporary theory and experiment in molecular biology.

Explanations in Molecular Biology

Molecular biology is not to be thought of only as molecular genetics,[10] nevertheless a number of the most important accomplishments of molecular biology have been made in this area. Furthermore, because of the centrality of the DNA code and the process of protein synthesis in the living cell, the *type* of explanations that are employed in the area of molecular genetics can be taken as paradigmatic of explanations in molecular biology. Studies in molecular biology outside the area of genetics, some of which will be referred to later, will support the theses about explanation to be developed in this paper.

Biochemical or molecular geneticists tend to concentrate on a relatively small number of organisms, but in so doing they have discovered a number of mechanisms that are of extremely general applicability. J. D. Watson has discussed the reasons for research concentration on *E. coli*, a fairly common bacterium,[11] and a recent series of memoirs also discusses the impact of extensive research on the bacteriophage viruses for molecular biology.[12] Such research, both with the organism and with cell-free systems prepared from the organism, has led to a considerably deepened understanding of the chemical mechanisms of cell maintenance, replication, growth, and differentiation. Happily, many of the discoveries, some of which will be discussed below, have nearly universal applicability to all living organisms.

From the point of view of the philosopher of science, the molecular biologist freely utilizes the theories and instruments of physics and chemistry in a reductionist program attempting to (*a*) identify biological entities with physico-chemical entities and systems, and (*b*) explain biological processes by reference to the identities cited in (*a*) and the laws and theories of physics and chemistry.[13,14]

Thus genes are identified with DNA nucleotide sequences, phenotypes with complexes of polypeptide chains, and the association between the identified genes and proteins is explained by reference to chemical laws and the chemical systems involved in protein synthesis. In pursuing his reductionist program, however, the molecular biologist cannot forget that he is, ultimately, working with a highly organized chemical system. The word "ultimately" is used here because it is possible to do significant experiments on isolable parts of the system — e.g., on the primary structure of protein as in V. Ingram's work on the hemoglobin molecule,[15] or, say, on cell free systems such as M. Nirenberg and J. H. Matthaei's breakthrough experiment on the genetic code.[16]

The systematization of chemical components that is involved in molecular biology is most important. To obtain a glimpse of the intricacy that is present in the mechanism of protein synthesis, consider the follwing recent observation:

> The requirements for protein synthesis include about 60 transfer RNA molecules, at least 20 activating enzymes which attach the amino acids to the t-RNAs, special enzymes for initiation, propagation, and termination of peptide synthesis, ribosomes containing three different structural RNAs, and as many as 50 different structural proteins, messenger RNA, ATP, GTP, and Mg^{+2}.[17]

Though considerable complexity and systematization have been discovered in the chemical mechanisms involved in living processes, no investigation in molecular biology has yielded any evidence that new laws, peculiar to living organisms, are required to explain biological processes. J. D. Watson's comment, though made several years ago, is still an accurate statement of the status of chemical laws in biology. Watson wrote:

> The growth and division of cells are based upon the same laws of chemistry that control the behavior of molecules outside of cells. Cells contain no atoms unique to the living state; they can synthesize no molecules which the chemist, with inspired, hard work, cannot some day make. Thus there is no special chemistry of living cells. A biochemist is not someone who studies unique types of chemical laws, but a chemist interested in learning about the behavior of molecules found within cells (biological molecules).[11]

Accordingly the molecular biologist or biochemist — the terms are not distinct — believes that biology is explicable by physics and chemistry. But consider what is involved in such physical and chemical explicability of biological processes. Clearly the laws of physics and chemistry are applied to the intricate biochemical machinery discussed in the previous paragraph. If one constructs a physico-chemical explanation for the synthesis of a hemoglobin molecule, say, included in the premises of the explanation must be statements which describe the elaborate chemical system for protein synthesis as already organized.

This should become quite clear if we consider the standard account of scientific explanation that is due to K. R. Popper[18] and C. G. Hempel and P. Oppenheim.[19] It could also be analyzed by working with standard accounts of reduction in the physical sciences and applying them to the reduction of biology.[20]

The PHO model is sometimes known as the "deductive-nomological" model of explanation, for it explicates scientific explanation as the deductive subsumption of an event to be explained under one or more scientific laws. In essence, one takes as premises for a deduction, one or more scientific laws $(L_1 \ldots L_n)$, conjoins sentences describing the appropriate initial conditions $(C_1 \ldots C_k)$, and deduces in accordance with the principles of logic and mathematics a sentence describing the event to be explained (E). In this schematic form this may be rendered:

$$L_1, \ldots L_n$$
$$\underline{C_1, \ldots C_k}$$
$$\therefore E$$

There are more sophisticated ways of articulating the model and its conditions that I need not go into here. It also need not be emphasized that there are other types of explanation, e.g., the statistical type, which require a different model. The existence of these alternatives does not, however, affect my thesis.

In terms of the Popper–Hempel–Oppenheim account as applied to physico-chemical explanation of a biological process like protein synthesis, the laws of physics and chemistry would not be sufficient, unless specific intricate structures, such as discussed above, were assumed as part of the initial conditions. These intricate structures or "machinery" are, for the molecular biologist, *nothing other than chemical systems*, but such systems are not the result of random physically and chemically characterizable initial conditions. The molecular biologist takes such organized machinery as given, as accepted initial conditions in his explanations.[21] More realistically, especially with respect to his research procedure, he assumes a very complex machinery exists, and focuses his attention on a *part* of the larger organized system. He then attempts to characterize that part in terms of its primary, secondary, and tertiary structures, and to understand in terms of chemical bonding, chemical kinetics, and other physical and chemical theories, how biological processes may be related to the physico-chemical organization and operation of that part. For example, like R. W. Holley, he may be concerned with the primary and secondary structure of tRNA and its role in the complex process of protein synthesis.[23] But whether the molecular biologist is concerned with the part or with the whole (for the molecular biologist the sum of the parts in all their physical and chemical relational aspects) he can only explain those most important initial conditions, the chemical systematization, by referring to another similarly organized chemical system. Usually this is the parent system(s), or perhaps the parent system's DNA and the relevant enzymes. Such a need to refer to previous organization holds not only at the structural level, but also at the control level of genetics. The complex development of a multicellular system will probably be traceable back to the chemical structure of the fertilized zygote and to the chemical control systems present therein. But going to the chemical level to explain the processes of embryological development, perhaps with the aid of Jacob and Monod's operon model[24] as complicated to include histone interaction by Bonner[25] and Zubay,[17] certainly does not obviate the need to refer to an intricately organized system. At every temporal point in the development and maintenance of the unicellular or multicellular organism, chemical systematization is present which cannot be accounted for chemically except by referring to a previous similarly organized system.

The difficulty for a reductionist protagonist is clear: though the laws and theories of physics and chemistry are adequate for molecular biology if elaborately organized chemical systematization is assumed, the molecular biologist cannot account for such systematization on the basis of non-organic initial conditions present in nature today.

(Incidentally, it might be inferred from such a claim that the "creation of life in a test tube" is impossible. Nothing could be further from the truth. If scientists succeed in forcing the link-up of the appropriate amino acids and nucleotides *in the right order*, life will have been created. At present, even in the most sophisticated of such experiments, forced link-ups require the use of a template provided by an already "living" organism.[26] Clearly, however, the use of such a template is not *a priori* necessary and the recent artificial synthesis of the enzyme RNase supports this claim.[27])

Such necessity as there is to refer the origin of chemical systematization to another similar chemical system (or to the active interference of the scientists directing the synthesis) *could* imply a weak vitalistic thesis. For what this necessity amounts to is the molecular version of the claim that "life only comes from life." A restriction of the molecular approach to current organisms is indeed possible, but it would expose a flank of the reductionist thesis to a serious vitalistic objection. This would be the claim that "entelechy," or perhaps "spirit," exercised an influence over molecules, nudging them together. Without such a *formative ordering force* living systems could not have originated. Such a claim would amount to a biological "Deism."

Though the idea of such a Deism is scientifically untenable, it does suggest an analogy which reveals important connections between current work in evolutionary theory, mutation theory, abiogenesis, and molecular biology.

The Need for a Molecular Evolutionary Theory

There is an interesting analogy which molecular biology displays with celestial mechanics, of either the Newtonian or Einsteinian variety. Celestial mechanics can be applied to the planets and other bodies in orbit to yield explanations and predictions of celestial phenomena. Such explanations require positions, moments, and gravitational masses as initial conditions to be used in conjunction with a general theory of mechanics. These mechanical explanations are good explanations, and in point of fact have often been taken as paradigms of scientific explanation.

But scientists have long been interested in the origin of any such initial conditions as might be used in celestial mechanical explanations, and in fact have developed a sophisticated branch of astrophysics, called cosmogony, to study such origins.

Molecular biologists are in a position which is analogous to the practitioners of celestial mechanics. Molecular biologists construct explanations which consist of general laws applied to given systems — systems, which to be sure, are considerably more complex than those which celestial mechanics studies — but systems which they contend are nothing more than chemical. Accordingly, molecular biologists can formulate legitimate explanations of living processes in physical and chemical terms but only, as noted above, by accepting complex chemical organizations as part of the initial conditions in the physico-chemical explicans.

A thoroughgoing reductionist program must account for the existence of such complex chemical organizations or systems or stand accused of leaving a gap in its account of living organisms — a gap which I suggested above might quickly be filled by entelechies and spirits. Now such complex systems as the molecular biologist works with cannot be explained on the basis of general physical and chemical conditions that are operative in the current environment. Nevertheless a naturalistic or materialistic explanation — or better an "explanation sketch"[28] — is available, and it is an explanation which is *expressible* in purely physico-chemical terms, though this tack has not generally been taken. This naturalistic explanation is *chemical evolution and chemically characterized Darwinian evolution.*

Thus the gap in the reductionist program can be naturalistically closed, but it must be closed by appealing to what we might term a "molecular biogony" that supplements the molecular biology studying contemporary organisms. Such a biogony studies the processes which began billions of years ago, perhaps in Haldane's "hot-dilute soup" of spontaneously forming organic compounds,[29] or perhaps in a region of volcanic activity, as Fox's experiments suggest.[30] In any case, what is sought and found in a number of varied experiments are chemically explicable prebiological systems.[31]

A considerable amount of research has been done concerning the possibility of "spontaneous generation" of these prebiological systems, and the results have been encouraging. A. I. Oparin has described the abiogenetic formation of small spheres of organic matter which he terms "coacervates" — and he argues plausibly and by reference to a number of experiments that such coacervates can grow, can alter their chemical structure, and can replicate as a result of being subjected to physical agitation.[5,32] S. W. Fox and his coworkers have produced proteinoid microspheres under conditions which are geologically plausible.[30] Fox has shown that such microspheres, which can occur naturally under primitive earth conditions, can replicate by a process of budding and agitation.[34] Fox believes that there are sufficient self-organizing *chemical* forces present in the amino acid polymerization process to account for the "limitations of diversification" found in living chemical systems. He also contends, however, that Darwinian natural selection played an important role in chemical systematization when replicability and mutatability appeared in chemical systems.[34,35]

Clearly what such research implies is that complex prebiological chemical systems could have been produced several billion years ago by natural conditions on the primitive earth. There is experimental evidence that such systems could have increased in complexity[33] — or at least altered their complexity — and when sufficiently accurate replicability and mutatability developed in such systems, that Darwinian evolution would have come into play.[36]

Darwinian natural selection can be characterized in terms of the physics and chemistry of chemical systems and their environment, though it certainly is not necessary to do this in all cases of evolutionary studies. Clearly one can and does discuss the process of natural selection in terms of phenotypic advantages without characterizing those phenotypic aspects in chemical language.[37] It may well be that in a number of cases giving a chemical characterization adds no new interesting information.[20] In certain cases, however, comparisons of the amino acid sequences does afford insights into the evolutionary process.[20,38] The major point that I wish to make in connection with Darwinian natural selection is that it can be understood at the level of the replicating and mutating chemical system — or perhaps better, at the level of "populations" of such chemical systems — interacting with a chemically characterizable environment.[39]

The study of molecular evolution is of considerable interest to many biologists. The results of the molecular geneticists' research on current organisms are being applied to evolutionary theory.[38] A number of analytical tools of the biochemists, such as comparison of aspects of organisms and species via the amino acid analysis of homologous polypeptide chains, has brought a clearer understanding of how evolution may operate through gene (DNA) duplication. Ingram's studies on chemical evolution of the myoglobin and hemoglobin molecules are cases in point.[15] Considerable research has been done on molecular mechanisms of mutation,[40] and this research indicates that the general account of genetic coding and protein synthesis referred to earlier is sufficient to account for the phenotypic variation which Darwinian selection requires.

Difficulties and Possible Applications of a Molecular Evolutionary Theory

The pursuit of a "synthetic" theory of *molecular* evolutionary biology which would draw on many different biochemical accomplishments should afford a number of basic insights into the origins of living organisms. There are, to be sure, great difficulties which must be surmounted before such a theory could receive the status of being more than a conjecture. The gathering of evolutionary evidence concerning genotypes, phenotypes, and environments at the molecular level must of necessity be extremely indirect and highly inferential: molecules may leave traces but they do not leave fossils in the usual sense. It seems, however, that we might look forward to the application of knowledge obtained from molecular mechanisms of contemporary organisms to the evolutionary record. Let us consider one possibility of the joint application of a theory of cell differentiation and a mutation theory.

George Wald has already used the rough "biogenetic law" to the effect that "ontogeny recapitulates phylogeny" as a useful tool and test for reconstructing possible evolutionary sequences at the molecular level.[41] It seems likely that this approach could be carried further by attempting to reconstruct the evolution of multicellular organisms in terms of the effect of likely mutations on the mechanisms of cell differentiation and embryological development.

James Bonner and G. L. Zubay, among others, have been pursuing research on the role of histones in genetic control and cell differentiation. Bonner's work[25] has been concerned with histone analysis and a "switching circuit logic" of differentiation and derepression involving the possible interaction of hormones, histones, and chromatin. G. L. Zubay has recently proposed a model of cell differentiation applicable to embryological development which postulates competition between specific repressors and histones during the S-phase of cell division, and which complements some of Bonner's views about the logic of cell differentiation.[42] Let us consider this model in some detail because of its possible connection with the thesis under consideration.

Zubay's model for stable differentiated cells would account in a relatively simple way for the fact that most of the genes in a cell remain unexpressed. He has proposed that specific repressors, such as are found in the Jacob–Monod model of the *lac* operon,[24] compete with general repressors, such as histone, for binding with the operator sites of genetic DNA. If histone binds with the operator site — this can happen only during the S-phase of cell duplication when both DNA and histone synthesis occur, since Zubay assumes that there are no inducers for histone — the histone-bound operator site and its associated structural genes are permanently repressed in that cell. If, however, a *specific* repressor has complexed with the operator site, the presence of a suitable inducer can render the operon functional. Zubay concludes on the basis of this model that:

> A new gene could only become inducible after S-phase, and then only if there were a buildup of the appropriate specific gene repressor prior to S-phase. The buildup of a new repressor would presumably result from the activation of one of the already inducible operons. This suggests a tight causal relationship between the activation of different groups of operons.

Should such a model of embryological development be corroborated,[43] the molecularly characterized ontogenetic sequences might well be used to reconstruct — with the aid of whatever paleontological information and molecular genetic theory is available — the possible molecular evolution of more and more complex chemical systems at the phylogenetic level. Such a molecular evolutionary theory would accordingly study the evolution of regulator genes as well as structural genes. Such a theory might well be able to show that a DNA based system of information storage and control was a necessary condition for the evolution of multicellular organisms. I do not conceive of such a theory as advocating a return to mutationism: the chemically characterizable phenotypes and environments would be as important as the genotype in such a molecular evolutionary theory.

Conclusion

On the basis of evidence from contemporary biology, and in the light of analyses of explanation and reduction in contemporary philosophy of science, it can be concluded that biology is autonomous only in the sense that it studies complex chemical systems, rather than the simpler systems which chemists who are not biochemists study. The origin of such complex systems are most likely explicable in terms of a molecular biogony or a theory of molecular evolution resulting from physico-chemical research in the areas of abiogenesis and evolutionary theory. G. G. Simpson anticipated one aspect of such a thesis when he wrote that:

> Life . . . is not . . . considered as nonphysical or nonmaterial. It is just that living beings have been affected for upward of two billion years by historical processes that are in themselves completely material but that do not affect nonliving matter or at least do not affect it in the same way.[1]

Simpson also believes that an organism's organization can and must be explained by evolution,[1] but he professes an antireductionist thesis when he asserts that biology utilizes principles of explanation which go beyond physics and chemistry,[1] and when he implies that a multilevel analysis is necessary and not merely pragmatic,[2] and hints that purely physico-chemical evolutionary studies are not adequate to explain current biological organization.[2]

Clearly it is possible to formulate good explanations of biological processes in chemical terms without necessarily referring to the evolutionary process. The *logos* or organizational pattern explanations utilized in molecular biology are autonomous with respect to the "gonetic" explanations that are necessary to explain the formation of such chemical patterns on the basis of general physico-chemical principles. Of course I do admit that reference to genesis is required for the reductionist protagonist, but the introduction of a historical dimension does not distinguish biology from physics. Celestial mechanics requires cosmogony in the same sense that molecular biology requires a molecular biogony.

REFERENCES AND NOTES

1. G. G. Simpson, *This View of Life* (Harcourt, Brace, and World, New York, 1964).

2.. G. G. Simpson, "Organisms and Molecules in Evolution," *Science, 148,* 1535 (1964).
3. E. Mayr, "Cause and Effect in Biology," *Science, 134,* 1501 (1961).
4. C. Grobstein, "Levels and Ontogeny," *American Scientist, 50,* 46 (1961).
5. A. I. Oparin, *Life: Its Nature, Origin, and Development* tr. A. Synge, (Academic Press, New York, 1962).
6. M. Polanyi, "Life's Irreducible Structure," *Science, 160,* 1308 (1968).
7. P. Weiss, "From Cell to Molecule," in *The Molecular Control of Cellular Activity,* J. M. Allen, Ed. (McGraw Hill, New York, 1961).
8. F. J. Ayala, "Biology as an Autonomous Science," *American Scientist, 56,* 207 (1968).
9. K. F. Schaffner, "Antireductionism and Molecular Biology," *Science, 157,* 644 (1967).
10. See D. E. Green and R. F. Goldberger's *Molecular Insights into the Living Process* (Academic Press, New York, 1967) for an introduction to the varied range of the molecular approach.
11. J. D. Watson, *Molecular Biology of the Gene* (W. A. Benjamin, New York, 1965).
12. J. Cairns, G. S. Stent, J. D. Watson, Eds., *Phage and the Origins of Molecular Biology* (Cold Spring Harbor Laboratory of Quantitative Biology, Cold Spring Harbor, New York, 1966).
13. E. Nagel, *The Structure of Science* (Harcourt, Brace, and World, New York, 1961).
14. K. F. Schaffner, "Approaches to Reduction," *Phil. Sci.,* 34, 137 (1967).
15. V. Ingram, *The Hemoglobins in Genetics and Evolution* (Columbia Univ. Press, New York, 1963).
16. M. Nirenberg and J. H. Matthaei, "The Dependence of Cell-free Protein Synthesis in *E. coli* upon Naturally Occurring or Synthetic Polyribonucleotides," *Proc. Natl. Acad. Sci. U.S.,* 47, 1588 (1961).
17. G. L. Zubay, *Papers in Biochemical Genetics* (Holt, Rinehart, and Winston, New York, 1968).
18. K. R. Popper, *The Logic of Scientific Discovery* (Harper and Row, New York, 1959).
19. C. G. Hempel and P. Oppenheim, "Studies in the Logic of Explanation," *Phil. Sci.,* 15, 135 (1948).
20. K. F. Schaffner, "Theories and Explanations in Biology," *J. Hist. Biol.,* 2, 19, 1969.
21. This thesis appears in my *Science* article, see Ref. 9, but is more extensively developed in references 20 and 22
22. K. F. Schaffner, "The Watson–Crick Model & Reductionism," *Brit. Jour. Phil. Sci.,* in press.
23. R. W. Holley, "Structure of a Ribonucleic Acid," *Science, 147,* 1462 (1965).
24. F. Jacob and J. Monod, "Genetic Regulatory Mechanisms in the Synthesis of Proteins," *J. Mol. Biol.,* 3, 318 (1961).
25. James Bonner, *The Molecular Biology of Development* (Clarendon Press, Oxford, 1965).
26. M. Goulian, A. Kornberg, and R. L. Sinsheimer, "Enzymatic Synthesis of DNA, XXIV: Synthesis of Infectious Phage ϕX174 DNA," *Proc. Natl. Acad. Sci. U.S.,* 58, 2321 (1967).
27. See the letters from C. B. Gutte and R. Merrifield and from R. G. Denkewalter, D. F. Veber, F. W. Holly, and R. F. Hirschmann in *J. Am. Chem. Soc.,* 91, 501 (1969).
28. C. G. Hempel, "The Function of General Laws in History," *J. Phil.,* 39, 35 (1942)
29. J. B. S. Haldane, "The Origin of Life," *Rationalist Annual* (1929).

30. S. W. Fox, "Thermal Polymerization of Amino-acids and Production of Formed Microparticles on Lava," *Nature*, *201*, 336 (1964).
31. See J. Koosian's *The Origin of Life*, 2nd. edit. (Reinhold, New York, 1968) for a review of such experiments.
32. A. I. Oparin, "The Pathways of the Primary Development of Metabolism and the Artificial Modeling of the Development in Coacervate Drops," in Fox, Ed., reference 33.
33. S. W. Fox, Ed., *The Origins of Prebiological Systems* (Academic Press, New York, 1965).
34. S. W. Fox, R. J. McCauley, and A. Wood, "A Model of Primitive Heterotrophic Proliferation," *Comp. Biochem. Physiol.*, *20*, 733 (1967).
35. S. W. Fox, "A Theory of Macromolecular and Cellular Origins," *Nature*, *205*, 328 (1965).
36. S. W. Fox, personal communication.
37. Consider Darwin's work, or more recently, R. A. Fisher's classic, *The Genetical Theory of Natural Selection*, 2nd. rev. edit., (Dover, New York, 1958), for nonchemical examples of evolutionary analyses.
38. T. H. Jukes, *Molecules and Evolution* (Columbia Univ. Press, New York, 1966).
39. G. G. Simpson would apparently disagree with this. See references 1 and 2 but also see Jukes reference 38 for a critique of reference 2.
40. See the section "The Molecular Basis of Mutagenesis" in Zubay reference 17 for examples and references.
41. G. Wald, "Phylogeny and Ontogeny at the Molecular Level," in *Evolutionary Biochemistry*, A. I. Oparin, Ed. (Pergamon, London, 1963).
42. What I term "Zubay's model" is developed by him in reference 17 pp. 466–467.
43. Zubay is currently engaged in research which will test some of the assumptions of this model. (Personal communication.)

20

JOHN BEATTY*

The Insights and Oversights of Molecular Genetics: The Place of the Evolutionary Perspective[1]

Nothing in biology makes sense except in the light of evolution. (Dobzhansky 1973)

1. Introduction

ALONG WITH THE notion that DNA is at the bottom of things biological, goes the notion that molecular genetics is on top. That is, along with the notion that DNA is the informational basis of life, goes the notion that the deepest explanations in biology are molecular-genetic. But even if the former notion is correct, the latter is not. More specifically, the molecular-genetic perspective alone is inappropriate for explaining those biological generalities that call out instead for an evolutionary account. Moreover, the molecular-genetic accounts that are brought to bear upon biological generalities are often themselves subject to evolutionary scrutiny in the final analysis. I cannot entirely rule out the possibility of an explanation of a biological generality that is, in the final analysis, molecular-genetic and nonevolutionary. I am, in this regard, in somewhat the position of the natural historian. "In natural history," as Gould and Lewontin put it, "all possible things happen sometimes; you generally do not support your favoured phenomenon by declaring

*From John Beatty, "The Insights and Oversights of Molecular Genetics: The Place of the Evolutionary Perspective," *PSA 1982*, P. Asquith and T. Nickles (eds.), (East Lansing, MI: Philosophy of Science Association, 1982). Vol. 1, pp. 341–355.

rivals impossible in theory. Rather, you acknowledge the rival, but circumscribe its domain of action so narrowly that it cannot have any importance in the affairs of nature. Then, you often congratulate yourself for being such an ecumenical chap." (1979, p. 585). Sometimes that is the best one can do in philosophy as well.

I will argue a general case about the insights and oversights of molecular genetics by arguing two specific cases: the first concerns the bearing of molecular genetics on Mendelian genetics, and the second concerns the bearing of molecular genetics on the replicability of the genetic material. As for the first case, I will argue that Mendel's law of segregation cannot be explained wholly in terms of molecular genetics —the law demands evolutionary scrutiny as well. As for the second case, I will argue that an account of the replicability of the genetic material in terms of molecular genetics is not entirely independent of evolutionary considerations, in the sense that it raises further evolutionary questions. The limitations of the molecular-genetic approach in these cases point to the limitations of that approach in general.[2]

2. Mendelian and Molecular Genetics

Consider Mendel's "law." In the simplest terms, Mendel's law describes the statistical outcome of the process by which gametes (sperm and egg cells) are formed from gamete-producing cells in the reproductive organs. This process is known as meiosis, and consequently Mendel's law is said to describe the outcome of "normal" meiosis (see diagram of stages of meiosis, p. 212). During normal meiosis, the total hereditary material of the gamete-producing cells is fractioned through a series of cell divisions in such a way that each resulting gamete receives a complementary half of the original hereditary material. To describe the process in a bit more detail, the hereditary material of the gamete-producing and other nongametic cells comes in pairs of morphologically similar, or "homologous" chromosomes. The genes reside linearly along each chromosome; the two genes that lie opposite one another when the homologous chromosomes pair during meiosis are said to occupy the same chromosomal "locus" (in the diagram, B and b occupy the same locus). During meiosis, the chromosome pairs double and then divide twice so that each of the four gametes formed receives one chromosome from each pair. Hence, each gamete receives one gene from each locus. Moreover, the probability that a gamete will receive one particular gene at a locus is one half. Thus, a common brief formulation of Mendel's law is:

> Each of the two genes at a locus has a probability of one half of being the single gene at that locus carried by a particular gamete. (Edwards 1977, p. 3).

On the basis of this law, and an additional assumption to the effect that gametes combine randomly with respect to their genetic types, one can calculate the familiar genotypic frequencies of offspring born to parents with specified single-locus gene combinations. For instance, a cross between two Bb-type parents will produce, in the long run, 1 BB:2Bb:1 bb ratio of offspring. If we add additional assumptions about the dominance and recessiveness of the genes in question, we can also calculate the phenotypic frequencies of the offspring. For instance, if B is completely dominant to b—i.e., if the phenotype of a Bb-type organism is the same in the relevant respects to the phenotype of a BB-type—then a cross between two Bb types will produce the familiar 3:1 phenotypic ratio.

But the purpose of this brief excursion into genetics is not so much to see what comes from normal meiosis, but to see where normal meiosis comes from. We have seen that normal meiosis is a cellular process. Now consider that this cellular process, like so many others, is *itself* genetically controlled by genes at specific locations along the chromosomes. This has become more and more clear with the discoveries of more and more genetic mutations that affect meiosis. Two kinds of genetic variation in the meiotic mechanism have received considerable attention: a variation known as "non-disjunction" and a variation known as "meiotic drive."

Non-disjunction occurs when homologous chromosomes do not disjoin or separate during meiosis. As a result, some of the gametes formed contain both or neither of the homologous chromosomes—meaning that those gametes contain both or neither of the genes at each locus of the non-disjoining chromosomes (see the bottom of the preceding diagram). This exceptional phenomenon was recognized microscopically as early as 1913, but its genetic basis was not established until 1933 when Gowen reported his experimental findings that, "the processes through which the chromosomes pass during the meiotic divisions are in part, at least, subject to the same specific gene regulation that guides other bodily development and inheritance." (1933, p. 83). Through breeding experiments, Gowen determined that cases of the non-disjunction of one or all four pairs of chromosomes of *Drosophila melanogaster* were influenced by a single gene locus that he located on the third chromosome pair (for more on the genetic basis of non-disjunction, see White 1973).

Meiotic drive, another acknowledged exception to normal meiosis, occurs when organisms with different genes at a locus (say, B and b) produce more than fifty percent of one kind of gamete (say B), on account of genetic differences between the homologous chromosomes. Perhaps the most well-known case of meiotic drive involves the so-called "t" gene in house mice, which is expressed in the form of taillessness. As much as ninety-five percent of the sperm of males of the Tt type may be of the t-type (Dunn 1953; for more on meiotic drive see White 1973).

So normal meiosis is a genetically controlled process—the phenotype of certain genotypes. Just as "blue eyes" is the phenotype of

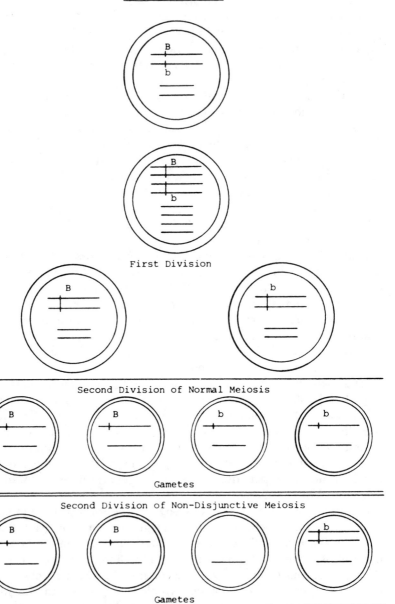

Stages of Meiosis

First Division

Second Division of Normal Meiosis

Gametes

Second Division of Non-Disjunctive Meiosis

Gametes

certain genotypes and "brown eyes" the phenotype of others, so too "normal meiosis" is the phenotype of certain genotypes and "abnormal meiosis" of others. The generality of 50:50 gamete ratios, as described by Mendel's law, is a consequence of the generality of the normal meiotic phenotype and its genotypes. Presumably, then, an explanation of Mendelian genetics in terms of molecular genetics would consist of a molecular account of the generality of this particular phenotype and its genotypes.

How can the generalizations of molecular genetics account for the generality of the normal meiotic phenotype and its genotype? First what are the generalizations of molecular genetics? I think a brief sketch will do. Among the generalizations of molecular genetics are:

(1) hypotheses concerning the structure of DNA, the various RNAs, and proteins;

(2) hypotheses about how DNA replicates; and

(3) hypotheses about how DNA carries the instructions for the formation of other building blocks of life and the direction of other life processes — i.e., how DNA carries the information for the formation of RNA, how RNA carries the information for the formation of proteins, etc.

Now how can such generalizations explain the generality of the normal meiotic phenotype and its genotype? The answer is that, alone, they cannot. But let us get clear about what we are asking. Following van Fraassen (1980, pp. 126 – 129), let us consider an explanation as an answer to a why question of the form,

Why P in contrast to X?

P is the phenomenon to be explained, the "presupposition" of the question. The question only arises, or is in order, on the assumption of P. X is the "contrast class," the set of possible alternatives to P. When we ask for an explanation of P we want to know why P is the case rather than some alternative to P.

With regard to the question at issue, P is the generality of normal meiosis, and X is the set of abnormal meiotic processes. We want to know,

Why is "normal" meiosis general while alternative forms of meiosis are not?

3. The Evolutionary Perspective

We may someday know the answer to that question, but the answer will not be strictly molecular-genetic. The generalizations of molecular genetics alone — generalizations about the structure of DNA and its replicative and informational properties — shed no more light on the generality of normal vs abnormal meiosis than they do on any single case of meiosis. And, alone, they no more explain any single case of normal vs abnormal meiosis than they explain why I have blue eyes vs brown. Normal meiosis vs abnormal, like blue eyes vs brown, requires the "right" kind of DNA sequence — not just any DNA sequence. So an account of the generality of normal vs abnormal meiosis in terms of molecular genetics would have to be supplemented by a claim concerning the generality of the right kinds of DNA sequences.

Now would any such supplemented explanation count toward the explanation of Mendelian in terms of molecular genetics? No, since the supplemental generalization is no more part of molecular genetics than are claims about the frequencies of any other kinds of DNA sequences. More importantly, however, the supplemented explanation is not the most appropriate explanation for understanding the phenomenon in question. We need a different kind of explanation. In Mayr's (1961) terms, an "ultimate" rather than a "proximate" explanation is called for.

Proximate explanations concern the processing of environmental and genetic information — the causal pathways from environment and genotype to phenotype. Ultimate explanations, on the other hand, concern the history of environmental and genetic information — the causal pathways that lead from frequencies of genotypes and their phenotypes in earlier generations to their frequencies in later generations. Physiology and developmental biology are proximate sciences, while evolutionary biology is an ultimate science. Molecular genetics is also a proximate science — population genetics would be an ultimate science.

It is especially important to recognize that proximate and ultimate explanations are generally not rival alternatives to the same questions; where one kind of explanation is appropriate, the other kind generally is not. In other words, some questions ask for proximate, others ultimate answers. Substantial confusion has resulted from confounding the roles of the two kinds of accounts. For instance, Mayr has made infamous the confusion of the early 20th-century physiologist Loeb, who complained that, "The earlier writers explained the growth of legs in the tadpole of the frog or toad as a case of adaptation to life on land. We know through Gudernatsch that the growth of the legs can be produced at any time even in the youngest tadpole, which is unable to live on land, by feeding the animal with the thyroid gland" (Loeb 1916, quoted in Mayr 1961, p. 1503). Thyroid hormone would be appropriately invoked to account for the growth of legs in an individual frog, but is not as appropriately invoked to account for the presence of legs as a general characteristic of frogs. An evolutionary account, presumably in terms of the differential reproductive success of legged vs nonlegged ancestors of frogs, would be more appropriate in the latter case.

Similarly, when we ask why normal vs abnormal meiosis is so general, we want an ultimate, evolutionary answer. As was explained earlier, the generality of normal meiosis rests upon the generality of the right kinds of DNA sequences. And the predominance of those sequences simply is not illuminated by a proximate explanation. Nothing (that we know of) that has to do with the processing of genetic information rules out the predominance of genotypes for alternative forms of meiosis. Evolutionary agents, on the other hand, could rule them out. For instance, it might be the case — we do not really know — that of the various forms of meiosis that have arisen through mutation, the normal form has always been selected. At least the latter explanation is an appropriate answer.

Molecular genetic considerations would be more appropriate for understanding why a certain kind of DNA sequence results in normal meiosis rather than abnormal. That question is no less interesting or important, but just a different question from why that kind of DNA sequence and its normal meiotic phenotype are so common. In order to differentiate more clearly these two types of questions, we need to focus on the "Why P" part of our general question form "Why P in contrast to X?" Depending on how we elaborate "Why P," the question may be one that calls for a molecular-genetic, or some other kind of proximate response, or one that calls for an ultimate, evolutionary response. The question about normal meiosis, as I have posed it, is an instance of a more general form:

(1) Why is phenotype Φ and/or genotype G general in contrast to alternatives?
(Why is/are normal meiosis and/or its underlying genotypes general in contrast to alternatives?)

Questions of form 1 call for ultimate, evolutionary answers — evolutionary considerations are most relevant to ruling out the generality of alternative phenotypes and genotypes. On the other hand, "Why P" questions of other forms call for proximate answers. Questions of form 2, for instance, are clearly more appropriately answered in molecular-genetic terms:

(2) Why, given genotype G and environment E does phenotype Φ result in contrast to alternatives?
(Why given a particular genotype does normal meiosis result in contrast to abnormal meiosis?)

In other words, molecular-genetic considerations may be relevant to ruling out alternative phenotypic outcomes. The point here is that the general question form "Why P in contrast to X?" does not automatically rule out the appropriateness of molecular-genetic responses. Whether such a response is appropriate depends on how one spells out "Why P?" The molecular-genetic approach has its place, to be sure, but its place is limited.

The case I have discussed — the bearing of evolutionary vs molecular perspectives on Mendelian genetics — is by no means an unusual case. In his recent, insightful analysis of theory structure in the biomedical sciences, Schaffner (1980) discusses four "representative" biomedical theories and the bearing of evolutionary theory on each. From these cases he generalizes the notion of "the metatheoretical function of evolutionary theory," which he summarizes as follows:

> We have seen in several of the examples mentioned above that evolutionary theory often functions as background providing a degree of intelligibility for components of a theory. Recall, for example, that the universality of the genetic code was explained evolutionarily, that the *lac* operon was conceived to have evolutionary advantages, that the "sharpness" of the clonal selection theory was justified evolutionarily, and that cell-mediated immunity was seen to confer important survival benefits as a defense against neoplasms in multicellular organisms. Evolutionary theory also allows us to understand why there is subtle variation in the organisms: Variation due to meiosis, mutation, and genetic drift, for example, predict that this type of variation should occur most frequently in evolving populations where strong selection pressures towards sharpness (involving lethal variations) are not present. Thus evolutionary theory at a very general level explains some of the specific and general features of other theories in the biomedical sciences. (Schaffner 1980, p. 76)

Schaffner is not explicitly concerned here with the issue of the appropriateness of evolutionary perspectives vs molecular-genetic, or of ultimate perspectives vs proximate. Nevertheless, a lesson to be learned from his analysis is that the extent of generality in the living world is a problem that calls for an ultimate, evolutionary solution rather than a proximate one.

4. More on the Evolutionary Perspective

So far I have acknowledged a restricted role for molecular genetics in the understanding of biological generalities. Molecular genetics is restricted to questions concerning the causal pathways from particular genotypes and environments to their phenotypes — i.e., to questions of form 2. But there is more to the restriction than that, and it is worth exploring further the nature of the restriction. To that end, let us consider a striking case in which a biological generality called for and received a molecular-genetic account. Having paid this powerful molecular-genetic account its due respects, I will then be in a better position to explain what I meant earlier when I said that the molecular-genetic accounts that are brought to bear upon biological generalities are often themselves subject to evolutionary scrutiny *in the final analysis.*

As early as 1922, the great geneticist Muller appealed for a molecular approach to genetics, and further suggested a criterion of adequacy that would have to be satisfied by a molecular account. That is, a satisfactory solution of the molecular constitution of the genetic material should illuminate the ability of the genetic material to "autocatalyze," or replicate, even after it had mutated. In other words, molecular genetics should address a question similar to form 2:

> Why, given the structure of the genetical material, is "mutable auto-catalysis" possible rather than not?

In Muller's own words:

> The fact that the genes have this autocatalytic power is in itself sufficiently striking, for they are undoubtedly complex substances, and it is difficult to understand by what strange coincidence of chemistry a gene can happen to have just that very special series of physico-chemical effects upon its surroundings which produces — of all possible end products — just this particular one, which is identical with its own complex structure. But the most remarkable feature of the situation is not this oft-noted autocatalytic action in itself — it is the fact that, when the structure of the gene becomes changed, through some "chance variation," the catalytic property of the gene may become correspondingly changed, in such a way as to leave it still *auto*catalytic. In other words, the change in gene structure — accidental though it was — has somehow resulted in a change of exactly *appropriate* nature in the catalytic reactions, so that the new reactions are now accurately adapted to produce more material just like that in the new changed gene itself.
>
> What sort of structure must the gene possess to permit it to mutate in this way? Since, through change after change in the gene, this same phenomenon persists, it is evident that it must depend upon some general feature of gene construction — common to all genes — which gives each one a *general* autocatalytic power — a "carte blanche" — to build material of whatever specific sort it itself happens to be composed of. . . . [This] question as to what the general principle of gene construction is, that permits this phenomenon of mutable autocatalysis, is the most fundamental question of genetics. (Muller 1922, pp. 34–35)

Muller's call for a molecular-genetic account of a biological (specifically genetic) generality was taken up by Watson and Crick, whose search for the structure of DNA was a search for a chemical structure with the appropriate biological properties. In other words, in Watson's and Crick's eyes, if DNA were really the genetic material, then a good chemical model of DNA would have to explain not only the data of DNA's physical and chemical analysis, but also general properties of the genetic material — like replicability even after mutation. As Watson later recollected in his textbook *Molecular Biology of the Gene*, the "excitement" of solving the structure of DNA "came not merely from the fact that the structure was solved, but also from the nature of the solution. Before the answer was known, there had always been the mild fear that it would turn out to be dull, and reveal nothing about how genes replicate and function. Fortunately, however, the answer was immensely exciting." (1965, p. 66).

In keeping with this goal, Watson and Crick triumphantly announced their discovery in 1953: "We wish to suggest a structure for the salt of deoxyribose nucleic acid (D.N.A.). This structure has novel features which are of considerable biological interest." (1953a, p. 737). As is now well known, their structure consisted of two intertwined helices, each of which consisted of a linear arrangement of the bases adenine, thymine, guanine, and cytosine. Moreover, the two helices were complementary in the sense that adenine on one helix was always paired with thymine on the other, and guanine was always paired with cytosine. After briefly describing the essentials of their structure, they concluded, "It has not escaped our notice that the specific pairing we have postulated immediately suggests a possible copying mechanism for the genetic material." (1953a, p. 737). Their idea was that the intertwined helices would, for purposes of replication, unwind and act as templates upon which complementary helices would be constructed. As they explained in their second article on DNA, published five weeks after the first:

> Now our model for deoxyribose nucleic acid is, in effect, a *pair* of templates, each of which is complementary to the other. We imagine that prior to duplication the hydrogen bonds [connecting the corresponding base pairs of the two helices] are broken, and the two chains unwind and separate. Each chain then acts as a template for the formation on to itself of a new companion chain, so that eventually, we shall have *two* pairs of chains where we only had one before. Moreover, the sequence of the pairs of bases will have been duplicated exactly. (1953b, p. 966).

Mutational permutations in the order of the bases are similarly replicated. Muller's question about the possibility of mutable autocatalysis was thus answered in the molecular terms he sought.

Here we have a case where a biological generality called for and received a strictly molecular-genetic account. The account *seems* to be entirely independent of evolutionary considerations, but in an interesting sense it is not. For the molecular account itself raises further evolutionary questions. Consider that such an account of the "mutable autocatalysis" of the genetic material in terms of the structure of DNA requires an identification of DNA as the genetic material. We might want to add that claim to the generalizations of molecular genetics, or we might not — that is not the point at issue. The point is that the identification of DNA as the genetic material is a biological generality that calls for an evolutionary perspective. That is, the fact, if it is a fact, that the genetic material is in all cases DNA is a fact that is best understood in terms of an ultimate account of the history of the genetic material, rather than wholly in terms of a proximate account of the processing of genetic information. Such an account would presumably concern the origin of life in a single common ancestor. At least, that would be an appropriate account.

We might conclude, then, that molecular-genetic accounts are appropriate for understanding generalizable relations between the genetic material and the other building blocks and processes of life, but the general occurrence of any particular kinds of genetic material, and any other building blocks and processes, ultimately calls for evolutionary understanding. In that sense, then, molecular-genetic generalities may account for biological generalities, but those molecular-genetic generalities may very well be subject to evolutionary scrutiny in the final analysis.

5. Difficulties and Conclusion

Exceptions to my general thesis are imaginable, however. As I excused myself earlier, I am in somewhat the position of the natural historian who recognizes that all possible things happen sometimes, but who emphasizes "only sometimes." Sometimes even questions of form 1 are better answered in terms of proximate rather than ultimate reasons. For instance, the question *why so many* trees of a certain population are charred and defoliated would be most appropriately answered in terms of circumstances that intervened in the processing of their environmental and genetic information, independently of any evolutionary considerations.

It would be nice if we could characterize this class of counterinstances generally and a priori instead of just citing instances of them a posteriori. It would be especially nice if we could tell simply from the vocabulary and/or grammar of a biological question whether it called for a proximate or an ultimate answer. I have suggested in effect that grammar and vocabulary are clues—that questions of form 1 should clue us to consider evolutionary answers, and questions of form 2 proximate answers. But to try to make the distinction completely in terms of language would be to overestimate seriously the power of philosophy. Whether a proximate or an ultimate explanation is most appropriate can only be determined by scientific investigation. For instance, there is no way to tell whether the generality of defoliation just discussed is the result of proximate or ultimate processes without "looking to see."

To be sure, it would be an error to overlook such exceptions as were just discussed. But it is a more common error, I think, to overlook the bearing of evolutionary perspectives on biological generalities, and to overemphasize the importance of the molecular-genetic perspective. We can at least say that the endeavor to explain all biological generalizations in molecular-genetic terms is a fundamentally misguided one, as is the endeavor to so explain any biological generalization if the sole rationale for trying to do so is the assumption that the deepest explanations in biology are molecular-genetic. Ultimately, evolution matters too.

NOTES

1. Thanks to Lindley Darden, Ernst Mayr, Nancy Maull, Alexander Rosenberg, Mary Williams, and especially Philip Kitcher for valuable help with earlier drafts of this paper.

2. See Schaffner (1969, 1974) and Ruse (1973, Chapter 10) on the possibility of explaining Mendelian genetics wholly in terms of molecular genetics. For an up-to-date review of other discussions concerning the bearing of molecular genetics on Mendelian genetics, see Hull (1982).

REFERENCES

Dobzhansky, T. (1973). "Nothing in Biology Makes Sense Except in the Light of Evolution." *American Biology Teacher* 35: 125–129.
Dunn, L. C. (1953). "Variations in the Segregation Ratio as Causes of Variations of Gene Frequency." *Acta Genetica et Statistica Medica* 4: 139–147.
———. (1957). "Evidence of Evolutionary Forces Leading to the Spread of Lethal Genes in Wild Populations of House Mice." *Proceedings of the National Academy of Science* 43: 158–163.
Edwards, A. W. F. (1977). *Foundations of Mathematical Genetics*. Cambridge: Cambridge University Press.
Gould, S. J. and Lewontin, R. C. (1979). "The Spandrels of San Marco and the Panglossian Paradigm: A Critique of the Adaptationist Programme." *Proceedings of the Royal Society of London* B205: 581–598.
Gowen, J. W. (1933). "Meiosis as a Genetic Character in *Drosophila Melanogaster*." *Journal of Experimental Zoology* 65: 83–106.
Hull, D. L. (1982). "Philosophy and Biology." In *Contemporary Philosophy: A New Survey*, Volume Two. Hague: Nijhoff. Pages 281–316.
Loeb, J. (1916). *The Organism as a Whole*. New York: Putnam.
Mayr, E. (1961). "Cause and Effect in Biology." *Science* 134: 1501–1506.
Muller, H. J. (1922). "Variation Due to Change in the Individual Gene." *American Naturalist* 56: 32–50.
Ruse, M. (1973). *The Philosophy of Biology*. London: Hutchinson.
Schaffner, K. F. (1969). "The Watson–Crick Model and Reductionism." *British Journal for the Philosophy of Science* 20: 325–348.
———. (1974). "The Peripherality of Reductionism in the Development of Molecular Biology." *Journal of the History of Biology* 7: 111–139.
———. (1980). "Theory Structure in the Biomedical Sciences." *Journal of Medicine and Philosophy* 5: 57–97.
van Fraassen, B. C. (1980). *The Scientific Image*. Oxford: Clarendon.
Watson, J. D. and Crick, F. H. C. (1953a). "Molecular Structure of Nucleic Acids." *Nature* 171: 737–738.
———. (1953b). "Genetical Implications of the Structure of Deoxyribonucleic Acid." *Nature* 171: 964–967.
———. (1965). *Molecular Biology of the Gene*. New York: Benjamin.
———. (1968). *The Double Helix*. New York: Atheneum.
White, M. J. D. (1973). *Animal Cytology and Evolution*. London: Clowes.

THE RECOMBINANT DNA DEBATE

21

JEREMY RIFKIN*

Creating the Efficient Gene

THE GREAT UNDERLYING myth of the Biotechnical Revolution is that it is possible to accelerate the production of more efficient living utilities without ever running out. This grand illusion will not be easy to dismiss or shunt aside. It has gained widespread currency over the past few decades because of the unfortunate choice of words we have come to use in distinguishing between fossil fuels and biological resources. We often refer to the former as non-renewable and to the latter as renewable. Therein lies the nub of the problem. In actuality, living resources are as depletable and finite as fossil fuels. Somewhere down the road, however, we have managed to confuse the idea of reproducible resources with the idea of perpetually inexhaustible resources. They are not the same. Living resources reproduce, but the life support systems that nourish them do not.

The genetic engineers refuse to come to grips with this underlying reality. Instead they continue to talk of the ever accelerating production of living things firmly convinced that the information coded in the genetic instructions provides the key to unlimited output. To illustrate this point, consider the possibility of genetically engineering new plants that could absorb greater sunlight and increase the rate of photosynthesis. While the benefit of such a procedure seems apparent, at first glance, a closer examination reveals the price that would have to be paid to achieve the desired results.

*From Jeremy Rifkin, *Declaration of a Heretic*, (London: Routledge & Kegan Paul, 1985), pp. 46–54.

222

Increased photosynthesis would require a greater use of soil nutrients, thus threatening the further depletion and erosion of an already endangered agricultural soil base. Soil depletion and erosion is one of the major problems in agriculture today. Up to one-third of some of our prime agricultural top-soil has been depleted in the past three decades, largely as a result of the accelerated production tempo of green revolution farming. Attempts to genetically engineer increases in speed of maturation and gross productivity will place additional burdens on an already over-taxed soil structure, thus posing the very real danger of inadequate nutrient reserves for sustaining future agricultural crops.

Genetic engineering will unquestionably result in the short term acceleration of biological materials into useful economic products, but at the expense of depleting the reservoir of life-support materials that are essential for maintaining the reproductive viability of living organisms in the future. In nature, there is no such thing as a free lunch. All biological and physical phenomena are subject to entropy and the Second Law of Thermodynamics. Being able to store and program the genetic instructions for living things is of little help if the biotic environment is bereft of the nutrients to sustain life.

All great technological revolutions secure the present by mortgaging the future. In this respect, genetic engineering represents the ultimate lien on the future. Genetic engineering raises the interest rates that will have to be paid by future generations beyond anything we've ever experienced in the long history of our attempts to control the forces of nature.

Everytime we choose to introduce a new genetically modified organism into the environment the ecological interest rate moves up a point. That's because every genetically engineered product presents a potential threat to the ecosystem it is released in. To appreciate why this is so, we need to be able to understand some of the defining characteristics of engineered organisms. The best way to do that is to contrast biotechnical products with petro-chemical products.

Genetically engineered products differ from petro-chemical products in several important ways. Because they are alive, genetically engineered products are inherently more unpredictable than petro-chemicals in the way they interact with other living things in the environment. Consequently, it is much more difficult to assess all of the potential impacts that a biotechnical product might have on the earth's ecosystems.

Genetically engineered products also reproduce. They grow and they migrate. Unlike petro-chemical products, it is impossible to constrain them within a given geographical locale. Finally, once released, it is virtually impossible to recall living products back to the laboratory, especially those products that are microscopic in nature. For all these reasons, genetically engineered products pose far greater long-term potential risks to the environment than petro-chemical substances.

Exactly how dangerous are genetically engineered products? Environmental scientists tell us that the risks in releasing biotechnical products into the biosphere are comparable to those we've encountered in introducing exotic organisms into the North American habitat. Over the past several hundred years thousands of non-native organisms have been brought to America from other regions of the globe. While most of these creatures have adapted to the ecosystem without severe dislocations, a small percentage of them have run wild, wreaking havoc on the flora and fauna of the continent. Gypsy moth, Kudzu vine, Dutch elm disease, chestnut blight, starlings, Mediterranean fruit flies come easily to mind. Each year the American continent is ravaged by these non-native organisms, with destruction to plant and animal life running into the tens of billions of dollars.

Whenever a genetically engineered organism is released there is always a small chance that it too will run amok because, like exotic organisms, it is not a naturally occurring life form. It has been artificially introduced into a complex environment that has developed a web of highly synchronized relationships over millions of years. Each new synthetic introduction is tantamount to playing ecological roulette. That is, while there is only a small chance of it triggering an environmental explosion, if it does, the consequences can be thunderous and irreversible.

For example, consider the first set of experiments in the US to release a genetically engineered organism into the open environment. Researchers at the University of California have modified a bacteria called P-syringae. This particular bacteria is found in its naturally occurring state in temperate regions all over the world. Its most unique attribute is its ability to nucleate ice crystals. In other words, it helps facilitate the formation of frost or ice. Using recombinant DNA technology, University of California researchers have found a way to delete the genetic instructions for making ice from the bacteria. This new genetically modified P-syringae microbe is called ice-minus.

Scientists are excited about the long-term commercial possibilities of ice-minus in agriculture. Frost damage has long been a major problem for American farmers. The chief culprit has been P-syringae which attaches itself to the plants, creating ice crystals. The American corporation financing this research hopes that by spraying massive concentrations of ice-minus P-syringae on agricultural crops, the naturally occurring P-syringae will be edged out, providing a protective blanket against frost damage. The benefits in introducing this genetically engineered organism appear impressive. It's only when one looks at the long-term ecological costs that problems begin to surface.

To begin with, the first question a good environmental scientist would ask is what role does the naturally occurring P-syringae play in nature? The experts that have studied this particular organism say that its ice-making capacity helps shape worldwide precipitation patterns and is a key determinant in establishing climatic conditions on the planet. The experts also contend that the P-syringae bacteria has played an important evolutionary role in enhancing the viability of frost resistant plants and insects in the northern areas of the globe.

Many of our agricultural crops, however, are tropical in origin and frost sensitive like citrus, corn, beans and tomatoes. Now consider the prospect of spraying ice-minus bacteria on millions of acres of frost-sensitive crops over several decades. While the crops will be protected against frost, the local flora and fauna will be at a disadvantage as the naturally occurring ice-nucleating bacteria which they have relied on for millions of years will have been edged out. Blanketing millions of acres of agricultural land with ice-minus also provides a protective coat of warmth allowing tropical based insects to begin migrating into colder regions. Then too, what will be the long-term effect on worldwide precipitation patterns and climate if ice-minus replaces the ice-making bacteria over millions of acres of land for a sustained period of time?

Introducing just this one genetically engineered product into the environment raises disturbing ecological questions. Yet, in the coming decades, industry is expected to introduce thousands of new genetically engineered products into the environment each year, just as industry introduced thousands of petro-chemical products into the environment each year. While many of these genetically engineered organisms will prove to be benign, sheer statistical probability suggests that a small percentage will prove to be dangerous and highly destructive to the environment.

For example, scientists are considering the possibility of producing a genetically engineered enzyme that could destroy lignin, an organic substance that makes wood rigid. They believe there might be great commercial advantage in using this genetically modified organism to clean up the effluent from paper mills or for decomposing biological material for energy. But if the enzyme were to migrate offsite and spread through forest land, it could well end up destroying millions of acres of woodland by eating away at the substance that provides trees with their rigidity.

Several years ago, General Electric developed and patented a micro-organism that eats up oil spills. This new microscopic creation has never been let out, probably for the reason that there is no way to guarantee that it won't get loose, reproduce in mass volumes and begin eating up oil reserves in gasoline storage tanks all over the planet. Environmental scientists also warn that new micro-organisms designed to consume toxic materials might develop an appetite for more valuable resources.

In fact, the long-term cumulative impact of thousands upon thousands of introductions of genetically modified organisms could well eclipse by a magnitude the damage that has resulted from the wholesale release of petro-chemical products into the earth's ecosystems. With these new biological based products, however, the damage is not containable, the destructive effects continue to reproduce, and the organisms can not be recalled, making the process irreversible.

Many people labor under the misguided assumption that genetic engineering has a good side and a bad side, and that steps can be taken to regulate potential abuses, assuring that only the beneficial aspects of the technology are employed. They fail to understand that it is the built-in assumptions of the technology, the inherent logic of the process, that creates the problem regardless of the good or bad intention of those using it. This can be seen quite clearly when looking at genetic engineering in agriculture and animal husbandry. The objective of genetic engineering technology is to improve the efficiency and productive output of plants and domestic animals. Efficiency and productivity, however, are cultural values not ecological rules.

Engineering efficiency and productivity into plants and animals means engineering sustainability out. Every breeder knows that attempts to streamline the productive efficiency of plants and animals results in more lucrative but less fit strains and breeds. It has long been acknowledged that over-breeding and over-hybridization result in monoculturing and loss of genetic variability. Reliance on a few super strains or breeds has proven to be very unwise, because it increases vulnerability to specific diseases or radical changes in the environment. Genetic diversity assures that each species will have enough variety to effectively adapt to changing environments. By eliminating all of the so called unprofitable strains and breeds, we undermine the adaptive capacity of each species. Farmers have witnessed, first hand, the problem that can arise from monoculturing. Not long ago the corn farmers were hit with a devastating blight. The corn strain they were all using was particularly vulnerable to the disease, resulting in massive losses. Had they planted a variety of corns, some of the strains would have been hearty enough to ward off the pest.

Animal breeding has posed similar problems. For example, many dairy farmers have chosen to breed only Holsteins because of their superior milk yield: Other less lucrative breeds have all but disappeared. The Holstein, while more lucrative, is less fit. It relies on specialized feeds, an array of technological support systems and continual monitoring, and cannot survive in pasture land over winter like other breeds.

Genetic engineering technology will dramatically accelerate the problems of monoculturing and loss of gene diversity. This technology allows scientists to more effectively increase short-term productivity by engineering efficiency directly into the genetic code of a species. At the same time, non-useful traits can be deleted directly from the hereditary blueprint further to increase productive output. There is even talk about introducing cloning techniques on a large scale in agriculture and animal husbandry over the next several decades. By reproducing millions of identical copies of a single superior strain or breed, agriculturists hope to increase efficiency and output dramatically. This kind of pure monoculturing is going to result in the almost complete loss of minor strains or breeds, as they will be considered uneconomical and uncompetitive in the open marketplace. The long-term environmental consequences could be profound. Imagine millions of exact cloned replicas of a particular cow being used throughout the country and the world. The spread of one disease to which that particular genotype is not immune could result in the wholesale destruction of entire herds and the collapse of much of the dairy industry. It could take years to search for any remaining minor breeds as replacements, and decades more to rebreed new herds from them.

Cloning livestock only begins to touch on the possibilities that lie ahead. Even more ambitious are current experiments being conducted by the US Department of Agriculture to insert human growth hormone genes into the permanent hereditary make-up of pigs, sheep and other domestic animals. Some scientists predict that within a very few years barnyard animals will double in size and develop to maturity in half the normal time. It is possible, say the experts, that with genetic engineering technology, they could produce a cow the size of a small elephant, producing over 45,000 pounds of milk products per year.

By transferring genes from one species into the biological codes of another species it is possible to change the essential character of domestic animals. These changes will not only revolutionize the business of animal husbandry, but also our concept of nature as well. As already mentioned, in accepting the notion of transferring genes from one species into another, we begin the process of eliminating species' borders from our thinking. Already researchers in the field of molecular biology are arguing that there is nothing particularly sacred about the concept of a species. As they see it, the important unit of life is no longer the organism, but rather the gene. They increasingly view life from the vantage point of the chemical composition at the genetic level. From this reductionist perspective, life is merely the aggregate representation of the chemicals that give rise to it and therefore they see no ethical problem whatsoever in transferring one, five or a hundred genes from one species into the hereditary blueprint of another species. For they truly believe that they are only transferring chemicals coded in the genes and not anything unique to a specific animal. By this kind of reasoning, all of life becomes desacralized. All of life becomes reduced to a chemical level and becomes available for manipulation.

Some ethicists and professional observers of science say they are not concerned about these first few experiments, but would be concerned with the transfer of more sophisticated genetic traits. Unfortunately, they fail to see that the blurring of species' borders begins the first moment a human gene is permanently implanted into the hereditary make-up of a mouse, pig or sheep. It is the first experiment that legitimizes the process. After all, if there is nothing particularly sacred about the human growth hormone gene, as researchers contend, then they might just as well argue that there is nothing particularly unique or special about all of the thousands of other individual genes that make-up the human gene pool. When it comes to more complex human traits (polygenic) that influence behavior and intellectual capacity, researchers will undoubtedly argue that they are not unique either, since they are merely a composite of the chemicals coded in the individual genes that make them up.

What, then, is unique about the human gene pool, or any other mammalian gene pool? Nothing, if you view each species as merely the sum total of the chemicals coded in the individual genes that make it up. It is this radical new concept of life that legitimizes the idea of crossing all species' barriers, and undermines the inviolability of discrete, recognizable species in nature.

Many scientists contend that it would be wrong to discontinue these kinds of experiments, because they broaden our field of knowledge. They rely on the rather clichéd argument that to halt such research would constitute a form of censorship. This is nonsense. Just because something can be done is no longer adequate justification for arguing that it should be done. The point is, it is a bit foolish to argue that every scientific experiment is worth pursuing. If certain types of scientific activity undermine the ethical principles and canons of civilization, we have an obligation to ourselves and future generations to be willing to say no. That doesn't make us guilty of stifling freedom of inquiry or "progress." It simply makes us responsible human beings.

Other proponents of this research argue that species have evolved, one from the other over the long period of history and, as such, the process of genetic transfer is merely a speed up of evolutionary development. On the other hand, it is also true that since *Homo sapiens* have populated the earth, we have never once recorded an event where one species has mutated into another species. Even accepting that these occurrences have taken place before human eyes could have ever recorded the events, we know little or nothing about how or why such changes might have occurred. In contrast, with the new genetic technologies we have the tools to "evolve" our own concept of life in a dramatically short span of historical time. Should we allow the cultural biases of a particular moment in human history to dictate basic changes in the biological blueprint of animals and humans? Should social criteria like efficiency, profits, productivity and national security determine which traits should be transferred between species? These are profound questions deserving long and prudent public debate. The time to discuss these questions is before the process unfolds, not after the technology has run its course.

22

STEPHEN P. STICH*

The Recombinant DNA Debate

THE DEBATE OVER recombinant DNA research is a unique event, perhaps a turning point, in the history of science. For the first time in modern history there has been widespread public discussion about whether and how a promising though potentially dangerous line of research shall be pursued. At root the debate is a moral debate and, like most such debates, requires proper assessment of the facts at crucial stages in the argument. A good deal of the controversy over recombinant DNA research arises because some of the facts simply are not yet known. There are many empirical questions we would like to have answered before coming to a decision — questions about the reliability of proposed containment facilities, about the viability of enfeebled strains of *E. coli*, about the ways in which pathogenic organisms do their unwelcome work, and much more. But all decisions cannot wait until the facts are available; some must be made now. It is to be expected that people with different hunches about what the facts will turn out to be will urge different decisions on how recombinant DNA research should be regulated. However, differing expectations about the facts have not been the only fuel for controversy. A significant part of the current debate can be traced to differences over moral principles. Also, unfortunately, there has been much unnecessary debate generated by care-

*Stephen P. Stich, "The Recombinant DNA Debate," *Philosophy & Public Affairs*, 7, no. 3 (Spring 1978). Copyright © 1978 by Princeton University Press. Reprinted with permission of Princeton University Press.

less moral reasoning and a failure to attend to the logical structure of some of the moral arguments that have been advanced.

In order to help sharpen our perception of the moral issues underlying the controversy over recombinant DNA research, I shall start by clearing away some frivolous arguments that have deflected attention from more serious issues. We may then examine the problems involved in deciding whether the potential benefits of recombinant DNA research justify pursuing it despite the risks that it poses.

I. Three Bad Arguments

My focus in this section will be on three untenable arguments, each of which has surfaced with considerable frequency in the public debate over recombinant DNA research.

The first argument on my list concludes that recombinant DNA research should not be controlled or restricted. The central premise of the argument is that scientists should have full and unqualified freedom to pursue whatever inquiries they may choose to pursue. This claim was stated repeatedly in petitions and letters to the editor during the height of the public debate over recombinant DNA research in the University of Michigan community.[1] The general moral principle which is the central premise of the argument plainly does entail that investigators using recombinant DNA technology should be allowed to pursue their research as they see fit. However, we need only consider a few examples to see that the principle invoked in this "freedom of inquiry" argument is utterly indefensible. No matter how sincere a researcher's interest may be in investigating the conjugal behavior of American university professors, few would be willing to grant him the right to pursue his research in my bedroom without my consent. No matter how interested a researcher may be in investigating the effects of massive doses of bomb-grade plutonium on preschool children, it is hard to imagine that anyone thinks he should be allowed to do so. Yet the "free inquiry" principle, if accepted, would allow both of these projects and countless other Dr. Strangelove projects as well. So plainly the simplistic "free inquiry" principle is indefensible. It would, however, be a

---

[1]For example, from a widely circulated petition signed by both faculty and community people: "The most important challenge may be a confrontation with one of our ancient assumptions—that there must be an absolute and unqualified freedom to pursue scientific inquiries. We will soon begin to wonder what meaning this freedom has if it leads to the destruction or demoralization of human beings, the only life forms able to exercise it." And from a letter to the editor written by a Professor of Engineering Humanities: "Is science beyond social and human controls, so that freedom of inquiry implies the absence of usual social restrictions which we all, as citizens, obey, respecting the social contract?"

It is interesting to note that the "freedom of inquiry" argument is rarely proposed by defenders of recombinant DNA research. Rather, it is proposed, then attacked, by those who are opposed to research involving recombinant molecules. Their motivation, it would seem, is to discredit the proponents of recombinant DNA research by attributing a foolish argument to them, then demonstrating that it is indeed a foolish argument.

mistake to conclude that freedom of inquiry ought not to be protected. A better conclusion is that the right of free inquiry is a qualified right and must sometimes yield to conflicting rights and to the demands of conflicting moral principles. Articulating an explicit and properly qualified principle of free inquiry is a task of no small difficulty. We will touch on this topic again toward the end of Section II.

The second argument I want to examine aims at establishing just the opposite conclusion from the first. The particular moral judgment being defended is that there should be a total ban on recombinant DNA research. The argument begins with the observation that even in so-called low-risk recombinant DNA experiments there is at least a possibility of catastrophic consequences. We are, after all, dealing with a relatively new and unexplored technology. Thus it is at least possible that a bacterial culture whose genetic makeup has been altered in the course of a recombinant DNA experiment may exhibit completely unexpected pathogenic characteristics. Indeed, it is not impossible that we could find ourselves confronted with a killer strain of, say, *E. coli* and, worse, a strain against which humans can marshal no natural defense. Now if this is possible — if we cannot say with assurance that the probability of it happening is zero — then, the argument continues, all recombinant DNA research should be halted. For the negative utility of the imagined catastrophe is so enormous, resulting as it would in the destruction of our society and perhaps even of our species, that no work which could possibly lead to this result would be worth the risk.

The argument just sketched, which might be called the "doomsday scenario" argument, begins with a premise which no informed person would be inclined to deny. It is indeed *possible* that even a low-risk recombinant DNA experiment might lead to totally catastrophic results. No ironclad guarantee can be offered that this will not happen. And while the probability of such an unanticipated catastrophe is surely not large, there is no serious argument that the probability is zero. Still, I think the argument is a sophistry. To go from the undeniable premise that recombinant DNA research might possibly result in unthinkable catastrophe to the conclusion that such research should be banned requires a moral principle stating that *all* endeavors that might possibly result in such a catastrophe should be prohibited. Once the principle has been stated, it is hard to believe that anyone would take it at all seriously. For the principle entails that, along with recombinant DNA research, almost all scientific research and many other commonplace activities having little to do with science should be prohibited. It is, after all, at least logically possible that the next new compound synthesized in an ongoing chemical research program will turn out to be an uncontainable carcinogen many orders of magnitude more dangerous than aerosol plutonium. And, to vary the example, there is a non-zero probability that experiments in artificial pollination will produce a weed that will, a decade from now, ruin the world's food grain harvest.[2]

I cannot resist noting that the principle invoked in the doomsday scenario argument is not new. Pascal used an entirely parallel argument to show that it is in our own best interests to believe in God. For though the probability of God's existence may be very low, if He nonetheless should happen to exist, the disutility that would accrue to the disbeliever would be catastrophic — an eternity in hell. But, as introductory philosophy students should all know, Pascal's argument only looks persuasive if we take our options to be just two: Christianity or atheism. A third possibility is belief in a jealous non-Christian God who will see to our damnation if and only if we *are* Christians. The probability of such a deity existing is again very small, but non-zero. So Pascal's argument is of no help in deciding whether or not to accept Christianity. For we may be damned if we do and damned if we don't.

I mention Pascal's difficulty because there is a direct parallel in the doomsday scenario argument against recombinant DNA research. Just as there is a non-zero probability that unforeseen consequences of recombinant DNA research will lead to disaster, so there is a non-zero probability that unforeseen consequences of *failing* to pursue the research will lead to disaster. There may, for example, come a time when, because of natural or man-induced climatic change, the capacity to alter quickly the genetic constitution of agricultural plants will be necessary to forestall catastrophic famine. And if we fail to pursue recombinant DNA research now, our lack of knowledge in the future may have consequences as dire as any foreseen in the doomsday scenario argument.

The third argument I want to consider provides a striking illustration of how important it is, in normative thinking, to make clear the moral *principles* being invoked. The argument I have in mind begins with a factual claim about recombinant DNA research and concludes that stringent restrictions, perhaps even a moratorium, should be imposed. However, advocates of the argument are generally silent on the normative principle(s) linking premise and conclusion. The gap thus created can be filled in a variety of ways, resulting in very different arguments. The empirical observation that begins the argument is that recombinant DNA methods enable scientists to move genes back and forth across natural barriers, "particularly the most fundamental such barrier, that which divides prokaryotes from eukaryotes. The results will be essentially new organisms, self-perpetuating and hence perma-

---

[2]Unfortunately, the doomsday scenario argument is *not* a straw man conjured only by those who would refute it. Consider, for example, the remarks of Anthony Mazzocchi, spokesman for the Oil, Chemical and Atomic Workers International Union, reported in *Science News*, 19 March 1977, p. 181: "When scientists argue over safe or unsafe, we ought to be very prudent. . . . If critics are correct and the Andromeda scenario has *even the smallest possibility* of occurring, we must assume it will occur on the basis of our experience" (emphasis added).

nent."[3] Because of this, it is concluded that severe restrictions are in order. Plainly this argument is an enthymeme; a central premise has been left unstated. What sort of moral principle is being tacitly assumed?

The principle that comes first to mind is simply that natural barriers should not be breached, or perhaps that "essentially new organisms" should not be created. The principle has an almost theological ring to it, and perhaps there are some people who would be prepared to defend it on theological grounds. But short of a theological argument, it is hard to see why anyone would hold the view that breaching natural barriers or creating new organisms is *intrinsically* wrong. For if a person were to advocate such a principle, he would have to condemn the creation of new bacterial strains capable of, say, synthesizing human clotting factor or insulin, *even if* creating the new organism generated *no unwelcome side effects*.

There is quite a different way of unraveling the "natural barriers" argument which avoids appeal to the dubious principles just discussed. As an alternative, this second reading of the argument ties premise to conclusion with a second factual claim and a quite different normative premise. The added factual claim is that at present our knowledge of the consequences of creating new forms of life is severely limited; thus we cannot know with any assurance that the probability of disastrous consequences is very low. The moral principle needed to mesh with the two factual premises would be something such as the following:

> If we do not know with considerable assurance that the probability of an activity leading to disastrous consequences is very low, then we should not allow the activity to continue.

Now this principle, unlike those marshaled in the first interpretation of the natural barriers argument, is not lightly dismissed. It is, to be sure, a conservative principle, and it has the odd feature of focusing entirely on the dangers an activity poses while ignoring its potential benefits.[4] Still, the principle may have a certain attraction in light of recent history, which has increasingly been marked by catastrophes attributable to technology's unanticipated side effects. I will not attempt a full scale evaluation of this principle just now. For the principle raises, albeit in a rather extreme way, the question of how risks and benefits are to be weighed against each other. In my opinion, that is the really crucial moral question raised by recombinant DNA research. It is a question which bristles with problems. In Section II I shall take a look at some of these problems and make a few tentative steps toward some

---

[3]The quotation is from George Wald, "The Case Against Genetic Engineering," *The Sciences*, September/October 1976; to be reprinted in David A. Jackson and Stephen P. Stich, eds., *The Recombinant DNA Debate*, forthcoming.

[4]It is important to note, however, that the principle is considerably less conservative, and correspondingly more plausible, than the principle invoked in the doomsday scenario argument. That latter principle would have us enjoin an activity if the probability of the activity leading to catastrophe is anything other than zero.

solutions. While picking our way through the problems we will have another opportunity to examine the principle just cited.

II. Risks and Benefits

At first glance it might be thought that the issue of risks and benefits is quite straightforward, at least in principle. What we want to know is whether the potential benefits of recombinant DNA research justify the risks involved. To find out we need only determine the probabilities of the various dangers and benefits. And while some of the empirical facts — the probabilities — may require considerable ingenuity and effort to uncover, the assessment poses no particularly difficult normative or conceptual problems. Unfortunately, this sanguine view does not survive much more than a first glance. A closer look at the task of balancing the risks and benefits of recombinant DNA research reveals a quagmire of sticky conceptual problems and simmering moral disputes. In the next few pages I will try to catalogue and comment on some of these moral disputes. I wish I could also promise solutions to all of them, but to do so would be false advertising.

Problems about Probabilities

In trying to assess costs and benefits, a familiar first step is to set down a list of possible actions and possible outcomes. Next, we assign some measure of desirability to each possible outcome, and for each action we estimate the conditional probability of each outcome given that the action is performed. In attempting to apply this decision-making strategy to the case of recombinant DNA research, the assignment of probabilities poses some perplexing problems. Some of the outcomes whose probabilities we want to know can be approached using standard empirical techniques. Thus, for example, we may want to know what the probability is of a specific enfeebled host *E. coli* strain surviving passage through the human intestinal system, should it be accidentally ingested. Or we may want to know what the probability is that a host organism will escape from a P-4 laboratory. In such cases, while there may be technical difficulties to be overcome, we have a reasonably clear idea of the sort of data needed to estimate the required probabilities. But there are other possible outcomes whose probabilities cannot be determined by experiment. It is important, for example, to know what the probability is of recombinant DNA research leading to a method for developing nitrogen-fixing strains of corn and wheat. And it is important to know how likely it is that recombinant DNA research will lead to techniques for effectively treating or preventing various types of cancer. Yet there is no experiment we can perform nor any data we can gather that will enable us to *empirically* estimate these probabilities. Nor are these the most problematic probabilities we may want to know. A possibility that weighs heavily on the minds of many who are worried about recombinant DNA research is that this research may lead to negative consequences for human health or for the environment *which have not yet even been thought of.* The history of technology during the last half-century surely demonstrates that this is not a quixotic concern. Yet here again there would appear to be no data we can gather that would help much in estimating the probability of such potential outcomes.

It should be stressed that the problems just sketched are not to be traced simply to a paucity of data. Rather, they are conceptual problems; it is doubtful whether there is *any clear empirical sense* to be made of objective probability assignments to contingencies like those we are considering.

Theorists in the Bayesian tradition may be unmoved by the difficulties we have noted. On their view all probability claims are reports of subjective probabilities.[5] And, a Bayesian might quite properly note, there is no special problem about assigning *subjective* probabilities to outcomes such as those that worried us. But even for the radical Bayesian, there remains the problem of *whose* subjective probabilities ought to be employed in making a *social* or *political* decision. The problem is a pressing one since the subjective probabilities assigned to potential dangers and benefits of recombinant DNA research would appear to vary considerably even among reasonably well informed members of the scientific community.

The difficulties we have been surveying are serious ones. Some might feel they are so serious that they render rational assessment of the risks and benefits of recombinant DNA research all but impossible. I am inclined to be rather more optimistic, however. Almost all of the perils posed by recombinant DNA research require the occurrence of a sequence of separate events. For a chimerical bacterial strain created in a recombinant DNA experiment to cause a serious epidemic, for example, at least the following events must occur:

(1) a pathogenic bacterium must be synthesized

(2) the chimerical bacteria must escape from the laboratory

(3) the strain must be viable in nature

(4) the strain must compete successfully with other micro-organisms which are themselves the product of intense natural selection.[6]

Since *all* of these must occur, the probability of the potential epidemic is the product of the probabilities of each individual contingency. And there are at least two items on the list, namely (2) and (3), whose

[5]For an elaboration of the Bayesian position, see Leonard J. Savage, *The Foundations of Statistics* (New York: John Wiley & Sons, 1954); also cf. Leonard J. Savage, "The Shifting Foundations of Statistics," in Robert G. Colodny, ed., *Logic, Laws and Life* (Pittsburgh: University of Pittsburgh Press, 1977).

[6]For an elaboration of this point, see Bernard D. Davis, "Evolution, Epidemiology and Recombinant DNA." *The Recombinant DNA Debate*, forthcoming.

probabilities are amenable to reasonably straightforward empirical assessment. Thus the product of these two individual probabilities places an upper limit on the probability of the epidemic. For the remaining two probabilities, we must rely on subjective probability assessments of informed scientists. No doubt there will be considerable variability. Yet even here the variability will be limited. In the case of (4), as an example, the available knowledge about microbial natural selection provides no precise way of estimating the probability that a chimerical strain of enfeebled *E. coli* will compete successfully outside the laboratory. But no serious scientist would urge that the probability is *high*. We can then use the highest responsible subjective estimate of the probabilities of (1) and (4) in calculating the "worst case" estimate of the risk of epidemic. If in using this highest "worst case" estimate, our assessment yields the result that benefits outweigh risks, then lower estimates of the same probabilities will, of course, yield the same conclusion. Thus it may well be the case that the problems about probabilities we have reviewed will not pose insuperable obstacles to a rational assessment of risks and benefits.

Weighing Harms and Benefits

A second cluster of problems that confronts us in assessing the risks and benefits of recombinant DNA research turns on the assignment of a measure of desirability to the various possible outcomes. Suppose that we have a list of the various harms and benefits that might possibly result from pursuing recombinant DNA research. The list will include such "benefits" as development of an inexpensive way to synthesize human clotting factor and development of a strain of nitrogen-fixing wheat; and such "harms" as release of a new antibiotic-resistant strain of pathogenic bacteria and release of a strain of *E. coli* carrying tumor viruses capable of causing cancer in man.

Plainly, it is possible that pursuing a given policy will result in more than one benefit and in more than one harm. Now if we are to assess the potential impact of various policies or courses of action, we must assign some index of desirability to the possible *total outcomes* of each policy, outcomes which may well include a mix of benefits and harms. To do this we must confront a tangle of normative problems that are as vexing and difficult as any we are likely to face. We must *compare* the moral desirabilities of various harms and benefits. The task is particularly troublesome when the harms and benefits to be compared are of different kinds. Thus, for example, some of the attractive potential benefits of recombinant DNA research are economic: we may learn to recover small amounts of valuable metals in an economically feasible way, or we may be able to synthesize insulin and other drugs inexpensively. By contrast, many of the risks of recombinant DNA research are risks to human life or health. So if we are to take the idea of cost–benefit analysis seriously, we must at some point decide how human lives are to be weighed against economic benefits.

There are those who contend that the need to make such decisions indicates the moral bankruptcy of attempting to employ risk–benefit analyses when human lives are at stake. On the critics' view, we cannot reckon the possible loss of a human life as just another negative outcome, albeit a grave and heavily weighted one. To do so, it is urged, is morally repugnant and reflects a callous lack of respect for the sacredness of human life.

On my view, this sort of critique of the very idea of using risk–benefit analyses is ultimately untenable. It is simply a fact about the human condition, lamentable as it is inescapable, that in many human activities we run the risk of inadvertently causing the death of a human being. We run such a risk each time we drive a car, allow a dam to be built, or allow a plane to take off. Moreover, in making social and individual decisions, we cannot escape weighing economic consequences against the risk to human life. A building code in the Midwest will typically mandate fewer precautions against earthquakes than a building code in certain parts of California. Yet earthquakes are not impossible in the Midwest. If we elect not to require precautions, then surely a major reason must be that it would simply be too expensive. In this judgment, as in countless others, there is no escaping the need to balance economic costs against possible loss of life. To deny that we must and do balance economic costs against risks to human life is to assume the posture of a moral ostrich.

I have been urging the point that it is not *morally objectionable* to try to balance economic concerns against risks to human life. But if such judgments are unobjectionable, indeed necessary, they also surely are among the most difficult any of us has to face. It is hard to imagine a morally sensitive person not feeling extremely uncomfortable when confronted with the need to put a dollar value on human lives. It might be thought that the moral dilemmas engendered by the need to balance such radically different costs and benefits pose insuperable practical obstacles for a rational resolution of the recombinant DNA debate. But here, as in the case of problems with probabilities, I am more sanguine. For while some of the risks and potential benefits of recombinant DNA research are all but morally incommensurable, the most salient risks and benefits are easier to compare. The major risks, as we have noted, are to human life and health. However, the major potential benefits are *also* to human life and health. The potential economic benefits of recombinant DNA research pale in significance when set against the potential for major breakthroughs in our understanding and ability to treat a broad range of conditions, from birth defects to cancer. Those of us, and I confess I am among them, who despair of deciding how lives and economic benefits are to be compared can nonetheless hope to settle our views about recombinant DNA research by comparing the potential risks to life and health with the potential benefits to life and health. Here we are comparing plainly commensurable outcomes. If the balance turns out to be favorable, then we need not worry about factoring in potential economic benefits.

There is a certain irony in the fact that we may well be able to ignore economic factors entirely in coming to a decision about recombinant DNA research. For I suspect that a good deal of the apprehension about recombinant DNA research on the part of the public at large is rooted in the fear that (once again) economic benefits will be weighed much too heavily and potential damage to health and the environment will be weighed much too lightly. The fear is hardly an irrational one. In case after well-publicized case, we have seen the squalid consequences of decisions in which private or corporate gain took precedence over clear and serious threats to health and to the environment. It is the profit motive that led a giant chemical firm to conceal the deadly consequences of the chemical which now threatens to poison the James River and perhaps all of Chesapeake Bay. For the same reason, the citizens of Duluth drank water laced with a known carcinogen. And the ozone layer that protects us all was eroded while regulatory agencies and legislators fussed over the loss of profits in the spray deodorant industry. Yet while public opinion about recombinant DNA research is colored by a growing awareness of these incidents and dozens of others, the case of recombinant DNA is fundamentally different in a crucial respect. The important projected benefits which must be set against the risks of recombinant DNA research are not economic at all, they are medical and environmental.

Problems about Principles

The third problem I want to consider focuses on the following question. Once we have assessed the potential harms and benefits of recombinant DNA research, how should we use this information in coming to a decision? It might be thought that the answer is trivially obvious. To assess the harms and benefits is, after all, just to compute for each of the various policies that we are considering, what might be called its *expected utility*. The expected utility of a given policy is found by first multiplying the desirability of each possible total outcome by the probability that the policy in question will lead to that total outcome, and then adding the numbers obtained. As we have seen, finding the needed probabilities and assigning the required desirabilities will not be easy. But once we know the expected utility of each policy, is it not obvious that we should choose the policy with the highest expected utility? The answer, unfortunately, is no, it is not at all obvious.

Let us call the principle that we should adopt the policy with the highest expected utility the *utilitarian principle*. The following example should make it clear that, far from being trivial or tautological the utilitarian principle is a substantive and controversial moral principle. Suppose that the decision which confronts us is whether or not to adopt policy A. What is more, suppose we know there is a probability close to

1 that 100,000 lives will be saved if we adopt A. However, we also know that there is a probability close to 1 that 1,000 will die as a direct result of our adopting policy A, and these people would survive if we did not adopt A. Finally, suppose that the other possible consequences of adopting A are relatively inconsequential and can be ignored. (For concreteness, we might take A to be the establishment of a mass vaccination program, using a relatively risky vaccine.) Now plainly if we take the moral desirability of saving a life to be exactly offset by the moral undesirability of causing a death, then the utilitarian principle dictates that we adopt policy A. But many people feel uncomfortable with this result, the discomfort increasing with the number of deaths that would result from A. If, to change the example, the choice that confronts us is saving 100,000 lives while causing the deaths of 50,000 others, a significant number of people are inclined to think that the morally right thing to do is to refrain from doing A, and "let nature take its course."

If we reject policy A, the likely reason is that we also reject the utilitarian principle. Perhaps the most plausible reason for rejecting the utilitarian principle is the view that our obligation to *avoid doing harm* is stronger than our obligation to do good. There are many examples, some considerably more compelling than the one we have been discussing, which seem to illustrate that in a broad range of cases we do feel that our obligation to avoid doing harm is greater than our obligation to do good.[7] Suppose, to take but one example, that my neighbor requests my help in paying off his gambling debts. He owes $5,000 to a certain bookmaker with underworld connections. Unless the neighbor pays the debt immediately, he will be shot. Here, I think we are all inclined to say, I have no strong obligation to give my neighbor the money he needs, and if I were to do so it would be a supererogatory gesture. By contrast, suppose a representative of my neighbor's bookmaker approaches me and requests that I shoot my neighbor. If I refuse, he will see to it that my new car, which cost $5,000, will be destroyed by a bomb while it sits unattended at the curb. In this case, surely, I have a strong obligation not to harm my neighbor, although not shooting him will cost me $5,000.

Suppose that this example and others convince us that we cannot adopt the utilitarian principle, at least not in its most general form, where it purports to be applicable to all moral decisions. What are the alternatives? One cluster of alternative principles would urge that in some or all cases we weigh the harm a contemplated action will cause more heavily than we weigh the good it will do. The extreme form of such a principle would dictate that we ignore the benefits entirely and

---

[7]For an interesting discussion of these cases, see J. O. Urmson, "Saints and Heros," in A. I. Melden, ed., *Essays In Moral Philosophy* (Seattle: University of Washington Press, 1958). Also see the discussion of positive and negative duties in Philippa Foot, "The Problem of Abortion and the Doctrine of Double Effect," *Oxford Review* 5 (1967). Reprinted in James Rachels, ed., *Moral Problems* (New York: Harper & Row) 1971.

opt for the action or policy that produces the *least* expected harm. (It is this principle, or a close relation, which emerged in the second reading of the "natural barriers" argument discussed in the third part of Section I above.) A more plausible variant would allow us to count both benefits and harms in our deliberations, but would specify how much more heavily harms were to count.

On my view, some moderate version of a "harm-weighted" principle is preferable to the utilitarian principle in a considerable range of cases. *However, the recombinant DNA issue is not one of these cases.* Indeed, when we try to apply a harm-weighted principle to the recombinant DNA case we run head on into a conceptual problem of considerable difficulty. The distinction between doing good and doing harm presupposes a notion of the normal or expectable course of events. Roughly, if my action causes you to be worse off than you would have been in the normal course of events, then I have harmed you; if my action causes you to be better off than in the normal course of events, then I have done you some good; and if my action leaves you just as you would be in the normal course of events, then I have done neither. In many cases, the normal course of events is intuitively quite obvious. Thus in the case of the neighbor and the bookmaker, in the expected course of events I would neither shoot my neighbor nor give him $5,000 to pay off his debts. Thus I am doing good if I give him the money and I am doing harm if I shoot him. But in other cases, including the recombinant DNA case, it is not at all obvious what constitutes the "expected course of events," and thus it is not at all obvious what to count as a harm. To see this, suppose that as a matter of fact many more deaths and illnesses will be prevented as a result of pursuing recombinant DNA research than will be caused by pursuing it. But suppose that there *will* be at least some people who become ill or die as a result of recombinant DNA research being pursued. If these are the facts, then who would be harmed by imposing a ban on recombinant DNA research? That depends on what we take to be the "normal course of events." Presumably, if we do not impose a ban, then the research will continue and the lives will be saved. If this is the normal course of events, then if we impose a ban we have *harmed* those people who would be saved. But it is equally natural to take as the normal course of events the situation in which recombinant DNA research is not pursued. And if *that* is the normal course of events then those who would have been saved are not harmed by a ban, for they are no worse off than they would be in the normal course of events. However, on this reading of "the normal course of events," if we *fail* to impose a ban, then we have harmed those people who will ultimately become ill or die as a result of recombinant DNA research, since as a result of not imposing a ban they are worse off than they would have been in the normal course of events. I conclude that, in the absence of a theory detailing how we are to recognize the normal course of events, harm-weighted principles have no clear application to the case of recombinant DNA research.

Harm-weighted principles are not the only alternatives to the utilitarian principle. There is another cluster of alternatives that take off in quite a different direction. These principles urge that in deciding which policy to pursue there is a strong presumption in favor of policies that adhere to certain formal moral principles (that is, principles which do not deal with the *consequences* of our policies). Thus, to take the example most directly relevant to the recombinant DNA case, it might be urged that there is a strong presumption in favor of a policy which preserves freedom of scientific inquiry. In its extreme form, this principle would protect freedom of inquiry *no matter what the consequences*; and as we saw in the first part of Section I, this extreme position is exceptionally implausible. A much more plausible principle would urge that freedom of inquiry be protected until the balance of negative over positive consequences reaches a certain specified amount, at which point we would revert to the utilitarian principle. On such a view, if the expected utility of banning recombinant DNA research is a bit higher than the expected utility of allowing it to continue, then we would nonetheless allow it to continue. But if the expected utility of a ban is enormously higher than the expected utility of continuation, banning is the policy to be preferred.[8]

III. Long Term Risks

Thus far in our discussion of risks and benefits, the risks that have occupied us have been what might be termed "short-term" risks, such as the release of a new pathogen. The negative effects of these events, though they might be long-lasting indeed, would be upon us relatively quickly. However, some of those who are concerned about recombinant DNA research think there are longer-term dangers that are at least as worrisome. The dangers they have in mind stem not from the accidental release of harmful substances in the course of recombinant DNA research, but rather from the unwise use of the *knowledge* we will likely gain in pursuing the research. The scenarios most often proposed are nightmarish variations on the theme of human genetic engineering. With the knowledge we acquire, it is conjectured, some future tyrant may have people built to order, perhaps creating a whole class of people who willingly and cheaply do the society's dirty or dangerous work, as in Huxley's *Brave New World.* Though the proposed scenarios clearly are science fiction, they are not to be lightly dismissed. For if the technology they conjure is not demonstrably achievable, neither is it demonstrably impossible. And if only a bit of the science fiction turns to fact, the dangers could be beyond reckoning.

Granting that potential misuse of the knowledge gained in recombi-

---

[8]Carl Cohen defends this sort of limited protection of the formal free inquiry principle over a straight application of the utilitarian principle in his interesting essay, "When May Research Be Stopped?" *New England Journal of Medicine* 296 (1977). To be reprinted in *The Recombinant DNA Debate.*

nant DNA research is a legitimate topic of concern, how ought we to guard ourselves against this misuse? One common proposal is to try to prevent the acquisition of such knowledge by banning or curtailing recombinant DNA research now. Let us cast this proposal in the form of an explicit moral argument. The conclusion is that recombinant DNA research should be curtailed, and the reason given for this conclusion is that such research could possibly produce knowledge which might be misused with disastrous consequences. To complete the argument we need a moral principle, and the one which seems to be needed is something such as this:

> If a line of research can lead to the discovery of knowledge which might be disastrously misused, then that line of research should be curtailed.

Once it has been made explicit, I think relatively few people would be willing to endorse this principle. For recombinant DNA research is hardly alone in potentially leading to knowledge that might be disastrously abused. Indeed, it is hard to think of an area of scientific research that could *not* lead to the discovery of potentially dangerous knowledge. So if the principle is accepted it would entail that almost all scientific research should be curtailed or abandoned.

It might be thought that we could avoid the extreme consequences just cited by retreating to a more moderate moral principle. The moderate principle would urge only that we should curtail those areas of research where the probability of producing dangerous knowledge is comparatively high. Unfortunately, this more moderate principle is of little help in avoiding the unwelcome consequences of the stronger principle. The problem is that the history of science is simply too unpredictable to enable us to say with any assurance which lines of research will produce which sorts of knowledge or technology. There is a convenient illustration of the point in the recent history of molecular genetics. The idea of recombining DNA molecules is one which has been around for some time. However, early efforts proved unsuccessful. As it happened, the crucial step in making recombinant DNA technology possible was provided by research on restriction enzymes, research that was undertaken with no thought of recombinant DNA technology. Indeed, until it was realized that restriction enzymes provided the key to recombining DNA molecules, the research on restriction enzymes was regarded as a rather unexciting (and certainly uncontroversial) scientific backwater.[9] In an entirely analogous way, crucial pieces of information that may one day enable us to manipulate the human genome may come from just about any branch of molecular biology. To guard against the discovery of that knowledge we should have to curtail not only recombinant DNA research but all of molecular biology.

[9] I am indebted to Prof. Ethel Jackson for both the argument and the illustration.

Before concluding, we would do well to note that there is a profound pessimism reflected in the attitude of those who would stop recombinant DNA research because it might lead to knowledge that could be abused. It is, after all, granted on all sides that the knowledge resulting from recombinant DNA research will have both good and evil potential uses. So it would seem the sensible strategy would be to try to prevent the improper uses of this knowledge rather than trying to prevent the knowledge from ever being uncovered. Those who would take the more extreme step of trying to stop the knowledge from being uncovered presumably feel that its improper use is all but inevitable, that our political and social institutions are incapable of preventing morally abhorrent applications of the knowledge while encouraging beneficial applications. On my view, this pessimism is unwarranted; indeed, it is all but inconsistent. The historical record gives us no reason to believe that what is technologically possible will be done, no matter what the moral price. Indeed, in the area of human genetic manipulation, the record points in quite the *opposite* direction. We have long known that the same techniques that work so successfully in animal breeding can be applied to humans as well. Yet there is no evidence of a "technological imperative" impelling our society to breed people as we breed dairy cattle, simply because we know that it can be done. Finally, it is odd that those who express no confidence in the ability of our institutions to forestall such monstrous applications of technology are not equally pessimistic about the ability of the same institutions to impose an effective ban on the uncovering of dangerous knowledge. If our institutions are incapable of restraining the application of technology when those applications would be plainly morally abhorrent, one would think they would be even more impotent in attempting to restrain a line of research which promises major gains in human welfare.

HUMAN SOCIOBIOLOGY

23

EDWARD O. WILSON*

ce

Heredity

ভ্ত

WE LIVE ON a planet of staggering organic diversity. Since Carolus Linnaeus began the process of formal classification in 1758, zoologists
have catalogued about one million species of animals and given each a
scientific name, a few paragraphs in a technical journal, and a small
space on the shelves of one museum or another around the world. Yet
despite this prodigious effort, the process of discovery has hardly
begun. In 1976 a specimen of an unknown form of giant shark, fourteen
feet long and weighing sixteen hundred pounds, was captured when it
tried to swallow the stabilizing anchor of a United States Naval vessel
near Hawaii. About the same time entomologists found an entirely new
category of parasitic flies that resemble large reddish spiders and live
exclusively in the nests of the native bats of New Zealand. Each year
museum curators sort out thousands of new kinds of insects, copepods,
wireworms, echinoderms, priapulids, pauropods, hypermastigotes, and
other creatures collected on expeditions around the world. Projections
based on intensive surveys of selected habitats indicate that the total
number of animal species is between three and ten million. Biology, as
the naturalist Howard Evans expressed it in the title of a recent book, is
the study of life "on a little known planet."

Thousands of these species are highly social. The most advanced

*Reprinted by permission of the publishers from *On Human Nature* by E. O. Wilson,
Cambridge, Massachusetts: Harvard University Press, Copyright © 1978 by the President and Fellows of Harvard College.

among them constitute what I have called the three pinnacles of social evolution in animals: the corals, bryozoans, and other colony-forming invertebrates; the social insects, including ants, wasps, bees, and termites; and the social fish, birds, and mammals. The communal beings of the three pinnacles are among the principal objects of the new discipline of sociobiology, defined as the systematic study of the biological basis of all forms of social behavior, in all kinds of organisms, including man. The enterprise has old roots. Much of its basic information and some of its most vital ideas have come from ethology, the study of whole patterns of behavior of organisms under natural conditions. Ethology was pioneered by Julian Huxley, Karl von Frisch, Konrad Lorenz, Nikolaas Tinbergen, and a few others and is now being pursued by a large new generation of innovative and productive investigators. It has remained most concerned with the particularity of the behavior patterns shown by each species, the ways these patterns adapt animals to the special challenges of their environments, and the steps by which one pattern gives rise to another as the species themselves undergo genetic evolution. Increasingly, modern ethology is being linked to studies of the nervous system and the effects of hormones on behavior. Its investigators have become deeply involved with developmental processes and even learning, formerly the nearly exclusive domain of psychology, and they have begun to include man among the species most closely scrutinized. The emphasis of ethology remains on the individual organism and the physiology of organisms.

Sociobiology, in contrast, is a more explicitly hybrid discipline that incorporates knowledge from ethology (the naturalistic study of whole patterns of behavior), ecology (the study of the relationships of organisms to their environment), and genetics in order to derive general principles concerning the biological properties of entire societies. What is truly new about sociobiology is the way it has extracted the most important facts about social organization from their traditional matrix of ethology and psychology and reassembled them on a foundation of ecology and genetics studied at the population level in order to show how social groups adapt to the environment by evolution. Only within the past few years have ecology and genetics themselves become sophisticated and strong enough to provide such a foundation.

Sociobiology is a subject based largely on comparisons of social species. Each living form can be viewed as an evolutionary experiment, a product of millions of years of interaction between genes and environment. By examining many such experiments closely, we have begun to construct and test the first general principles of genetic social evolution. It is now within our reach to apply this broad knowledge to the study of human beings.

Sociobiologists consider man as though seen through the front end of a telescope, at a greater than usual distance and temporarily diminished in size, in order to view him simultaneously with an array of other social experiments. They attempt to place humankind in its proper place in a catalog of the social species on Earth. They agree with Rousseau that "One needs to look near at hand in order to study men, but to study man one must look from afar."

This macroscopic view has certain advantages over the traditional anthropocentrism of the social sciences. In fact, no intellectual vice is more crippling than defiantly self-indulgent anthropocentrism. I am reminded of the clever way Robert Nozick makes this point when he constructs an argument in favor of vegetarianism. Human beings, he notes, justify the eating of meat on the grounds that the animals we kill are too far below us in sensitivity and intelligence to bear comparison. It follows that if representatives of a truly superior extraterrestrial species were to visit Earth and apply the same criterion, they could proceed to eat us in good conscience. By the same token, scientists among these aliens might find human beings uninteresting, our intelligence weak, our passions unsurprising, our social organization of a kind already frequently encountered on other planets. To our chagrin they might then focus on the ants, because these little creatures, with their haplodiploid form of sex determination and bizarre female caste systems, are the truly novel productions of the Earth with reference to the Galaxy. We can imagine the log declaring, "A scientific breakthrough has occurred; we have finally discovered haplodiploid social organisms in the one- to ten-millimeter range." Then the visitors might inflict the ultimate indignity: in order to be sure they had not underestimated us, they would simulate human beings in the laboratory. Like chemists testing the structural characterization of a problematic organic compound by assembling it from simpler components, the alien biologists would need to synthesize a hominoid or two.

This scenario from science fiction has implications for the definition of man. The impressive recent advances by computer scientists in the design of artificial intelligence suggests the following test of humanity: that which behaves like man *is* man. Human behavior is something that can be defined with fair precision, because the evolutionary pathways open to it have not all been equally negotiable. Evolution has not made culture all-powerful. It is a misconception among many of the more traditional Marxists, some learning theorists, and a still surprising proportion of anthropologists and sociologists that social behavior can be shaped into virtually any form. Ultra-environmentalists start with the premise that man is the creation of his own culture: "culture makes man," the formula might go, "makes culture makes man." Theirs is only a half truth. Each person is molded by an interaction of his environment, especially his cultural environment, with the genes that affect social behavior. Although the hundreds of the world's cultures seem enormously variable to those of us who stand in their midst, all versions of human social behavior together form only a tiny fraction of the realized organizations of social species on this planet and a still smaller fraction of those that can be readily imagined with the aid of sociobiological theory.

The question of interest is no longer whether human social behavior is genetically determined; it is to what extent. The accumulated evidence for a large hereditary component is more detailed and compelling than most persons, including even geneticists, realize. I will go further: it already is decisive.

That being said, let me provide an exact definition of a genetically determined trait. It is a trait that differs from other traits at least in part as a result of the presence of one or more distinctive genes. The important point is that the objective estimate of genetic influence requires comparison of two or more states of the same feature. To say that blue eyes are inherited is not meaningful without further qualification, because blue eyes are the product of an interaction between genes and the largely physiological environment that brought final coloration to the irises. But to say that the *difference* between blue and brown eyes is based wholly or partly on differences in genes is a meaningful statement because it can be tested and translated into the laws of genetics. Additional information is then sought: What are the eye colors of the parents, siblings, children, and more distant relatives? These data are compared to the very simplest model of Mendelian heredity, which, based on our understanding of cell multiplication and sexual reproduction, entails the action of only two genes. If the data fit, the differences are interpreted as being based on two genes. If not, increasingly complicated schemes are applied. Progressively larger numbers of genes and more complicated modes of interaction are assumed until a reasonably close fit can be made. In the example just cited, the main differences between blue and brown eyes are in fact based on two genes, although complicated modifications exist that make them less than an ideal textbook example. In the case of the most complex traits, hundreds of genes are sometimes involved, and their degree of influence can ordinarily be measured only crudely and with the aid of sophisticated mathematical techniques. Nevertheless, when the analysis is properly performed it leaves little doubt as to the presence and approximate magnitude of the genetic influence.

Human social behavior can be evaluated in essentially the same way, first by comparison with the behavior of other species and then, with far greater difficulty and ambiguity, by studies of variation among and within human populations. The picture of genetic determinism emerges most sharply when we compare selected major categories of animals with the human species. Certain general human traits are shared with a majority of the great apes and monkeys of Africa and Asia, which on grounds of anatomy and biochemistry are our closest living evolutionary relatives:

• Our intimate social groupings contain on the order of ten to one hundred adults, never just two, as in most birds and marmosets, or up to thousands, as in many kinds of fishes and insects.

• Males are larger than females. This is a characteristic of considerable significance within the Old World monkeys and apes and many other kinds of mammals. The average number of females consorting with successful males closely corresponds to the size gap between males and females when many species are considered together. The rule makes sense: the greater the competition among males for females the greater the advantage of large size and the less influential are any disadvantages accruing to bigness. Men are not very much larger than women; we are similar to chimpanzees in this regard. When the sexual size difference in human beings is plotted on the curve based on other kinds of mammals, the predicted average number of females per successful male turns out to be greater than one but less than three. The prediction is close to reality; we know we are a mildly polygynous species.

• The young are molded by a long period of social training, first by closest associations with the mother, then to an increasing degree with other children of the same age and sex.

• Social play is a strongly developed activity featuring role practice, mock aggression, sex practice, and exploration.

These and other properties together identify the taxonomic group consisting of Old World monkeys, the great apes, and human beings. It is inconceivable that human beings could be socialized into the radically different repertories of other groups such as fishes, birds, antelopes, or rodents. Human beings might self-consciously *imitate* such arrangements, but it would be a fiction played out on a stage, would run counter to deep emotional responses and have no chance of persisting through as much as a single generation. To adopt with serious intent, even in broad outline, the social system of a nonprimate species would be insanity in the literal sense. Personalities would quickly dissolve, relationships disintegrate, and reproduction cease.

At the next, finer level of classification, our species is distinct from the Old World monkeys and apes in ways that can be explained only as a result of a unique set of human genes. Of course, that is a point quickly conceded by even the most ardent environmentalists. They are willing to agree with the great geneticist Theodosius Dobzhansky that "in a sense, human genes have surrendered their primacy in human evolution to an entirely new, nonbiological or superorganic agent, culture. However, it should not be forgotten that this agent is entirely dependent on the human genotype." But the matter is much deeper and more interesting than that. There are social traits occurring through all cultures which upon close examination are as diagnostic of mankind as are distinguishing characteristics of other animal species—as true to the human type, say, as wing tessellation is to a fritillary butterfly or a complicated spring melody to a wood thrush. In 1945 the American anthropologist George P. Murdock listed the following characteristics that have been recorded in every culture known to history and ethnography:

Age-grading, athletic sports, bodily adornment, calendar, cleanliness
training, community organization, cooking, cooperative labor, cosmol-
ogy, courtship, dancing, decorative art, divination, division of labor,
dream interpretation, education, eschatology, ethics, ethnobotany, etiquette, faith healing, family feasting, fire making, folklore, food
taboos, funeral rites, games, gestures, gift giving, government, greetings, hair styles, hospitality, housing, hygiene, incest taboos, inheritance rules, joking, kin groups, kinship nomenclature, language, law,
luck superstitions, magic, marriage, mealtimes, medicine, obstetrics,
penal sanctions, personal names, population policy, postnatal care,
pregnancy usages, property rights, propitiation of supernatural
beings, puberty customs, religious ritual, residence rules, sexual restrictions, soul concepts, status differentiation, surgery, tool making,
trade, visiting, weaving, and weather control.

Few of these unifying properties can be interpreted as the inevitable
outcome of either advanced social life or high intelligence. It is easy to
imagine nonhuman societies whose members are even more intelligent
and complexly organized than ourselves, yet lack a majority of the
qualities just listed. Consider the possibilities inherent in the insect
societies. The sterile workers are already more cooperative and altruistic than people and they have a more pronounced tendency toward
caste systems and division of labor. If ants were to be endowed in
addition with rationalizing brains equal to our own, they could be our
peers. Their societies would display the following peculiarities:

Age-grading, antennal rites, body licking, calendar, cannibalism, caste
determination, caste laws, colony-foundation rules, colony organization, cleanliness training, communal nurseries, cooperative labor, cos-
mology, courtship, division of labor, drone control, education, eschatology, ethics, etiquette, euthanasia, fire making, food taboos, gift
giving, government, greetings, grooming rituals, hospitality, housing,
hygiene, incest taboos, language, larval care, law, medicine, metamor-
phosis rites, mutual regurgitation, nursing castes, nuptial flights, nutrient eggs, population policy, queen obeisance, residence rules, sex
determination, soldier castes, sisterhoods, status differentiation, sterile workers, surgery, symbiont care, tool making, trade, visiting,
weather control,

and still other activities so alien as to make mere description by our
language difficult. If in addition they were programmed to eliminate
strife between colonies and to conserve the natural environment they
would have greater staying power than people, and in a broad sense
theirs would be the higher morality.

Civilization is not intrinsically limited to hominoids. Only by accident was it linked to the anatomy of bare-skinned, bipedal mammals
and the peculiar qualities of human nature.

Freud said that God has been guilty of a shoddy and uneven piece of
work. That is true to a degree greater than he intended: human nature
is just one hodgepodge out of many conceivable. Yet if even a small
fraction of the diagnostic human traits were stripped away, the result
would probably be a disabling chaos. Human beings could not bear to

simulate the behavior of even our closest relative among the Old World primates. If by perverse mutual agreement a human group attempted to imitate in detail the distinctive social arrangements of chimpanzees or gorillas, their effort would soon collapse and they would revert to fully human behavior.

It is also interesting to speculate that if people were somehow raised from birth in an environment devoid of most cultural influence, they would construct basic elements of human social life *ab initio*. In short time new elements of language would be invented and their culture enriched. Robin Fox, an anthropologist and pioneer in human socio-biology, has expressed this hypothesis in its strongest possible terms. Suppose, he conjectured, that we performed the cruel experiment linked in legend to the Pharaoh Psammetichus and King James IV of Scotland, who were said to have reared children by remote control, in total social isolation from their elders. Would the children learn to speak to one another?

> I do not doubt that they *could* speak and that, theoretically, given time, they or their offspring would invent and develop a language despite their never having been taught one. Furthermore, this language, although totally different from any known to us, would be analyzable to linguists on the same basis as other languages and translatable into all known languages. But I would push this further. If our new Adam and Eve could survive and breed — still in total isolation from any cultural influences — then eventually they would produce a society which would have laws about property, rules about incest and marriage, customs of taboo and avoidance, methods of settling disputes with a minimum of bloodshed, beliefs about the supernatural and practices relating to it, a system of social status and methods of indicating it, initiation ceremonies for young men, courtship practices including the adornment of females, systems of symbolic body adornment generally, certain activities and associations set aside for men from which women were excluded, gambling of some kind, a tool- and weapon-making industry, myths and legends, dancing, adultery, and various doses of homicide, suicide, homosexuality, schizophrenia, psychosis and neuroses, and various practitioners to take advantage of or cure these, depending on how they are viewed.

Not only are the basic features of human social behavior stubbornly idiosyncratic, but to the limited extent that they can be compared with those of animals they resemble most of all the repertories of other mammals and especially other primates. A few of the signals used to organize the behavior can be logically derived from the ancestral modes still shown by the Old World monkeys and great apes. The grimace of fear, the smile, and even laughter have parallels in the facial expressions of chimpanzees. This broad similarity is precisely the pattern to be expected if the human species descended from Old World primate ancestors, a demonstrable fact, and if the development of human social behavior retains even a small degree of genetic constraint, the broader hypothesis now under consideration.

24

STEPHEN JAY GOULD*

Sociobiology and the Theory of Natural Selection

Natural Selection as Storytelling

LUDWIG VON BERTALANFFY, a founder of general systems theory and a holdout against the neo-Darwinian tide, often argued that natural selection must fail as a comprehensive theory because it explains *too much*—a paradoxical, but perceptive statement. He wrote (1969:24, 11):

> If selection is taken as an axiomatic and *a priori* principle, it is always possible to imagine auxiliary hypotheses—unproved and by nature unprovable—to make it work in any special case . . . Some adaptive value . . . can always be construed or imagined.
>
> I think the fact that a theory so vague, so insufficiently verifiable and so far from the criteria otherwise applied in "hard" science, has become a dogma, can only be explained on sociological grounds. Society and science have been so steeped in the ideas of mechanism, utilitarianism, and the economic concept of free competition, that instead of God, Selection was enthroned as ultimate reality.

Similarly, the arguments of Christian fundamentalism used to frustrate me until I realized that there are, in principle, no counter cases and that, on this ground alone, literal bibliolatry is bankrupt. The theory of natural selection is, fortunately, in much better straits. It could be invalidated as a general cause of evolutionary change. (If, for

*From Stephen Jay Gould, "Sociobiology and the Theory of Natural Selection," *Sociobiology: Beyond Nature/Nurture?* G. W. Barlow and J. Silverberg, eds. (Boulder, Colorado: Westview Press, Inc., 1980), pp. 257–269. Reprinted by permission of the author.

example, Lamarckian inheritance were true and general, then adaptation would arise so rapidly in the Lamarckian mode that natural selection would be powerless to create and would operate only to eliminate.) Moreover, its action and efficacy have been demonstrated experimentally by 60 years of manipulation within *Drosophila* bottles —not to mention several thousand years of success by plant and animal breeders.

Yet in one area, unfortunately a very large part of evolutionary theory and practice, natural selection has operated like the fundamentalist's God—he who maketh *all* things. Rudyard Kipling asked how the leopard got its spots, the rhino its wrinkled skin. He called his answers "just-so stories." When evolutionists try to explain form and behavior, they also tell just-so stories—and the agent is natural selection. Virtuosity in invention replaces testability as the criterion for acceptance. This is the procedure that inspired von Bertalanffy's complaint. It is also the practice that has given evolutionary biology a bad name among many experimental scientists in other disciplines. We should heed their disquiet, not dismiss it with a claim that they understand neither natural selection nor the special procedures of historical science.

This style of storytelling might yield acceptable answers if we could be sure of two things: 1) that all bits of morphology and behavior arise as direct results of natural selection, and 2) that only one selective explanation exists for each bit. But, as Darwin insisted vociferously, and contrary to the mythology about him, there is much more to evolution than natural selection. (Darwin was a consistent pluralist who viewed natural selection as the most important agent of evolutionary change, but who accepted a range of other agents and specified the conditions of their presumed effectiveness. In Chapter 7 of the *Origin* (6th ed.), for example, he attributed the cryptic coloration of a flat fish's upper surface to natural selection and the migration of its eyes to inheritance of acquired characters. He continually insisted that he wrote his 2-volume *Variation of Animals and Plants Under Domestication* (1868), with its Lamarckian hypothesis of pangenesis, primarily to illustrate the effect of evolutionary factors other than natural selection. In a letter to *Nature* in 1880, he used the sharpest and most waspish language of his life to castigate Sir Wyville Thomson for caricaturing his theory by ascribing all evolutionary change to natural selection.)

Since God can be bent to support all theories, and since Darwin ranks closest to deification among evolutionary biologists, panselectionists of the modern synthesis tended to remake Darwin in their image. But we now reject this rigid version of natural selection and grant a major role to other evolutionary agents—genetic drift, fixation of neutral mutations, for example. We must also recognize that many features arise indirectly as developmental consequences of other features subject to natural selection—see classic (Huxley 1932) and modern (Gould 1966 and 1975; Cock 1966) work on allometry and the developmental consequences of size increase. Moreover, and perhaps most importantly, there are a multitude of potential selective explanations for each feature. There is no such thing in nature as a self-evident and unambiguous story.

When we examine the history of favored stories for any particular adaptation, we do not trace a tale of increasing truth as one story replaces the last, but rather a chronicle of shifting fads and fashions. When Newtonian mechanical explanations were riding high, G. G. Simpson wrote (1961:1686):

> The problem of the pelycosaur dorsal fin . . . seems essentially solved by Romer's demonstration that the regression relationship of fin area to body volume is appropriate to the functioning of the fin as a temperature regulating mechanism.

Simpson's firmness seems almost amusing since now — a mere 15 years later with behavioral stories in vogue — most paleontologists feel equally sure that the sail was primarily a device for sexual display. (Yes, I know the litany: It might have performed both functions. But this too is a story.)

On the other side of the same shift in fashion, a recent article on functional endothermy in some large beetles had this to say about the why of it all (Bartholomew and Casey 1977:883):

> It is possible that the increased power and speed of terrestrial locomotion associated with a modest elevation of body temperatures may offer reproductive advantages by increasing the effectiveness of intraspecific aggressive behavior, particularly between males.

This conjecture reflects no evidence drawn from the beetles themselves, only the current fashion in selective stories. We may be confident that the same data, collected 15 years ago, would have inspired a speculation about improved design and mechanical advantage.

Sociobiological Stories

Most work in sociobiology has been done in the mode of adaptive storytelling based upon the optimizing character and pervasive power of natural selection. As such, its weaknesses of methodology are those that have plagued so much of evolutionary theory for more than a century. Sociobiologists have anchored their stories in the basic Darwinian notion of selection as individual reproductive success. Though previously underemphasized by students of behavior, this insistence on selection as individual success is fundamental to Darwinism. It arises directly from Darwin's construction of natural selection as a conscious analog to the laissez-faire economics of Adam Smith with its central notion that order and harmony arise from the natural interaction of individuals pursuing their own advantages (see Schweber 1977).

Sociobiologists have broadened their range of selective stories by invoking concepts of inclusive fitness and kin selection to solve (successfully I think) the vexatious problem of altruism—previously the greatest stumbling block to a Darwinian theory of social behavior. Altruistic acts are the cement of stable societies. Until we could explain apparent acts of self-sacrifice as potentially beneficial to the genetic fitness of sacrificers themselves—propagation of genes through enhanced survival of kin, for example—the prevalence of altruism blocked any Darwinian theory of social behavior.

Thus, kin selection has broadened the range of permissible stories, but it has not alleviated any methodological difficulties in the process of storytelling itself. Von Bertalanffy's objections still apply, if anything with greater force because behavior is generally more plastic and more difficult to specify and homologize than morphology. Sociobiologists are still telling speculative stories, still hitching without evidence to one potential star among many, still using mere consistency with natural selection as a criterion of acceptance.

David Barash (1976), for example, tells the following story about mountain bluebirds. (It is, by the way, a perfectly plausible story that may well be true. I only wish to criticize its assertion without evidence or test, using consistency with natural selection as the sole criterion for useful speculation.) Barash reasoned that a male bird might be more sensitive to intrusion of other males before eggs are laid than after (when he can be certain that his genes are inside). So Barash studied two nests, making three observations at 10-day intervals, the first before the eggs were laid, the last two after. For each period of observation, he mounted a stuffed male near the nest while the male occupant was out foraging. When the male returned he counted aggressive encounters with both model and female. At time one, males in both nests were aggressive toward the model and less, but still substantially, aggressive toward the female as well. At time two, after eggs had been laid, males were less aggressive to models and scarcely aggressive to females at all. At time three, males were still less aggressive toward models, and not aggressive at all toward females. Barash concludes that he has established consistency with natural selection and need do no more (1976:1099–1100):

> These results are consistent with the expectations of evolutionary theory. Thus aggression toward an intruding male (the model) would clearly be especially advantageous early in the breeding season, when territories and nests are normally defended . . . The initial, aggressive response to the mated female is also adaptive in that, given a situation suggesting a high probability of adultery (i.e., the presence of the model near the female) and assuming that replacement females are available, obtaining a new mate would enhance the fitness of males . . . The decline in male–female aggressiveness during incubation and fledgling stages could be attributed to the impossibility of being cuckolded after the eggs have been laid . . . The results are consistent with an evolutionary interpretation. In addition, the term "adultery" is unblushingly employed in this letter without quotation marks, as I believe it reflects a true analogy to the human concept, in

the sense of Lorenz. It may also be prophesied that continued application of a similar evolutionary approach will eventually shed considerable light on various human foibles as well.

Consistent, yes. But what about the obvious alternative, dismissed without test in a line by Barash: male returns at times two and three, approaches the model a few times, encounters no reaction, mutters to himself the avian equivalent of "it's that damned stuffed bird again," and ceases to bother. And why not the evident test: expose a male to the model for the *first* time *after* the eggs are laid.

We have been deluged in recent years with sociobiological stories. Some, like Barash's are plausible, if unsupported. For many others, I can only confess my intuition of extreme unlikeliness, to say the least —for adaptive and genetic arguments about why fellatio and cunnilingus are more common among the upper classes (Weinrich 1977), or why male panhandlers are more successful with females and people who are eating than with males and people who are not eating (Lockard et al. 1976).

Not all of sociobiology proceeds in the mode of storytelling for individual cases. It rests on firmer methodological ground when it seeks broad correlations across taxonomic lines, as between reproductive strategy and distribution of resources, for example (Wilson 1975), or when it can make testable, quantitative predictions as in Trivers and Hare's work on haplodiploidy and eusociality in Hymenoptera (Trivers and Hare 1976). Here sociobiology has had and will continue to have success. And here I wish it well. For it represents an extension of basic Darwinism to a realm where it should apply.

Special Problems for Human Sociobiology

Sociobiological explanations of human behavior encounter two major difficulties, suggesting that a Darwinian model may be generally inapplicable in this case.

Limited Evidence and Political Clout

We have little direct evidence about the genetics of behavior in humans; and we do not know how to obtain it for the specific behaviors that figure most prominently in sociobiological speculation — aggression, conformity, etc. With our long generations, it is difficult to amass much data on heritability. More important, we cannot (ethically, that is) perform the kind of breeding experiments, in standardized environments, that would yield the required information. Thus, in dealing with humans, sociobiologists rely even more heavily than usual upon speculative storytelling.

At this point, the political debate engendered by sociobiology comes appropriately to the fore. For these speculative stories about human behavior have broad implications and proscriptions for social policy — and this is true quite apart from the intent or personal politics of the storyteller. Intent and usage are different things; the latter marks political and social influence, the former is gossip or, at best, sociology.

The common political character and effect of these stories lies in the direction historically taken by innatist arguments about human behavior and capabilities—a defense of existing social arrangements as part of our biology.

In raising this point, I do not act to suppress truth for fear of its political consequences. Truth, as we understand it, must always be our primary criterion. We live, because we must, with all manner of unpleasant biological truth—death being the most pervasive and ineluctable. I complain because sociobiological stories are not truth but unsupported speculations with political clout (again, I must emphasize, quite apart from the intent of the storyteller). All science is embedded in cultural contexts, and the lower the ratio of data to social importance, the more science reflects the context.

In stating that there is politics in sociobiology, I do not criticize the scientists involved in it by claiming that an unconscious politics has intruded into a supposedly objective enterprise. For they are behaving like all good scientists—as human beings in a cultural context. I only ask for a more explicit recognition of the context and, specifically, for more attention to the evident impact of speculative sociobiological stories. For example, when the *New York Times* runs a weeklong front page series on women and their rising achievements and expectations, spends the first four days documenting progress toward social equality, devotes the last day to potential limits upon this progress, and advances sociobiological stories as the only argument for potential limits—then we know that these are stories with consequences:

> Sociologists believe that women will continue for some years to achieve greater parity with men, both in the work place and in the home. But an uneasy sense of frustration and pessimism is growing among some advocates of full female equality in the face of mounting conservative opposition. Moreover, even some staunch feminists are reluctantly reaching the conclusion that women's aspirations may ultimately be limited by inherent biological differences that will forever leave men the dominant sex (New York Times, Nov. 30, 1977).

The article then quotes two social scientists, each with a story.

> If you define dominance as who occupies formal roles of responsibility, then there is no society where males are not dominant. When something is so universal, the probability is—as reluctant as I am to say it—that there is some quality of the organism that leads to this condition.

> It may mean that there never will be full parity in jobs, that women will always predominate in the caring tasks like teaching and social work and in the life sciences, while men will prevail in those requiring more aggression—business and politics, for example—and in the 'dead' sciences like physics.

Adaptation in Humans Need Not Be Genetic and Darwinian

The standard foundation of Darwinian just-so stories does not apply to humans. That foundation is the implication: if adaptive, then genetic — for the inference of adaptation is usually the only basis of a genetic story, and Darwinism is a theory of genetic change and variation in populations.

Much of human behavior is clearly adaptive, but the problem for sociobiology is that humans have so far surpassed all other species in developing an alternative, non-genetic system to support and transmit adaptive behavior — cultural evolution. As adaptive behavior does not require genetic input and Darwinian selection for its origin and maintenance in humans; it may arise by trial and error in a few individuals that do not differ genetically from their groupmates, spread by learning and imitation, and stabilize across generations by value, custom and tradition. Moreover, cultural transmission is far more powerful in potential speed and spread than natural selection — for cultural evolution operates in the "Lamarckian" mode by inheritance through custom, writing and technology of characteristics acquired by human activity in each generation.

Thus, the existence of adaptive behavior in humans says nothing about the probability of a genetic basis for it, or about the operation of natural selection. Take, for example, Trivers' (1971) concept of "reciprocal altruism." The phenomenon exists, to be sure, and it is clearly adaptive. In honest moments, we all acknowledge that many of our "altruistic" acts are performed in the hope and expectation of future reward. Can anyone imagine a stable society without bonds of reciprocal obligation? But structural necessities do not imply direct genetic coding. (All human behaviors are, of course, part of the potential range permitted by our genotype — but sociobiological speculations posit direct natural selection for specific behavioral traits.) As Benjamin Franklin said: "We must all hang together, or assuredly we shall all hang separately."

Failure of the Research Program for Human Sociobiology

The grandest goal — I do not say the only goal — of human sociobiology must fail in the face of these difficulties. That goal is no less than the reduction of the behavioral (indeed most of the social) sciences to Darwinian theory. Wilson (1975) presents a vision of the human sciences shrinking in their independent domain, absorbed on one side by neurobiology and on the other by sociobiology.

But this vision cannot be fulfilled, for the reasons cited above. Although we can identify adaptive behavior in humans, we cannot tell thereby if it is genetically based (while much of it must arise by fairly pure cultural evolution). Yet the reduction of the human sciences to Darwinism requires the genetic argument, for Darwinism is a theory about genetic change in populations. All else is analogy and metaphor.

My crystal ball shows the human sociobiologists retreating to a fall-back position—indeed it is happening already. They will argue that this fallback is as powerful as their original position, though it actually represents the unravelling of their fondest hopes. They will argue: yes, indeed, we cannot tell whether an adaptive behavior is genetically coded or not. But it doesn't matter. The same adaptive constraints apply whether the behavior evolved by cultural or Darwinism routes, and biologists have identified and explicated the adaptive constraints. (Steve Emlen reports, for example, that some Indian peoples gather food in accordance with predictions of optimal foraging strategy, a theory developed by ecologists. This is an exciting and promising result within an anthropological domain—for it establishes a fruitful path of analogical illumination between biological theory and non-genetic cultural adaptation. But it prevents the assimilation of one discipline by the other and frustrates any hope of incorporating the human sciences under the Darwinian paradigm.)

But it does matter. It makes all the difference in the world whether human behaviors develop and stabilize by cultural evolution or by direct Darwinian selection for genes influencing specific adaptive actions. Cultural and Darwinian evolution differ profoundly in the three major areas that embody what evolution, at least as a quantitative science, is all about:

1. Rate. Cultural evolution, as a "Lamarckian" process, can proceed orders of magnitude more rapidly than Darwinian evolution. Natural selection continues its work within *Homo sapiens*, probably at characteristic rates for change in large, fairly stable populations, but the power of cultural evolution has dwarfed its influence (alteration in frequency of the sickling gene vs. changes in modes of communication and transportation). Consider what we have done with ourselves in the past 3000 years, all without the slightest evidence for any biological change in the size or power of the human brain.

2. Modifiability. Complex traits of cultural evolution can be altered profoundly all at once (social revolution, for example). Darwinian change is much slower and more piecemeal.

3. Diffusibility. Since traits of cultural evolution can be transmitted by imitation and inculcation, evolutionary patterns include frequent and complex anastomosis among branches. Darwinian evolution in sexually reproducing animals is a process of continuous divergence and ramification with few opportunities for coming together (hybridization or parallel modification of the same genes in independent groups).

I believe that the future will bring mutual illumination between two vigorous, independent disciplines—Darwinian theory and cultural history. This is a good thing, joyously to be welcomed. But there will be no reduction of the human sciences to Darwinian theory and the research program of human sociobiology will fail. The name, of course, may survive. It is an irony of history that movements are judged successful if their label sticks, even though the emerging content of a discipline may lie closer to what opponents originally advocated. Modern geology, for example, is an even blend of Lyell's strict uniformitarianism and the claims of catastrophists (Rudwick 1972; Gould 1977). But we call the hybrid doctrine by Lyell's name and he has become the conventional hero of geology.

I welcome the coming failure of reductionistic hopes because it will lead us to recognize human complexity at its proper level. For consumption of *Time*'s millions, my colleague Bob Trivers maintained: "Sooner or later, political science, law, economics, psychology, psychiatry, and anthropology will all be branches of sociobiology" (*Time*, Aug. 1, 1977:54). It is one thing to conjecture, as I would allow, that common features among independently developed legal systems might reflect adaptive constraints and might be explicated usefully with some biological analogies. It is quite another to state, as Trivers did, that the mores of the entire legal profession will be subsumed, along with a motley group of other disciplines, as mere epiphenomena of Darwinian processes.

I read Trivers' statement the day after I had sung in a full production of Berlioz' *Requiem*. And I remembered the visceral reaction I had experienced upon hearing the 4 brass choirs, finally amalgamated with the 10 tympani in the massive din preceding the great *Tuba mirum* — the spine tingling and the involuntary tears that almost prevented me from singing. I tried to analyze it in the terms of Wilson's conjecture — reduction of behavior to neurobiology on the one hand and sociobiology on the other. And I realized that this conjecture might apply to my experience. My reaction had been physiological and, as a good mechanist, I do not doubt that its neurological foundation can be ascertained. I will also not be surprised to learn that the reaction has something to do with adaptation (emotional overwhelming to cement group coherence in the face of danger, to tell a story). But I also realized that these explanations, however "true," could never capture anything of importance about the meaning of that experience.

And I say this not to espouse mysticism or incomprehensibility, but merely to assert that the world of human behavior is too complex and multifarious to be unlocked by any simple key. I say this to maintain that this richness — if anything — is both our hope and our essence.

Summary

Ever since Darwin proposed it, the theory of natural selection has been marred by an uncritical style of speculative application to the study of individual adaptations: one simply constructs a story to explain how a shape, function, or behavior might benefit its possessor. Virtuosity in invention replaces testability and mere consistency with evolutionary theory becomes the primary criterion of acceptance. Although this dubious procedure has been used throughout evolutionary biology, it has recently become the primary style of explanation in sociobiology.

Human sociobiology presents two major problems related to this tradition. First, evidence is so poor or lacking that speculative story-telling assumes even greater importance than usual. Secondly, the exis-tence of behavioral adaptation does not imply the operation of Darwin-ian processes at all — for non-genetic cultural evolution, working in the Lamarckian mode, dwarfs by its rapidity the importance of slower Darwinian change. The sociobiological vision of a reduction of the human sciences to biology via Darwinism and natural selection will fail. Instead, I anticipate fruitful, mutual illumination by analogy between independent theories of the human and biological sciences.

LITERATURE CITED

Barash, D. 1976. Male response to apparent female adultery in the mountain bluebird (*Sialia currocoides*): An evolutionary interpretation. *American Naturalist* 110:1097–1101.

Bartholomew, G. A. and T. M. Casey. 1977. Endothermy during terrestrial activity in large beetles. *Science* 195:882–883.

Bertalanffy, L. von. 1969. Chance or law. *In* A. Koestler (ed.). *Beyond reduc-tionism.* Hutchinson, London.

Cock, A. G. 1966. Genetical aspects of metrical growth and form in animals. *Quarterly Review of Biology* 41:131–190.

Darwin, C. 1868. *The variation of animals and plants under domestication.* John Murray, London.

———. 1880. Sir Wyville Thomson and natural selection. *Nature* 23:32.

Gould, S. J. 1966. Allometry and size in ontogeny and phylogeny. *Biological Reviews* 41:587–640.

———. 1975. Allometry in primates, with emphasis on scaling and the evolu-tion of the brain. *In* Approaches to primate paleobiology. *Contributions to Primatology* 5:244–292.

———. 1977. Eternal metaphors of paleontology. *In* A. Hallam (ed.). *Patterns of evolution.* Elsevier, Amsterdam, pp. 1–26.

Huxley, J. 1932. *Problems of relative growth.* MacVeagh, London.

Lockard, J. S., L. L. McDonald, D. A. Clifford, and R. Martinez. 1976. Panhan-dling: Sharing of resources. *Science* 191:406–408.

Rudwick, M. J. S. 1972. *The meaning of fossils.* Macdonald, London.

Schweber, S. S. 1977. The origin of the *Origin* revisited. *Journal of the History of Biology* 10:229–316.

Simpson, G. G. 1961. Some problems of vertebrate paleontology. *Science* 133:1679–1689.

Trivers, R. 1971. The evolution of reciprocal altruism. *Quarterly Review of Biology* 46:35–57.

Trivers, R. and H. Hare. 1976. Haplodiploidy and the evolution of the social insects. *Science* 191:249–263.

Weinrich, J. D. 1977. Human sociobiology: Pair-bonding and resource predictability (effects of social class and race). *Behavioral Ecology and Sociobiology* 2:91–118.

Wilson, E. O. 1975. *Sociobiology: The New Synthesis*. Harvard University Press, Cambridge, Massachusetts.

EDITORIAL NOTE

In this article Gould refers to one of the very greatest triumphs of recent biological understanding, the solution to a puzzle that had troubled evolutionists from the time of Darwin, namely, the evolution of sterile worker castes in the social insects, especially the ants, the bees, and the wasps (the "hymenoptera"). The answer to this problem—an impetus for the whole sociobiological movement—came to the English biologist William Hamilton in the 1960s, when he realized that selection can work by proxy, as it were. As long as copies of one's genes are passed on, it matters not if one does it directly oneself or indirectly through bearers of similar genes, namely close relatives. Indeed, if relatives can do the job better, then selection may promote the evolution of "altruistic" features, where one is positively inclined to help. (Such an evolutionary process is now known generally as "kin selection.") In the particular case of the hymenoptera, Hamilton seized on their peculiar method of reproduction. They are "Haplodiploid." Females are like other sexual organisms, having father and mother, receiving (as is normal) a half set of chromosomes from each parent. Males have only mothers, receiving just one half set of chromosomes. As a result—unlike the normal situation, where the relationship is symmetrical—females are more closely related to sisters than to daughters. Hence, biologically it pays females to forego personal reproduction, even to the point of sterility, and raise fertile sisters.

This and other examples are superbly discussed in John Maynard Smith's "The evolution of behavior," *Scientific American*, 239, September 1978, 176–192.

25

ARTHUR L. CAPLAN*

Say It Just Ain't So: Adaptational Stories and Sociobiological Explanations of Social Behavior

A KEY PROBLEM with many sociobiological accounts, according to Stephen Jay Gould, is that they amount to no more than the weaving of plausible stories. Gould alleges that many sociobiologists commit the fallacy of what he terms—borrowing a phrase from Rudyard Kipling—the "just-so story." He cites approvingly Ludwig von Bertalanffy's indictment of the modern synthetic theory of evolution as just so much fantastic myth-making: "If selection is taken as an axiomatic and *a priori* principle, it is always possible to imagine auxiliary hypotheses— unproved and by nature unprovable—to make it work in any special case. . . . Some adaptive value . . . can always be construed or imagined."[1] Gould notes that von Bertalanffy's objection applies with a vengeance to sociobiological explanations of behavior: "Von Bertalanffy's objections still apply, if anything with greater force, because behavior is generally more plastic and more difficult to specify and homologize than morphology. Sociobiologists are still telling speculative stories, still hitching without evidence to one potential star among many, still using mere consistency with natural selection as a criterion of acceptance."[2]

A strange sense of *déjà vu* surrounds the just-so story complaint against sociobiology. For, as Gould rightly observes, it is precisely this complaint that has been levelled against the adequacy of Darwinian

*First published in *the Philosophical Forum,* Vol. 13, nos. 2–3, Winter–Spring 1981–1982. By permission of the publisher.

accounts of evolution for much of the twentieth century. Evolutionary theory has been charged time and again with the sins of excessive malleability, *ad hocness*, and unfalsifiability.[3]

It is precisely these concerns which led Karl Popper and various other philosophers[4] to relegate Darwinism to the dustbin of metaphysics. The charge lingers on to the present day, particularly in paleontological and taxonomic circles. Biologists in these fields have shown an astounding capacity for self-flagellation as they bemoan the theoretical docility of many of their peers who are unable to see that the modern synthetic theory of evolution rests upon sandy mythological foundations.[5]

The problem with Gould's complaint, as with von Bertalanffy's before him, is that it throws the scientific baby of evolutionary science out with a vast amount of naive historicist bathwater. Of course it is true that myth-making and deliberate *ad hocness* have no place in evolutionary biology — sociobiological or otherwise. But the appeal to remove story-telling from evolutionary biology conflates important differences between science, history, stories, and myths.

Popper and those critics of evolutionary theory such as von Bertalanffy, who were inspired by Popper's criticisms of historicist laws, fallaciously view evolutionary theory as a maxim in search of biological facts to explain. Natural selection, on the Popperian account, is a principle invoked to explain each and every twist and turn in the history of organic life on this planet. If giraffes have long necks, it is, presumably, as a result of natural selection. If a sub-group of giraffes is found possessing short necks, this too is presumably a consequence of natural selection. The accordion-like ability of natural selection in explaining disparate facts reminds Popper and other kindred spirits of two other allegedly similar and suspect explanatory outlooks — Marxism and Freudianism.[6] Popper smells historicism in Darwinism and argues, since the events of human and animal history are unique, they cannot be subsumed under a principle such as natural selection which purports to explain all of organic history.[7]

The primary problem with this line of attack against Darwinism and its descendant, the modern synthetic theory of evolution, is that it totally and utterly misconstrues the status of natural selection. Natural selection is not a maxim of any sort, *a priori* or otherwise. It is, as Darwin stated repeatedly, a metaphor for describing both the processes and outcomes of biological evolution. Natural selection *per se* explains nothing. Recent misguided efforts by a variety of authors[8] to the contrary notwithstanding, natural selection cannot and should not be reified to the status of a nomological principle. Rather, natural selection is a useful label for referring to an extraordinarily complex array of causal interactions occurring at the level of genes, genotypes, phenotypes, and environments. It is the laws and generalizations of genetics, development, ecology, and demography which ultimately are invoked by biologists to explain change and descent in the history of life. Natural selection is simply a covering term or place-holder for describing the various processes involved in producing evolutionary change, or the products of such processes.[9]

Perhaps the easiest way of seeing the emptiness of the charge that natural selection is utilized in all evolutionary accounts as an *a priori* explanatory law is by reflecting upon a rather peculiar episode in the recent history of the philosophy of biology. During the 1950s and early 1960s a number of articles were written by some of the leading advocates of the modern synthetic theory denying the existence of laws in evolutionary biology. For example, Ernst Mayr wrote: "Uniqueness is particularly characteristic for evolutionary biology. It is quite impossible to have, for unique phenomena, general laws like those existing in classical mechanics."[10] George G. Simpson argued that: "History does not correspond with possible mechanistic models, such as some in the physical sciences. That history is not simple and tidy is unfortunate, perhaps, but it is true. . . . the human desire for neat and unequivocal conclusions explains the long and necessarily futile search for simple, absolute, deterministic laws of evolution."[11]

Ironically, numerous philosophers in the 1970s have argued quite strenuously that evolutionary biology does indeed have laws. For example, Michael Ruse has argued that population genetics surely has identifiable laws, e.g., the Hardy–Weinberg law, and since population genetics is at the core of evolutionary theory, then evolutionary theory must surely have laws.[12]

This juxtaposition of opposing views about the existence of laws in evolutionary theory appears, at first glance, to be utterly bizarre. Defenders of the scientific status of contemporary evolutionary theory, such as Ruse, claim to find laws in nearly all evolutionary explanations, while leading proponents of the theory, such as Mayr and Simpson, argue that the theory has no laws, can never have laws, and that this state of affairs confers a distinctive status upon evolutionary theorizing in comparison with other branches of natural science!

However, the contradictory nature of these points of view concerning laws in evolutionary theory evaporates once it is seen that the biologists and philosophers involved were really talking at cross-purposes. Surely Mayr and Simpson are as aware as Ruse and other philosophers of biology of the vast edifice of laws, theorems, generalizations, models, and principles erected in the past seventy-five years in the fields of population genetics, ecology, and demography. Their aim in arguing against the existence of laws or maxims in evolutionary theory was to debunk the existence of a particular type of law — historical, purposive, or directional laws.[13] Mayr, Simpson, and other architects of the modern synthetic theory were concerned to refute the views of Berg, Schindewolf, Teilhard de Chardin, and others who argued for various orthogenetic explanations of evolution — that evolution can only proceed through fixed, predetermined cycles or stages of development.[14]

The arguments of biologists such as Mayr and Simpson against the existence of laws in evolutionary biology are best understood as arguments against historicist interpretations of the history of life — the very point of such concern to Popper, von Bertalanffy and, most recently, Gould. Darwinian evolutionary biologists, from Darwin himself to contemporary exponents of Neo-Darwinism, have been adament opponents of all forms of historicism in explaining the history of life. At the heart of the Darwinian analysis of evolution is the belief that historical phenomena in the organic world can only be explained by ahistorical, mechanistic laws. Thus, philosophers of biology, such as Ruse, are not, as they have often thought, really at odds with contemporary evolutionists over the nature of laws in evolutionary biology. Vagaries of language have simply obscured the fact that Darwinians, Popperians, and devotees of the received view of scientific theories, such as Ruse, are all in agreement about the nomological character of laws in evolutionary theory. There are no distinctive laws of evolution in the sense of historicist or directional laws. There are no evolutionary laws which can subsume distinct events in the history of life in order to explain such events. Rather, non-historical laws can be applied to particular events or occurrences in the history of life in order to explain subsequent changes and developments. Evolutionists only have available the mechanistic, nomological generalizations and models of population genetics, demography, ecology, molecular biology, and now sociobiology for explaining events in the history of life.[15]

If my analysis of the misunderstanding that has arisen over the character of laws in evolutionary explanations is correct, then the inappropriateness of the charge that evolutionary accounts are based upon a blind adherence to the law of natural selection becomes apparent. Darwin and the Neo-Darwinists who follow in his scientific footsteps have no tolerance for the type of metaphysical invocation of explanatory principle so anathematized by Popper, von Bertalanffy, and Gould. While it is indeed possible to challenge the belief that historical or directional laws of evolution do not exist or that such laws are conceptually incoherent,[16] the fact is that the modern synthetic theory has no truck with this type of law. The essence of the Darwinian approach to the explanation of biological evolution over time is that such changes can only be explained by means of laws, principles and models that make no essential reference to time or history as subsumed variables.

Sociobiologists, if they can be fairly characterized as anything, can surely be characterized as Darwinian in explanatory outlook. Their models of kin selection, parental investment, reciprocal altruism, and the like, are meant to explain the evolution of social behavior by means of the interactions that obtain between genotypes, phenotypes, and environments. Natural selection for sociobiologists, as for any evolutionary biologist committed to a Darwinian understanding of evolution, can never serve as an *a priori* maxim, unfalsifiable nomic principle, or *ad hoc* explanatory device. It is simply a phrase that acts as a capsule summary for the complex set of causal interactions that, acting over time, eventuate in the myriad forms of life, traits, and behaviors we refer to as the end-products of evolution. Whatever the sins of sociobiologists may be, and they may, as Gould and others have noted, be numerous, reification of natural selection into an *a priori* law or principle is not one of them. Sociobiologists may believe that natural selection should be at the heart of any explanation of evolutionary change in animals and humans, past or present, living or dead. But this belief is merely a belief in the power and scope of evolutionary theory to explain all aspects of organic change in every species. While this belief may be (and probably is) false, its falseness does not arise as a consequence of the invocation of natural selection as some sort of all powerful, untestable law of nature. This red herring derives its fishy smell from a failure to perceive the single-minded devotion to anti-historicist views of history that permeates the Darwinian, Neo-Darwinian, and sociobiological view of life.

The criticism that sociobiologists adopt an explanatory strategy toward all social behavior, such that they view it solely as the result of natural selection and thereby make their accounts untestable or *a priori*, seriously misconstrues the meaning of natural selection. I know of no single instance where any sociobiologist has argued for a metaphysical interpretation of natural selection *sensu* Gould or von Bertalanffy. What sociobiologists do is construct hypotheses about the evolution of various social behaviors based upon an emended version of Neo-Darwinism. The emendations they make to Darwinian theory involve (1) the recognition that similar environmental forces act similarly on similar genotypes, (2) that certain forms of behavior can be mutually beneficial to interacting organisms, and (3) that parents are in direct competition with their own offspring for environmental resources.[17] These emendations are still part and parcel of the modern synthetic theory which posits various laws governing the interaction of genes, phenotypes, environments, and isolating factors. The request that sociobiologists cease spinning selectionist stories about social behavior is equivalent to the request that they not extend the scope of the modern synthetic theory to social behavior — which of course is the central aim of the sociobiological enterprise. Sociobiological accounts are not, as Gould suggests, "consistent" with evolutionary theory — they are *derived* directly from an emended version of that theory.

Numerous philosophers of history[18] have claimed that there are a number of additional criteria that distinguish stories and myths from history. For example, there is near unanimity of opinion about the claim that among the properties possessed by history, as opposed to stories and myths, are internal consistency, the avowed intention to produce a "factual" account of past events, and the willingness of historians to test their accounts against publicly available forms of evidence. Stories normally lack all of these characteristics.

If such criteria can be utilized to distinguish history from stories, myths, and fables, then surely sociobiological accounts count as history, not stories. Sociobiological explanations of the incest taboo, homosexuality, panhandling among humans, and inheritance patterns among persons in various cultural settings are constructed so as to be grounded in an established theory (an emended version of the modern synthetic theory of evolution), to be "factual," and to be testable by publicly available evidence. Indeed, the evidence for the adequacy of sociobiological accounts regarding these phenomena seems to refute many of these hypotheses. But the real point at issue is that many sociobiological accounts do approximate the classificatory standards for being understood as history (perhaps false history but still history) operative in the social sciences and human history, which is probably all that can reasonably be asked of sociobiological hypotheses on methodological grounds at this point in time. While sociobiological accounts of the origins of social behavior may indeed be slap-dash or false, they are patently not fictions or fables.

ACKNOWLEDGMENTS

I would like to thank Walter Bock, Janet Caplan, and Bruce Jennings for their willingness to engage in numerous discussions with me about the ideas presented in this paper.

FOOTNOTES

1. S. J. Gould, "Sociobiology and the Theory of Natural Selection," in G. Barlow and J. Silverberg, eds. *Sociobiology: Beyond Nature/Nurture?* (Boulder: Westview, 1980), p. 257.
2. *Ibid.*, p. 260.
3. For an excellent review of different versions of these criticisms, see M. Ruse, *The Philosophy of Biology* (London: Hutchinson, 1973), Chapters II and III. Also, see A. Caplan, "Darwinism and Deductivist Models of Theory Structure," *Studies in History and Philosophy of Science*, 10 (1979), 341–353.
4. See especially K. R. Popper, *The Poverty of Historicism*, 3rd ed. (London: Routledge & Kegan Paul, 1961), Chapter IV.
5. See the review of Ernst Mayr's *Evolution and the Diversity of Life* by N. Platnick in *Systematic Zoology*, 26 (1977), 224–228; D. E. Rosen and D. G. Buth, "Empirical Evolutionary Research Versus Neo-Darwinian Speculation," *Systematic Zoology*, 29 (1980), 300–308; and Niles Eldredge and Joel Cracraft, *Phylogenetic Patterns and the Evolutionary Process* (New York: Columbia, 1980).
6. Popper, *The Poverty of Historicism*, pp. 73–83. Also, see K. Popper, *Objective Knowledge* (Oxford: Oxford University Press, 1972).
7. An excellent critique of Popper's views can be found in M. Ruse, "Karl

Popper's Philosophy of Biology," *Philosophy of Science*, 44 (1977), 638–661.

8. M. B. Williams, "Deducing the Consequences of Evolution," *Journal of Theoretical Biology*, 29 (1970), 342–85, and E. S. Reed, "The Lawfulness of Natural Selection," *The American Naturalist*, 118 (1981), 61–71.

9. A. L. Caplan, "Testability, Disreputability and the Structure of the Modern Synthetic Theory of Evolution," *Erkenntnis*, 13 (1978), 261–278.

10. E. Mayr, "Cause and Effect in Biology," rpt. in R. Munson, ed., *Man and Nature* (New York: Dell, 1971), p. 114.

11. G. G. Simpson, "Evolutionary Determinism," rpt. in R. Munson, ed., *Man and Nature* (New York: Dell, 1971), p. 210.

12. See Ruse, *The Philosophy of Biology*, and Chapter II of his *Sociobiology: Sense or Nonsense?* (Dordrecht: Reidel, 1979).

13. A clear explanation of the difference between functional or mechanistic laws, and historical or directional laws has been given by Maurice Mandelbaum:

> [a functional law] would only enable us to predict immediately subsequent events, and each further prediction would have to rest upon knowledge of the initial and boundary conditions obtaining at that time. The second type of law [a directional law] would not demand a knowledge of subsequent initial conditions. . . . For if there were a law of directional change which could be discovered in any segment of history, we could extrapolate to the past and to the future without needing to gather knowledge of the initial conditions obtaining at each successive point in the historical process.

M. Mandelbaum, "Societal Laws," In W. H. Dray, ed., *Philosophical Analysis and History* (New York: Harper & Row, 1966), p. 334.
    Karl Popper denies the possibility of this type of directional law as abject historicism in the *Poverty of Historicism, op. cit.*, Note 15. See also Chapter 7 of Mandelbaum's *History, Man and Reason* (Baltimore: Johns Hopkins, 1971), for a superb discussion of the problems facing proponents of directional laws in history and the biological sciences.

14. For a defense of the orthogenetic approach to evolution, see M. Grene, "Two Evolutionary Theories," *British Journal for the Philosophy of Science*, 9 (1959), 110–127.

15. However, Stephen J. Gould has recently done some plumping for historicism in evolutionary theory. See his "The Promise of Paleobiology as a Nomothetic Evolutionary Discipline," *Paleobiology*, 6 (1980), 96–118.

16. P. Urbach, "Is Any of Popper's Arguments Against Historicism Valid?," *British Journal for the Philosophy of Science*, 29 (1978), 117–130, and A. Olding, "A Defense of Evolutionary Laws," *British Journal for the Philosophy of Science*, 29 (1978), 131–143.

17. See R. Dawkins, *The Selfish Gene* (Oxford: Oxford University Press, 1976).

18. See, for example, W. B. Gallie's discussion of stories and histories in *Philosophy and the Historical Understanding*, 2nd ed. (New York: Shocken, 1968), Chapters II and III.

EXTRATERRESTRIALS?

26

ROBERT BIERI*

Huminoids on Other Planets?

"The probability of evolving some living system was likely high. That evolution would go in a particular direction is a different matter. Thus the *a priori* probability of evolving man must have been extremely small—for there were an almost infinite number of other possibilities. Even the probability of an organism evolving with a nervous system like ours, was, I think, extremely small because of the enormous number of alternatives. I am therefore not at all hopeful that we will ever establish communication with living things on other planets, even though there may well be many such on many planets."[1]

I THINK THAT at least a few biologists would disagree with all but the first sentence of this statement. Because of continued interest in possible interplanetary communication, I would like to present arguments for the opposite view, namely, that if life has evolved on other planets in other solar systems and if some population has reached the level of conceptual thought, it is highly probable that the organisms so endowed will bear a strong resemblance to *Homo sapiens*.

Essentially this argument is based on the premise that the physical properties of the elements, the forms of energy available, and the environmental conditions which would allow life to arise and evolve are such that severe limitations are imposed on the number of routes available to evolving forms. The number of alternative possibilities is by no means infinite; on the contrary, the number is quite limited. This limited number of available routes has led to the innumerable cases of convergent evolution in plants and animals.

*From Robert Bieri, "Huminoids on Other Planets?" *American Scientist,* Vol. 52 (New Haven, CT: Scientific Research Society, 1964), pp. 452–458. By permission of the publisher.

Man is the result of a series of evolutionary "breakthroughs"; he is the result of the solution of a number of major evolutionary problems. Only a few of these are discussed here in order to illustrate the validity of the premise of limited solutions or routes available at each turn in the long evolution from abiogenetically produced organic matter to conceptualizing being.

In this discussion, life is defined as a system of matter having heritability and mutability. Such a system of matter requires periodic or continual capture of matter and energy. As long ago as 1913, Henderson[2] advanced the argument that, of all the elements known, only carbon fulfills the many requirements necessary for the existence of a duplicating, living system. He further argues that such a system could only develop in water. Urey[3] has recently reaffirmed this idea. This argument is now so generally accepted by scientists that it is seldom emphasized that it is a major limit on the number of evolutionary pathways available to life.[4] More recently, biochemists have argued that a living system based on carbon must also be based on the high energy phosphate bond. The fact that the ATP – ADP system is common to all living systems illustrates the limited range of biochemical alternatives.[5] The construction of living systems on a carbon framework in water means that we are limited to a temperature range not widely different from that of our earth.

Because of the limited time available, on the order of five to ten billion years, and because of other limiting factors, it is most probable that a living system will arise within a liquid medium after chemical contributions have been made from the solid – liquid and liquid – gas interfaces. The emergence of life from a solid or gaseous system, although perhaps not impossible, certainly is highly improbable.

If a living system is to evolve to any significant degree of complexity, some form of autotrophic organism must evolve to support the primary heterotrophic forms. Because of the greater energy capture possible, photosynthesis and an oxidizing atmosphere will eventually supersede the many possible chemosynthetic autotrophic energy systems.

In the long evolution within a sea that must precede invasion of the land, herbivores and carnivores will have reached a high degree of complexity and specialization. Both types of organisms will be bilaterally symmetrical because the viscosity of the medium demands streamlining for speed of pursuit, speed of escape, and the efficient use of captured energy. There are many radially symmetrical animals in the earth's oceans. However, nearly all of these have evolved from bilaterally symmetrical ancestors after adopting a sessile mode of life. The Radiolaria among the protozoa and the majority of the coelenterates represent animals that probably show primary radial symmetry. They are good examples of the low level of organization attained by radially symmetrical animals. Probably the best examples of secondarily derived radial symmetry are found in the annelids. Many modifications accompany sessile living but the most important for the present discussion are the reduction of sensory structures and degeneration of the nervous system. Only an active mode of life, the pursuit of prey and escape from predators, has led to larger, more complex nervous systems.[6]

Active search for food and pursuit of prey have resulted in a bilaterally symmetrical animal with an anterior mouth and a posterior anus.[7] Invariably the mouth is surrounded by sensory and grasping organs. There may be small sense organs scattered along the body and at the trailing end, but in essentially all cases known in the animal kingdom the leading surface of the bilateral animal has the largest and greatest variety of sensing devices.[8] This is not because of some accident in the dim past, only one of many possibilities. The anterior mouth with surrounding sense and grasping organs has been independently evolved in group after group, again and again. It is not surprising therefore to find the largest ganglion or the brain at the front end in close proximity to the major sensing organs.

An organism with brain and sense organs at the front end has several major adaptive advantages. First, it takes time to send sensed data over nerves and the successful animal is not one that sends the data to its tail and then back to the grasping or feeding organs. Secondly, the ganglion or brain is the integrating center that evaluates various incoming signals and sends out the command signal. To save space, to reduce the chance of damage, and to reduce interference and noise, it is more efficient to have short sense conductors and long command conductors. Thus, the brain and major sense organs are close together. Even in highly efficient two-directional animals such as the squids, the major sense organs are grouped at one end of the body next to the brain and near the mouth. The major sense organs do not migrate to distant parts of the body nor does the brain.

Thus, we can be more than reasonably sure that advanced animals will be bilaterally symmetrical, and will have a large brain at the front end near the mouth in close proximity to the largest and most diverse sense organs.

Given that life will first arise in a liquid medium, water, we can be sure there will be a long evolution in the liquid evironment before any population of organisms is able to invade and occupy the solid–gas interface. It is conceivable that an aquatic population might leave its ocean milieu by way of a subterranean route or directly penetrate the air from the ocean surface. However, if we recall that an invading population needs an energy source, it becomes apparent that both of these possibilities are very remote. Also, we can probably rule out the possibility of flying plants or burrowing chemosynthetic autotrophs. Permanent invasion and extensive utilization of the land–air interface require terrestrial plants, autotrophs, decomposing heterotrophs, and terrestrial animals, heterotrophs, evolving together in interdependent communities. The air, in all probability, can only be occupied after the land–air interface and by way of the land–air interface. Although some fish and squids are able to make short journeys into the air, all true fliers have evolved from the land. The aquatic insects have secondarily invaded the water from the land.

On the earth, the evolution of the largest brain occurred at the solid–air interface. We now know that at least five different chemical synaptic transmitters are used in various nerve tissues. It seems highly probable that the higher forms of terrestrial life have adopted the most efficient of these and have achieved the greatest miniaturization of nerve cells and conductive tissue possible. If this supposition is correct, large integrating centers, brains, of a size similar to ours will be necessary for conceptual thought.[9] If this assumption is accepted, we can be reasonably certain that a large brain will not develop underground or in the aerial habitat.

Recent work on the dolphins[10] indicates that we cannot be as certain that the largest brain on other earths would not develop in an aquatic medium. However, I think that this is highly unlikely for the following reasons. It is now reasonably well established that man's large brain evolved concurrently with social behavior, the use of tools, and speech.[11] If any one of these had been suppressed, *Homo* would never have reached the level of development of conceptual thought. Although social behavior and speech or at least communication would have and have had great adaptive and survival value in the sea, the development of tools in a liquid medium is highly improbable because of the density and viscosity of water. The sea otter rises to the surface to break his fragile, sea urchin food on his chest while floating on his back. It is very difficult to slam one's fist against one's chest while completely submerged in water. Using a lever to pry rocks or food from the bottom while completely submerged is also difficult. The simplest tools, thrown rocks and sticks, would be useless in the sea. Thus, on the basis that tools would be much less effective under water than on land it is highly probable that the largest-brained animal will evolve at the air–land interface.[12] Similarly the use of tools while flying is a very remote possibility.

This then raises the question of locomotion on land. There are many answers and solutions to this problem including sliding on slime, wriggling, and walking with legs. There can be no doubt that the latter method is superior to all others when we consider the friction involved, the energy required, and the speed and maneuverability obtainable. It is quite significant that wheels, one of man's greatest achievements, have never been evolved in any living organism. The primary reason for this is that living tissue just is not suitable for the high pressures that must be sustained at the bearing of a wheel. On land there is the further limiting factor of an inadequate amount of flat surface on which to use wheels. Thus, again, the number of possible structures is quite limited. Legs will be required.

We come then to the question of how many legs, one, two, three, four, six, ten, one thousand? The fossil record shows that there are strong selective pressures in both the marine and terrestrial environments that lead to the reduction of multiple appendages. This is so well demonstrated that it is given the distinction of a "law," Williston's Law. We can probably eliminate the odd number of appendages approached in many different animal groups but never really achieved. There are the three "pointed" kangaroos, the seven "legged" Collembola or spring tails, the five "armed" platyrrhine monkeys and many other examples. All of these can, for various reasons, be eliminated as real challengers to the paired appendage body plan. Although they are all numbers of paired appendaged organisms living on the earth, all but the vertebrates with two pairs and the insects with three pairs are essentially back water fugitives of only moderate or less success (with apologies to the acarologist and their eight-legged friends). Just why the higher vertebrates ended up with four legs is another tale that takes us back to the liquid medium of their early evolution. As higher and higher speed swimmers evolved, stabilizing and steering appendages of various sorts were developed.[13] The details of why four came out on top is only briefly discussed in vertebrate zoology treatises although this surely was a fundamental necessity for the vertebrate invasion of the land.[14]

Just why the insects have six legs is not immediately apparent but is perhaps related to their small size and exoskeleton of chitin. At any rate, it seems most probable that our extraterrestrial huminoid will have either two or three sets of paired appendages. I'm willing to bet on the smaller number.

Multiple sense organs with diverse functions have often been postulated for extraterrestrial animals. Almost certainly, outer space huminoids will have sensors for light (vision), sensors for chemicals in solution or dispersed in air (taste and smell), and for pressure changes (hearing). Whether they will have sensors for magnetic fields is less certain.[15] Strong arguments can be advanced for the presence of only two eyes set for binocular vision and two ears for binaural hearing. We can also argue strongly for the smell sensor being directly over the mouth. Its primary function in animals with high visual acuity, and probably in most others, is to test the nature of materials about to enter the mouth, is it poisonous or edible?

Could the conceptualizing organism have additional sense organs based on other parts of the electromagnetic spectrum besides the visual portion? Wald,[16] on the basis of biochemical limitations and energy requirements, argues no. Certainly, the ability to detect infrared at night would have great survival value. An acoustical ranging system similar to that of bats might be present, in which case the ears would be proportionately larger than ours. However, in this case the eyes might be reduced somewhat. On the other hand, any marked reduction in the visual sensors would be a serious impediment to the evolution of tools and the associated evolution of a large brain.

What about the dactyls? Although the human hand is considered to be a relatively unspecialized vertebrate structure, it is hard to visualize a more effective arrangement. Three, four, six, or seven fingers might work as well, but certainly an arrangement of claws, fingers, tweezers, knives, and so forth would not work as well. If we accept two as the minimum number and ten as the maximum, then four, five, or six dactyls would seem most probable and our extraterrestrial huminoid will count by eights, tens, or twelves.

We might consider the nature of the outer covering. Will it have scales, feathers, hair, foam, a cellophane wrapper? I would suspect that a competent vertebrate zoologist consulting with a biochemist could make a strong argument for skin and hair.

No doubt many other possibilities have occurred to the reader and the task of answering them is endless. I have considered and rejected green skin, reproductive organs on the chest, antenna systems, a ventral nerve cord, different temperature regimens, stronger and weaker gravitational systems and many other possibilities.

Those who wish to suggest alternatives not considered here, should realize that, if their suggested modifications of the extraterrestrial huminoids are to be considered seriously, a reasonable set of evolutionary steps must be set up to explain the final structures derived.

To restate the argument, a conceptualizing population of living organisms can only develop by the process of evolution. Given the ninety-two known, naturally occurring elements, the forms of energy available, and limited time, the number of alternative solutions to the major steps leading to a conceptual organism are strictly limited. The phenomenon of convergent evolution is so widespread in both the plant and animal kingdoms that it needs no special elucidation here. Suffice it to say that the evidence shows that, again and again, animals and plants have independently evolved not only similar structures but also similar biochemical systems and similar behavioral patterns as solutions to the same fundamental problems.

If we fail to communicate with conceptualizing beings on other planets it will not be because they are fundamentally different from us but because they have either far surpassed our state of technology and have no interest in communicating with us or have not yet reached our state of advancement and are thus unable to do so. This probably is the fundamental factor which would greatly reduce the number of such populations attempting to communicate with us.

If we ever succeed in communicating with conceptualizing beings in outer space, they won't be spheres, pyramids, cubes, or pancakes. In all probability they will look an awful lot like us.

ACKNOWLEDGMENT

I wish to thank Stephen Gould, Martin Murie, Stephen Rothbaum, Edmund Samuel, and Albert Stewart for their valuable criticism of the ideas expressed in this paper.

REFERENCES

1. Beadle, G. W., 1959. The place of genetics in modern biology. Eleventh Annual A.D. Little Mem. Lecture, Mass. Inst. Tech. (One of the better, more concise statements of this position.) G. G. Simpson, 1964. *Sci.*, *143* (3608): 769–775 has recently presented a similar argument.
2. Henderson, L. J., 1927. The fitness of the environment. Macmillan Co., N.Y., 317 pp. (copyright 1913).
3. Urey, H. C., 1963. The origin of organic molecules. In "The Nature of biological diversity," J. M. Allen (Ed.), McGraw-Hill, N.Y., 1–13.
4. Calvin, M., 1961. Chemical evolution. Condon Lectures, Univ. Oregon Press, Eugene, 41 pp.
5. Baldwin, E., 1963. Biochemistry and evolution. In "The nature of biological diversity," J. M. Allen (Ed.), McGraw-Hill, N.Y., 45–68, discusses the problem of common ancestorship versus convergence in the evolution of biochemical systems. See also Pantin, C. F. A., 1951. Organic design. *Advanc. Sci.*, *8*, 138–150; and Calvin, M., and G. J. Calvin, 1964. Atom to Adam, *Am. Sci.*, *52*, 163–186.
6. Ramsay, J. A., 1952. A physiological approach to the lower animals. Cambr. Univ. Press, 148 pp.
Carthy J. D., 1957. An introduction to the behavior of invertebrates. Macmillan Co., N.Y., 380 pp.
7. In the flatworms, which do not have an anus, the mouth is in the middle of the animal's body on the belly. Despite this exception and a few others, there is no doubt that the best solution to the problems of ingestion and excretion in an active animal is an anterior mouth and posterior anus.
8. Interesting exceptions to this, aside from the lowest invertebrates such as the ctenophores, coelenterates, and flatworms, might be the crabs which run sideways and squids which pursue their prey backwards.
9. Jerison, H., 1963. Interpreting the evolution of the brain. *Human Biol.*, *35* (3), 263–291.
10. Kellog, W. N., 1961. Porpoises and sonar. Univ. Chicago Press, 177 pp.
11. Washburn, S. L., and F. C. Howell, 1960. Human evolution and culture. In S. Tax (Ed.). The evolution of Man. Univ. Chicago Press, 33–56.
12. In the snapping shrimps, such as *Crangon* we see the development of a hydraulic pistol for stunning prey. Further modification of this system to propel missiles is conceivable but the weight and size required for effective operation would probably lead to a sedentary mode of life, and, hence, a small brain. The poison darts of the gastropod genus, *Conus*, represent another possible beginning of marine missiles; however, they are entirely productions of their own bodies analogous to the stinging bodies of the coelenterates and hence not tools.
13. Gregory, W. K., and H. C. Raven, 1941. Studies on the origin and early evolution of paired fins and limbs. *Annals New York Acad. Sci.*, *42* (3), 273–360.
14. Gunter, G., 1956. The origin of the tetrapod limb. *Sci.*, *123* (3195), 495–496. Szarski, H., 1962. The origin of the Amphibia. *Quart. Rev. Biol.*, *37* (3), 189–241.
15. Brown, F. A., Jr., 1962. Response of the planarian, *Dugesia*, and the protozoan, *Paramecium*, to very weak horizontal magnetic fields. *Biol. Bull.*, *123*, 264–281.
16. Wald, G., 1959. Life and light. *Sci. Amer.*, *201* (4), 92–108.

27

ERNST MAYR*

The Probability of Extraterrestrial Intelligent Life

A NUMBER OF very different problems are often confused during discussions of the SETI [Search for Extraterrestrial Intelligence] project: (1) the probability of the existence of "life" elsewhere in the universe, (2) the probability of intelligent extraterrestrial life, and (3) the chances of being able to communicate with such life, if it should exist.

At the present time we have no positive evidence whatsoever that life exists elsewhere, and thus, of course, also of intelligent life. The probabilities of either can be guessed at only by highly indirect inferences.

Life in the Universe

When the Mars missions were being prepared, the astronomer Donald Menzel and I had a $5 bet as to whether or not "life as on earth" [as was our precise designation] would be discovered on Mars. The physical scientist Menzel said yes, the evolutionary biologist Mayr said no. Who was right is on record. By now it is quite evident that none of the other planets in this solar system is suitable for life.

One negative instance, of course, proves nothing. If all suns in the universe have planets (actually a rather dubious assumption), we would have hundreds of millions of planets. Surely, it is argued, some of these

*From Ernst Mayr, "The Probability of Extraterrestrial Intelligent Life," in *Extraterrestrials: Science and Alien Intelligence,* edited by Edward Regis, Jr. (New York: Cambridge University Press, 1985), pp. 23–30. Reprinted with the permission of Cambridge University Press.

should have spawned life. And I agree, the probability for a multiple origin of a self-replicating nucleic acid–protein aggregate is indeed high.

It has been known for some time that smaller organic molecules, like amino-acids, purines, and pyrimidines, can arise spontaneously in the universe, and that such processes can be duplicated in the laboratory. Nevertheless, for a long time it seemed impossible to explain how the amino-acids (and peptides) could get together with nucleic acids to form truly replicating, i.e. living, macromolecules. Through the researches of Eigen and his school (Küppers 1983) there seems to be no longer a difficulty of principle. What is particularly interesting is the important role played by natural selection, even during the pre-biotic phase. The probability of the repeated origin of macromolecular systems with an ability for information storage and replication can no longer be doubted.

What is still entirely uncertain is how often this has happened, where it has happened, and how much evolution might have occurred subsequent to the origin of such life. We who live on the earth do not fully appreciate what an inhospitable place most planets must be. To be able to support life they must be just the right distance from their sun, have the right temperature, a sufficient amount of water, a sufficient density to be able to hold an atmosphere, a protection against damaging ultraviolet radiation, and so forth. Furthermore, every planet changes in the course of its history, and the sequence of changes has to be just right. If, for instance, there were too much free oxygen at an early stage, it would destroy life. The total set of prerequisites for the origin and maintenance of life drastically reduces the number of planets that would have been suitable for the origin of life. There is, indeed, the probability that the combination and sequence of conditions that permitted the origin of life on earth was not duplicated on a single other planet in the universe. I do not make such a claim, and it would not be science if I did, since it would be impossible ever to refute it. However, measured by the possibility of refutation, the claims of the proponents of extraterrestrial life and intelligence are equally outside the bounds of science. The only thing we know for sure is that of the nine planets of the solar system the earth is the only one that has produced life. Let us assume, however, for the sake of the argument that life has originated on some of the supposedly hundreds of millions of planets in the universe. Since we do not know how many suns have planets, the mentioned figure might be a gross overestimation.

The Existence of Extraterrestrial Intelligence

It is interesting and rather characteristic that almost all the promoters of the thesis of extraterrestrial intelligence are physical scientists. They are joined by a number of molecular and microbial biologists, and by a handful of romantic organismic biologists.

Why are those biologists, who have the greatest expertise on evolutionary probabilities, so almost unanimously skeptical of the probability of extraterrestrial intelligence? It seems to me that this is to a large extent due to the tendency of physical scientists to think deterministically, while organismic biologists know how opportunistic and unpredictable evolution is.

Some 20 years ago when I argued a great deal with the astronomer Donald Menzel about life on Mars, I was forever astonished how certain he was that if life had ever originated on Mars (or been transported to Mars), this would inevitably lead to intelligent humanoids. The production of man was for him like the end product of a chemical reaction chain where the end product can be predicted once you know with what chemicals you had started. He took it virtually for granted that if there was life on a planet it would in due time give rise to intelligent life: "Our own Milky Way might contain up to a million planets [favorable to the development of life], all inhabited by intelligent life" (Menzel 1965, p. 218).

Everybody knows, of course, that determinism is no longer the fashion in modern physics, and yet in conversation with physical scientists I have discovered again and again how strongly they still think along deterministic lines. If organic evolution on earth culminated in intelligence, why should it not have resulted in intelligence on all planets on which life had originated?

By contrast an evolutionist is impressed by the incredible improbability of intelligent life ever to have evolved, even on earth. To demonstrate this, let us look at the history of life on earth.

Date of origin of kinds of organisms if age of earth (4.5 billion years) is made equivalent to a calendar year:

Origin of Earth = 1 January

Life (Prokaryotes) = 27 February

Eukaryotes = 28 October

Chordates = 17 November

Vertebrates = 21 November

Mammals = 12 December

Primates = 26 December

Anthropoids = 30 December, at 01:00 a.m.

Hominid line = 31 December 10:00 a.m.

Homo sapiens = 31 December 11.562$\frac{1}{2}$ p.m. (= 3$\frac{1}{2}$
minutes before year's end).

Let us look at the chronology of major evolutionary events on earth. (All cited figures are rough estimates, the upward or downward revision of which would have no effect on the argument; the order of magnitude, however, is right.) Let us assume the earth originated 4.5 billion years ago. There is evidence that life began only about 700 million years (*my*) later. Definite early prokaryote fossils are known from 3.5 billion years ago. What is most remarkable is that for about 3000 *my* nothing very spectacular happened as far as life on earth was concerned. There was apparently a rich diversification of prokaryotes, but these—although quite successful in their way—are poor potential as progenitors of intelligent life. Nevertheless, they displayed remarkable metabolic diversification, the blue-green bacteria even became phototropic and produced oxygen. Up to that time the earth's atmosphere had been reducing.

Sometime, between 800 and 1000 *my* ago a most improbable event took place. According to the most likely explanation, a symbiosis was established between two (or more) kinds of prokaryotes, one of them supplying cytoplasmic organelles, the other one the nucleus of an entirely new type of organism, the first eukaryote. This was apparently such a successful combination that within a period of about 100 *my* (estimate) four new kingdoms evolved, the protists (one-celled animals and plants), fungi, plants, and animals. All higher organisms are eukaryotes, characterized by the possession of a well-organized nucleus and chromosomes in each cell.

We can see that from the origin of life to the origin of the eukaryotes about two-thirds of the age of the earth had passed by without any noticeable events except for diversification within the prokaryotes. But once the eukaryotes had been "invented" an almost explosive innovative diversification took place. Within each of the four mentioned kingdoms scores of separate evolutionary lines originated, many of them strikingly different from each other. However, in none of these kingdoms, except that of the animals, was there even the beginning of any evolutionary trends toward intelligence.

What about evolution of intelligence among the animals? After the animalian "type" had been invented, different structural types originated with such fertility that one could probably recognize at least 40 different phyla of animals in the Cambrian, including unique types in the Ediacaran and surviving Burgess shale formations. Many of these became extinct rather quickly and there is no good evidence for the origin of any new phylum after the end of the Cambrian (500 *my* ago). However, the surviving phyla experienced a continuing abundant proliferation into classes, orders, families, and lower taxa. Of the 40 or so original phyla of animals only one, that of the chordates, eventually gave rise to intelligent life, but the world still had to wait some 500 *my* before this happened. At first, still in Paleozoic, the vertebrates appeared in exceedingly diverse types, formerly all lumped together under the name "Fishes," but it is now realized how different the early vertebrates were from each other. Among this multitude of types only one gave rise to the amphibians, and among the various types of amphibians only one to the reptiles. What are called the reptiles are again a highly diverse group of vertebrates including such different organisms as turtles, lizards, snakes, crocodiles, and numerous extinct lineages as ichthyosaurs, plesiosaurs, pterodactyls, and dinosaurs. Among these numerous types of reptiles, only two, the pseudosuchians (ancestors of birds) and the therapsids (ancestors of mammals) gave rise to descendants to some of whom a reasonable degree of intelligence can be attributed. But with all my bias in favor of birds, I would not say that a raven or parrot has the amount and kind of intelligence to found a civilization. So we have to continue with the mammalian class. It contains such unusual types as the monotremes (e.g. platypus) and marsupials, as well as a rich assortment of placental orders, some still living, many others having become extinct in the course of the Tertiary. Forms with a rather high development of the central nervous system and a good deal of intelligence are quite common among the mammals, but only one of these many orders led to the development of a truly superior intelligent life, the primates. The primates, however, are a rather diversified group, with prosimians (lemurs, etc.), New World monkeys, and Old World monkeys, but only the anthropoid apes produced intelligence that clearly surpasses other mammals. Only after 18 of the 25 *my* of the existence of the anthropoid apes, and after a splitting of this major lineage into a number of minor lineages, like the gibbons (and relatives), the orang utan (and relatives), the African apes (chimpanzee and gorilla), and a considerable number of extinct lineages, did the lineage emerge which eventually, less than one-third of a million years ago, led to *Homo sapiens*.

The reason why I have buried you under this mass of tedious detail is to make one point, but an all-important one. In conflict with the thinking of those who see a straight line from the origin of life to intelligent man, I have shown that at each level of this pathway there were scores, if not hundreds, of branching points and separately evolving phyletic lines, with only a single one in each case forming the ancestral lineage that ultimately gave rise to Man.

If evolutionists have learned anything from a detailed analysis of evolution, it is the lesson that the origin of new taxa is largely a chance event. Ninety-nine of 100 newly arising species probably become extinct without giving rise to descendant taxa. And the characteristic of any new taxon is to a large extent determined by such chance factors as the genetic composition of the founding population, the special internal structure of its genotype, and the physical as well as biotic environment that supplies the selection forces of the new species population.

My argument based on the incredibly low probability of life ever having originated on earth, must not be misunderstood. I do not claim in the least that an extraterrestrial intelligent "life" must have the slightest anatomical similarity to man. I already mentioned that we get a certain amount of intelligence in other mammals and even in birds. Indeed, an ability to make use of previous experience in subsequent actions, in other words a rudimentary kind of intelligence, is widely distributed in the animal kingdom. Intelligence, on another planet, might reside in a being inconceivably different from any living being on earth. Any devotee of science fiction will have no trouble in coming up with possibilities.

The point I am making is the incredible improbability of genuine intelligence emerging. There were probably more than a billion species of animals on earth, belonging to many millions of separate phyletic lines, all living on this planet earth which is hospitable to intelligence, and yet only a single one of them succeeded in producing intelligence.

Proponents of extraterrestrial intelligence have mentioned the convergent evolution of two such "highly improbable" organs as the eye of the cephalopods and of the vertebrates as an analog to the presumably equally improbable but not at all impossible convergent evolution of intelligence. Those who have thus argued, unfortunately, do not know their biology. The case of the convergent evolution of eyes is, indeed, of decisive importance for the estimation of the probability of convergent evolution of intelligence. The crucial point is that the convergent evolution of eyes in different phyletic lines of animals is not at all improbable. In fact, eyes evolved whenever they were of selective advantage in the animal kingdom. As Salvini-Plawen and I have shown, eyes evolved independently no less than at least 40 times in different groups of animals (Salvini-Plawan & Mayr 1977). This shows that a highly complicated organ can evolve repeatedly and convergently when advantageous, provided such evolution is at all probable. For genuine intelligence this is evidently not the case, as the history of life on earth has shown.

One additional improbability must be mentioned. Somehow, the supporters of SETI naively assume that "intelligence" means developing a technology capable of intragalactic or even intergalactic communication. But such a development is highly improbable. For instance, Neanderthal Man, living 100,000 years ago, had a brain as big as ours. Yet, his "civilization" was utterly rudimentary. The wonderful civilizations of the Greeks, the Chinese, the Mayas or the Renaissance, although they were created by people who were for all intents and purposes physically identical with us, never developed such a technology, and neither did we until a few years ago. The assumption that any intelligent extraterrestrial life must have the technology and mode of thinking of late twentieth-century Man is unbelievably naive.

Civilizations, as human history demonstrates, are fleeting moments in the history of an intelligent species. For two civilizations to communicate with each other, it is necessary that they flourish simultaneously. Let me illustrate the importance of this point by a little fable. Let me assume there is another high technology civilization in our galaxy. By some extraordinary instrumentation their inhabitants were able to discover the origin of the earth 4.5 billion years ago. At once they began to send signals to the earth and continued to do so for 4.5 billion years. Finally at the time of the birth of Christ they decided that they would terminate their program after another 1900 years, if they had not received any answer by then. When they abandoned their program in the year 1900, they had proven to their own satisfaction that there was no other intelligent life in our galaxy.

I am trying to demonstrate by this fable that even if there were intelligent extraterrestrial life, and even if it had developed a highly sophisticated technology [although if they were truly intelligent they would probably carefully avoid this], the timing of their efforts and those of our engineers would have to coincide to an altogether improbable degree, considering the amounts of astronomical time available. Every aspect of "extraterrestrial intelligence" that we consider confronts us with astronomically low probabilities. If one multiplies these with each other, one comes out so close to zero, that it is zero for all practical purposes. This was already pointed out by Simpson in 1964. Those biologists who doubt the probability of ever establishing contact with extraterrestrial intelligent life if it should exist, do not "deny categorically the possibility of extraterrestrial intelligence," as they have been accused. How could they? There are no facts that would permit such a categorical denial. Nor have I seen a published statement of such a "categorical denial." All they claim is that the probabilities are close to zero. This is why evolutionary biologists, as a group, are so skeptical of the existence of extraterrestrial intelligence, and even more so of any possibility of communicating with it, if it exists.

For all these reasons I conclude that the SETI program is a deplorable waste of taxpayers' money, money that could be spent far more usefully for other purposes.[1]

NOTE

1. After completing the manuscript, I reread Simpson's classical paper (1964) on the subject and was struck by how similar his analysis was to mine. Perhaps subconsciously I still remembered his arguments. Be that as it may, I warmly recommend reading his famous essay on the "Nonprevalence of humanoids." It is as pertinent today as it was 20 years ago.

REFERENCES

Küppers, B.-O. (1983). *Molecular Theory of Evolution*. Berlin, Heidelberg, New York: Springer.
Menzel, D. H. (1965). Life in the Universe. *The Graduate Journal*, 7, 195–219.
Salvini-Plawen, L. v. & Mayr, E. (1977). On the evolution of photoreceptors and eyes. *Evolutionary Biology*, 10, 207–63.
Simpson, G. G. (1964). The nonprevalence of humanoids. In *This View of Life*, pp. 253–71. New York: Harcourt, Brace, and World.

EVOLUTION AND ETHICS

28

WILLIAM GRAHAM SUMNER*

The Challenge of Facts

SOCIALISM IS NO new thing. In one form or another it is to be found throughout all history. It arises from an observation of certain harsh facts in the lot of man on earth, the concrete expression of which is poverty and misery. These facts challenge us. It is folly to try to shut our eyes to them. We have first to notice what they are, and then to face them squarely.

Man is born under the necessity of sustaining the existence he has received by an onerous struggle against nature, both to win what is essential to his life and to ward off what is prejudicial to it. He is born under a burden and a necessity. Nature holds what is essential to him, but she offers nothing gratuitously. He may win for his use what she holds, if he can. Only the most meager and inadequate supply for human needs can be obtained directly from nature. There are trees which may be used for fuel and for dwellings, but labor is required to fit them for this use. There are ores in the ground, but labor is necessary to get out the metals and make tools or weapons. For any real satisfaction, labor is necessary to fit the products of nature for human use. In this struggle every individual is under the pressure of the necessities for food, clothing, shelter, fuel, and every individual brings with him more or less energy for the conflict necessary to supply his needs. The relation, therefore, between each man's needs and each man's energy, or "individualism," is the first fact of human life.

*From William Graham Sumner, *The Challenge of Facts and Other Essays*, A. S. Kelle, ed. (New Haven: Yale University Press, 1914), pp. 17–30.

It is not without reason, however, that we speak of a "man" as the individual in question, for women (mothers) and children have special disabilities for the struggle with nature, and these disabilities grow greater and last longer as civilization advances. The perpetuation of the race in health and vigor, and its success as a whole in its struggle to expand and develop human life on earth, therefore, require that the head of the family shall, by his energy, be able to supply not only his own needs, but those of the organisms which are dependent upon him. The history of the human race shows a great variety of experiments in the relation of the sexes and in the organization of the family. These experiments have been controlled by economic circumstances, but, as man has gained more and more control over economic circumstances, monogamy and the family education of children have been more and more sharply developed. If there is one thing in regard to which the student of history and sociology can affirm with confidence that social institutions have made "progress" or grown "better," it is in this arrangement of marriage and the family. All experience proves that monogamy, pure and strict, is the sex relation which conduces most to the vigor and intelligence of the race, and that the family education of children is the institution by which the race as a whole advances most rapidly, from generation to generation, in the struggle with nature. Love of man and wife, as we understand it, is a modern sentiment. The devotion and sacrifice of parents for children is a sentiment which has been developed steadily and is now more intense and far more widely practiced throughout society than in earlier times. The relation is also coming to be regarded in a light quite different from that in which it was formerly viewed. It used to be believed that the parent had unlimited claims on the child and rights over him. In a truer view of the matter, we are coming to see that the rights are on the side of the child and the duties on the side of the parent. Existence is not a boon for which the child owes all subjection to the parent. It is a responsibility assumed by the parent towards the child without the child's consent, and the consequence of it is that the parent owes all possible devotion to the child to enable him to make his existence happy and successful.

The value and importance of the family sentiments, from a social point of view, cannot be exaggerated. They impose self-control and prudence in their most important social bearings, and tend more than any other forces to hold the individual up to the virtues which make the sound man and the valuable member of society. The race is bound, from generation to generation, in an unbroken chain of vice and penalty, virtue and reward. The sins of the fathers are visited upon the children, while, on the other hand, health, vigor, talent, genius, and skill are, so far as we can discover, the results of high physical vigor and wise early training. The popular language bears witness to the universal observation of these facts, although general social and political dogmas have come into fashion which contradict or ignore them. There is no other such punishment for a life of vice and self-indulgence as to see children grow up cursed with the penalties of it, and no such reward for self-denial and virtue as to see children born and grow up vigorous in mind and body. It is time that the true import of these observations for moral and educational purposes was developed, and it may well be questioned whether we do not go too far in our reticence in regard to all these matters when we leave it to romances and poems to do almost all the educational work that is done in the way of spreading ideas about them. The defense of marriage and family, if their sociological value were better understood, would be not only instinctive but rational. The struggle for existence with which we have to deal must be understood, then, to be that of a man for himself, his wife, and his children.

The next great fact we have to notice in regard to the struggle of human life is that labor which is spent in a direct struggle with nature is severe in the extreme and is but slightly productive. To subjugate nature, man needs weapons and tools. These, however, cannot be won unless the food and clothing and other prime and direct necessities are supplied in such amount that they can be consumed while tools and weapons are being made, for the tools and weapons themselves satisfy no needs directly. A man who tills the ground with his fingers or with a pointed stick picked up without labor will get a small crop. To fashion even the rudest spade or hoe will cost time, during which the laborer must still eat and drink and wear, but the tool, when obtained, will multiply immensely the power to produce. Such products of labor, used to assist production, have a function so peculiar in the nature of things that we need to distinguish them. We call them capital. A lever is capital, and the advantage of lifting a weight with a lever over lifting it by direct exertion is only a feeble illustration of the power of capital in production. The origin of capital lies in the darkness before history, and it is probably impossible for us to imagine the slow and painful steps by which the race began the formation of it. Since then it has gone on rising to higher and higher powers by a ceaseless involution, if I may use a mathematical expression. Capital is labor raised to a higher power by being constantly multiplied into itself. Nature has been more and more subjugated by the human race through the power of capital, and every human being now living shares the improved status of the race to a degree which neither he nor any one else can measure, and for which he pays nothing.

Let us understand this point, because our subject will require future reference to it. It is the most short-sighted ignorance not to see that, in a civilized community, all the advantage of capital except a small fraction is gratuitously enjoyed by the community. For instance, suppose the case of a man utterly destitute of tools, who is trying to till the ground with a pointed stick. He could get something out of it. If now he should obtain a spade with which to till the ground, let us suppose, for illustration, that he could get twenty times as great a product. Could, then, the owner of a spade in a civilized state demand, as its price, from the man who had no spade, nineteen-twentieths of the product which could be produced by the use of it? Certainly not. The price of a spade is fixed by the supply and demand of products in the community. A spade is bought for a dollar and the gain from the use of it is an inheritance of knowledge, experience, and skill which every man who lives in a civilized state gets for nothing. What we pay for steam transportation is no trifle, but imagine, if you can, eastern Massachusetts cut off from steam connection with the rest of the world, turnpikes and sailing vessels remaining. The cost of food would rise so high that a quarter of the population would starve to death and another quarter would have to emigrate. To-day every man here gets an enormous advantage from the status of a society on a level of steam transportation, telegraph, and machinery, for which he pays nothing.

So far as I have yet spoken, we have before us the struggle of man with nature, but the social problems, strictly speaking, arise at the next step. Each man carries on the struggle to win his support for himself, but there are others by his side engaged in the same struggle. If the stores of nature were unlimited, or if the last unit of the supply she offers could be won as easily as the first, there would be no social problem. If a square mile of land could support an indefinite number of human beings, or if it cost only twice as much labor to get forty bushels of wheat from an acre as to get twenty, we should have no social problem. If a square mile of land could support millions, no one would ever emigrate and there would be no trade or commerce. If it cost only twice as much labor to get forty bushels as twenty, there would be no advance in the arts. The fact is far otherwise. So long as the population is low in proportion to the amount of land, on a given stage of the arts, life is easy and the competition of man with man is weak. When more persons are trying to live on a square mile than it can support, on the existing stage of the arts, life is hard and the competition of man with man is intense. In the former case, industry and prudence may be on a low grade; the penalties are not severe, or certain, or speedy. In the latter case, each individual needs to exert on his own behalf every force, original or acquired, which he can command. In the former case, the average condition will be one of comfort and the population will be all nearly on the average. In the latter case, the average condition will not be one of comfort, but the population will cover wide extremes of comfort and misery. Each will find his place according to his ability and his effort. The former society will be democratic; the latter will be aristocratic.

The constant tendency of population to outstrip the means of subsistence is the force which has distributed population over the world, and produced all advance in civilization. To this day the two means of escape for an overpopulated country are emigration and an advance in the arts. The former wins more land for the same people; the latter makes the same land support more persons. If, however, either of these means opens a chance for an increase of population, it is evident that the advantage so won may be speedily exhausted if the increase takes place. The social difficulty has only undergone a temporary amelioration, and when the conditions of pressure and competition are renewed, misery and poverty reappear. The victims of them are those who have inherited disease and depraved appetites, or have been brought up in vice and ignorance, or have themselves yielded to vice, extravagance, idleness, and imprudence. In the last analysis, therefore, we come back to vice, in its original and hereditary forms, as the correlative of misery and poverty.

The condition for the complete and regular action of the force of competition is liberty. Liberty means the security given to each man that, if he employs his energies to sustain the struggle on behalf of himself and those he cares for, he shall dispose of the product exclusively as he chooses. It is impossible to know whence any definition or criterion of justice can be derived, if it is not deduced from this view of things; or if it is not the definition of justice that each shall enjoy the fruit of his own labor and self-denial, and of injustice that the idle and the industrious, the self-indulgent and the self-denying, shall share equally in the product. Aside from the *a priori* speculations of philosophers who have tried to make equality an essential element in justice, the human race has recognized, from the earliest times, the above conception of justice as the true one, and has founded upon it the right of property. The right of property, with marriage and the family, gives the right of bequest.

Monogamic marriage, however, is the most exclusive of social institutions. It contains, as essential principles, preference, superiority, selection, devotion. It would not be at all what it is if it were not for these characteristic traits, and it always degenerates when these traits are not present. For instance, if a man should not have a distinct preference for the woman he married, and if he did not select her as superior to others, the marriage would be an imperfect one according to the standard of true monogamic marriage. The family under monogamy, also, is a closed group, having special interests and estimating privacy and reserve as valuable advantages for family development. We grant high prerogatives, in our society, to parents, although our observation teaches us that thousands of human beings are unfit to be parents or to be entrusted with the care of children. It follows, therefore, from the organization of marriage and the family, under monogamy, that great inequalities must exist in a society based on those institutions. The son of wise parents cannot start on a level with the son of foolish ones, and the man who has had no home discipline cannot be equal to the man who has had home discipline. If the contrary were true, we could rid ourselves at once of the wearing labor of inculcating sound morals and manners in our children.

Private property, also, which we have seen to be a feature of society organized in accordance with the natural conditions of the struggle for existence produces inequalities between men. The struggle for existence is aimed against nature. It is from her niggardly hand that we have to wrest the satisfactions for our needs, but our fellow-men are our competitors for the meager supply. Competition, therefore, is a law of nature. Nature is entirely neutral; she submits to him who most energetically and resolutely assails her. She grants her rewards to the fittest, therefore, without regard to other considerations of any kind. If, then, there be liberty, men get from her just in proportion to their works, and their having and enjoying are just in proportion to their being and their doing. Such is the system of nature. If we do not like it, and if we try to amend it, there is only one way in which we can do it. We can take from the better and give to the worse. We can deflect the penalties of those who have done ill and throw them on those who have done better. We can take the rewards from those who have done better and give them to those who have done worse. We shall thus lessen the inequalities. We shall favor the survival of the unfittest, and we shall accomplish this by destroying liberty. Let it be understood that we cannot go outside of this alternative: liberty, inequality, survival of the fittest; not-liberty, equality, survival of the unfittest. The former carries society forward and favors all its best members; the latter carries society downwards and favors all its worst members.

For three hundred years now men have been trying to understand and realize liberty. Liberty is not the right or chance to do what we choose; there is no such liberty as that on earth. No man can do as he chooses: the autocrat of Russia or the King of Dahomey has limits to his arbitrary will; the savage in the wilderness, whom some people think free, is the slave of routine, tradition, and superstitious fears; the civilized man must earn his living, or take care of his property, or concede his own will to the rights and claims of his parents, his wife, his children, and all the persons with whom he is connected by the ties and contracts of civilized life.

What we mean by liberty is civil liberty, or liberty under law; and this means the guarantees of law that a man shall not be interfered with while using his own powers for his own welfare. It is, therefore, a civil and political status; and that nation has the freest institutions in which the guarantees of peace for the laborer and security for the capitalist are the highest. Liberty, therefore, does not by any means do away with the struggle for existence. We might as well try to do away with the need of eating, for that would, in effect, be the same thing. What civil liberty does is to turn the competition of man with man from violence and brute force into an industrial competition under which men vie with one another for the acquisition of material goods by industry, energy, skill, frugality, prudence, temperance, and other industrial virtues. Under this changed order of things the inequalities are not done away with. Nature still grants her rewards of having and enjoying, according to our being and doing, but it is now the man of the highest training and not the man of the heaviest fist who gains the highest reward. It is impossible that the man with capital and the man without capital should be equal. To affirm that they are equal would be to say that a man who has no tool can get as much food out of the ground as the man who has a spade or a plough; or that the man who has no weapon can defend himself as well against hostile beasts or hostile men as the man who has a weapon. If that were so, none of us would work any more. We work and deny ourselves to get capital just because, other things being equal, the man who has it is superior, for attaining all the ends of life, to the man who has it not. Considering the eagerness with which we all seek capital and the estimate we put upon it, either in cherishing it if we have it, or envying others who have it while we have it not, it is very strange what platitudes pass current about it in our society so soon as we begin to generalize about it. If our young people really believed some of the teachings they hear, it would not be amiss to preach them a sermon once in a while to reassure them, setting forth that it is not wicked to be rich, nay even, that it is not wicked to be richer than your neighbor.

It follows from what we have observed that it is the utmost folly to denounce capital. To do so is to undermine civilization, for capital is the first requisite of every social gain, educational, ecclesiastical, political, aesthetic, or other.

It must also be noticed that the popular antithesis between persons and capital is very fallacious. Every law or institution which protects persons at the expense of capital makes it easier for persons to live and to increase the number of consumers of capital while lowering all the motives to prudence and frugality by which capital is created. Hence every such law or institution tends to produce a large population, sunk in misery. All poor laws and all eleemosynary institutions and expenditures have this tendency. On the contrary, all laws and institutions which give security to capital against the interests of other persons than its owners, restrict numbers while preserving the means of subsistence. Hence every such law or institution tends to produce a small society on a high stage of comfort and well-being. It follows that the antithesis commonly thought to exist between the protection of persons and the protection of property is in reality only an antithesis between numbers and quality.

I must stop to notice, in passing, one other fallacy which is rather scientific than popular. The notion is attributed to certain economists that economic forces are self-correcting. I do not know of any economists who hold this view, but what is intended probably is that many economists, of whom I venture to be one, hold that economic forces act compensatingly, and that whenever economic forces have so acted as to produce an unfavorable situation, other economic forces are brought into action which correct the evil and restore the equilibrium. For instance, in Ireland overpopulation and exclusive devotion to agriculture, both of which are plainly traceable to unwise statesmanship in the past, have produced a situation of distress. Steam navigation on the ocean has introduced the competition of cheaper land with Irish agriculture. The result is a social and industrial crisis. There are, however, millions of acres of fertile land on earth which are unoccupied and which are open to the Irish, and the economic forces are compelling the direct corrective of the old evils, in the way of emigration or recourse to urban occupations by unskilled labor. Any number of economic and legal nostrums have been proposed for this situation, all of which propose to leave the original causes untouched. We are told that economic causes do not correct themselves. That is true. We are told that when an economic situation becomes very grave it goes on from worse to worse and that there is no cycle through which it returns. That is not true, without further limitation. We are told that moral forces alone can elevate any such people again. But it is plain that a people which has sunk below the reach of the economic forces of self-interest has certainly sunk below the reach of moral forces, and that this objection is superficial and short-sighted. What is true is that economic forces always go before moral forces. Men feel self-interest long before they feel prudence, self-control, and temperance. They lose the moral forces long before they lose the economic forces. If they can be regenerated at all, it must be first by distress appealing to self-interest and forcing recourse to some expedient for relief. Emigration is certainly an economic force for the relief of Irish distress. It is a palliative only, when considered in itself, but the virtue of it is that it gives the non-emigrating population a chance to rise to a level on which the moral forces can act upon them. Now it is terribly true that only the better ones emigrate, and only the better ones among those who remain are capable of having their ambition and energy awakened, but for the rest the solution is famine and death, with a social regeneration through decay and the elimination of that part of the society which is not capable of being restored to health and life. As Mr. Huxley once said, the method of nature is not even a word and a blow, with the blow first. No explanation is vouchsafed. We are left to find out for ourselves why our ears are boxed. If we do not find out, and find out correctly, what the error is for which we are being punished, the blow is repeated and poverty, distress, disease, and death finally remove the incorrigible ones. It behooves us men to study these terrible illustrations of the penalties which follow on bad statesmanship, and of the sanctions by which social laws are enforced. The economic cycle does complete itself; it must do so, unless the social group is to sink in permanent barbarism. A law may be passed which shall force somebody to support the hopelessly degenerate members of a society, but such a law can only perpetuate the evil and entail it on future generations with new accumulations of distress.

The economic forces work with moral forces and are their handmaidens, but the economic forces are far more primitive, original, and universal. The glib generalities in which we sometimes hear people talk, as if you could set moral and economic forces separate from and in antithesis to each other, and discard the one to accept and work by the other, gravely misconstrue the realities of the social order.

We have now before us the facts of human life out of which the social problem springs. These facts are in many respects hard and stern. It is by strenuous exertion only that each one of us can sustain himself against the destructive forces and the ever recurring needs of life; and the higher the degree to which we seek to carry our development the greater is the proportionate cost of every step. For help in the struggle we can only look back to those in the previous generation who are responsible for our existence. In the competition of life the son of wise and prudent ancestors has immense advantages over the son of vicious and imprudent ones. The man who has capital possesses immeasurable advantages for the struggle of life over him who has none. The more we break down privileges of class, or industry, and establish liberty, the greater will be the inequalities and the more exclusively will the vicious bear the penalties. Poverty and misery will exist in society just so long as vice exists in human nature.

29

THOMAS HENRY HUXLEY*

Evolution and Ethics

MODERN THOUGHT IS making a fresh start from the base whence Indian and Greek philosophy set out; and, the human mind being very much what it was six-and-twenty centuries ago, there is no ground for wonder if it presents indications of a tendency to move along the old lines to the same results.

We are more than sufficiently familiar with modern pessimism, at least as a speculation; for I cannot call to mind that any of its present votaries have sealed their faith by assuming the rags and the bowl of the mendicant Bhikku, or the cloak and the wallet of the Cynic. The obstacles placed in the way of sturdy vagrancy by an unphilosophical police have, perhaps, proved too formidable for philosophical consistency. We also know modern speculative optimism, with its perfectibility of the species, reign of peace, and lion and lamb transformation scenes; but one does not hear so much of it as one did forty years ago; indeed, I imagine it is to be met with more commonly at the tables of the healthy and wealthy, than in the congregations of the wise. The majority of us, I apprehend, profess neither pessimism nor optimism. We hold that the world is neither so good, nor so bad, as it conceivably might be; and, as most of us have reason, now and again, to discover that it can be. Those who have failed to experience the joys that make life worth living are, probably, in as small a minority as those who have never known the

*From T. H. Huxley, "Evolution and ethics" (The Romanes Lecture for 1893). In *Collected Essays*, 9 (London: Macmillan, 1893).

griefs that rob existence of its savour and turn its richest fruits into mere dust and ashes.

Further, I think I do not err in assuming that, however, diverse their views on philosophical and religious matters, most men are agreed that the proportion of good and evil in life may be very sensibly affected by human action. I never heard anybody doubt that the evil may be thus increased, or diminished; and it would seem to follow that good must be similarly susceptible of addition or subtraction. Finally, to my knowledge, nobody professes to doubt that, so far forth as we possess a power of bettering things, it is our paramount duty to use it and to train all our intellect and energy to this supreme service of our kind.

Hence the pressing interest of the question, to what extent modern progress in natural knowledge, and, more especially, the general outcome of that progress in the doctrine of evolution, is competent to help us in the great work of helping one another?

The propounders of what are called the "ethics of evolution," when the "evolution of ethics" would usually better express the object of their speculations, adduce a number of more or less interesting facts and more or less sound arguments, in favour of the origin of the moral sentiments, in the same way as other natural phenomena, by a process of evolution. I have little doubt, for my own part, that they are on the right track; but as the immoral sentiments have no less been evolved, there is, so far, as much natural sanction for the one as the other. The thief and the murderer follow nature just as much as the philanthropist. Cosmic evolution may teach us how the good and the evil tendencies of man may have come about; but, in itself, it is incompetent to furnish any better reason why what we call good is preferable to what we call evil than we had before. Some day, I doubt not, we shall arrive at an understanding of the evolution of the aesthetic faculty; but all the understanding in the world will neither increase nor diminish the force of the intuition that this is beautiful and that is ugly.

There is another fallacy which appears to me to pervade the so-called "ethics of evolution." It is the notion that because, on the whole, animals and plants have advanced in perfection or organization by means of the struggle for existence and the consequent "survival of the fittest"; therefore men in society, men as ethical beings, must look to the same process to help them towards perfection. I suspect that this fallacy has arisen out of the unfortunate ambiguity of the phrase "survival of the fittest." "Fittest" has a connotation of "best"; and about "best" there hangs a moral flavour. In cosmic nature, however, what is "fittest" depends upon the conditions. Long since,[1] I ventured to point out that if our hemisphere were to cool again, the survival of the fittest might bring about, in the vegetable kingdom, a population of more and more stunted and humbler and humbler organisms, until the "fittest" that survived might be nothing but lichens, diatoms, and such microscopic organisms as those which give red snow its colour; while, if it became hotter, the pleasant valleys of the Thames and Isis might be uninhabitable by any animated beings save those that flourish in a tropical jungle. They, as the fittest, the best adapted to the changed conditions, would survive.

Men in society are undoubtedly subject to the cosmic process. As among other animals, multiplication goes on without cessation, and involves severe competition for the means of support. The struggle for existence tends to eliminate those less fitted to adapt themselves to the circumstances of their existence. The strongest, the most self-assertive, tend to tread down the weaker. But the influence of the cosmic process on the evolution of society is the greater the more rudimentary its civilization. Social progress means a checking of the cosmic process at every step and the substitution for it of another, which may be called the ethical process; the end of which is not the survival of those who may happen to be the fittest, in respect of the whole of the conditions which obtain, but of those who are ethically the best.[2]

As I have already urged, the practice of that which is ethically best —what we call goodness or virtue—involves a course of conduct which, in all respects, is opposed to that which leads to success in the cosmic struggle for existence. In place of ruthless self-assertion it demands self-restraint; in place of thrusting aside, or treading down, all competitors, it requires that the individual shall not merely respect, but shall help his fellows; its influence is directed, not so much to the survival of the fittest, as to the fitting of as many as possible to survive. It repudiates the gladiatorial theory of existence. It demands that each man who enters into the enjoyment of the advantages of a polity shall be mindful of his debt to those who have laboriously constructed it; and shall take heed that no act of his weakens the fabric in which he has been permitted to live. Laws and moral precepts are directed to the end of curbing the cosmic process and reminding the individual of his duty to the community, to the protection and influence of which he owes, if not existence itself, at least the life of something better than a brutal savage.

It is from neglect of these plain considerations that the fanatical individualism[3] of our time attempts to apply the analogy of cosmic nature to society. Once more we have a misapplication of the stoical injunction to follow nature; the duties of the individual to the State are forgotten, and his tendencies to self-assertion are dignified by the name of rights. It is seriously debated whether the members of a community are justified in using their combined strength to constrain one of their number to contribute his share to the maintenance of it; or even to prevent him from doing his best to destroy it. The struggle for existence, which has done such admirable work in cosmic nature, must, it appears, be equally beneficent in the ethical sphere. Yet if that which I have insisted upon is true; if the cosmic process has no sort of relation to moral ends; if the imitation of it by man is inconsistent with the first principles of ethics; what becomes of this surprising theory?

Let us understand, once for all, that the ethical progress of society depends, not on imitating the cosmic process, still less in running away from it, but in combating it. It may seem an audacious proposal thus to pit the microcosm against the macrocosm and to set man to subdue nature to his higher ends; but I venture to think that the great intellectual difference between the ancient times with which we have been occupied and our day, lies in the solid foundation we have acquired for the hope that such an enterprise may meet with a certain measure of success.

The history of civilization details the steps by which men have succeeded in building up an artificial world within the cosmos. Fragile reed as he may be, man, as Pascal says, is a thinking reed:[4] there lies within him a fund of energy, operating intelligently and so far akin to that which pervades the universe, that it is competent to influence and modify the cosmic process. In virtue of his intelligence, the dwarf bends the Titan to his will. In every family, in every polity that has been established, the cosmic process in man has been restrained and otherwise modified by law and custom; in surrounding nature, it has been similarly influenced by the art of the shepherd, the agriculturist, the artisan. As civilization has advanced, so has the extent of this interference increased; until the organized and highly developed sciences and arts of the present day have endowed man with a command over the course of nonhuman nature greater than that once attributed to the magicians. The most impressive, I might say startling, of these changes have been brought about in the course of the last two centuries; while a right comprehension of the process of life and of the means of influencing its manifestations is only just dawning upon us. We do not yet see our way beyond generalities; and we are befogged by the obtrusion of false analogies and crude anticipations. But Astronomy, Physics, Chemistry, have all had to pass through similar phrases, before they reached the stage at which their influence became an important factor in human affairs. Physiology, Psychology, Ethics, Political Science, must submit to the same ordeal. Yet it seems to me irrational to doubt that, at no distant period, they will work as great a revolution in the sphere of practice.

The theory of evolution encourages no millennial anticipations. If, for millions of years, our globe has taken the upward road, yet, some time, the summit will be reached and the downward route will be commenced. The most daring imagination will hardly venture upon the suggestion that the power and the intelligence of man can ever arrest the procession of the great year.

Moreover, the cosmic nature born with us and, to a large extent, necessary for our maintenance, is the outcome of millions of years of severe training, and it would be folly to imagine that a few centuries will suffice to subdue its masterfulness to purely ethical ends. Ethical nature may count upon having to reckon with a tenacious and powerful enemy as long as the world lasts. But, on the other hand, I see no limit to the extent to which intelligence and will, guided by sound principles of investigation, and organized in common effort, may modify the conditions of existence, for a period longer than that now covered by history. And much may be done to change the nature of man himself.[5] The intelligence which has converted the brother of the wolf into the faithful guardian of the flock ought to be able to do something towards curbing the instincts of savagery in civilized men.

But if we may permit ourselves a larger hope of abatement of the essential evil of the world than was possible to those who, in the infancy of exact knowledge, faced the problem of existence more than a score of centuries ago, I deem it as essential condition of the realization of that hope that we should cast aside the notion that the escape from pain and sorrow is the proper object of life.

We have long since emerged from the heroic childhood of our race, when good and evil could be met with the same "frolic welcome"; the attempts to escape from evil, whether Indian or Greek, have ended in flight from the battle-field; it remains to us to throw aside the youthful over-confidence and the no less youthful discouragement of nonage. We are grown men, and must play the man

> strong in will
> To strive, to seek to find, and not to yield,

cherishing the good that falls in our way, and bearing the evil, in and around us, with stout hearts set on diminishing it. So far, we all may strive in one faith towards one hope:

> It may be that the gulfs will wash us down,
> It may be we shall touch the Happy Isles,
> . . . . but something ere the end,
> Some work of noble note may yet be done.[6]

NOTES

1. "Criticisms on the Origin of Species," 1864. *Collected Essays*, vol. ii, p. 91. [1894.]

2. Of course, strictly speaking, social life, and the ethical process in virtue of which it advances towards perfection, are part and parcel of the general process of evolution, just as the gregarious habit of innumerable plants and animals, which has been of immense advantage to them, is so. A hive of bees is an organic polity, a society in which the part played by each member is determined by organic necessities. Queens, workers, and drones are, so to speak, castes, divided from one another by marked physical barriers. Among birds and mammals, societies are formed, of which the bond in many cases seems to be purely psychological; that is to say, it appears to depend upon the liking of the individuals for one another's company. The tendency of individuals to over self-assertion is kept down by fighting. Even in these rudimentary forms of society, love and fear come into play, and enforce a greater or less renunciation of self-will. To this extent the general cosmic process begins to be checked by a rudimentary ethical process, which is, strictly speaking, part of the former,

just as the "governor" in a steam-engine is part of the mechanism of the engine.

3. See "Government: Anarchy or Regimentation," *Collected Essays*, vol. i, pp. 413–418. It is this form of political philosophy to which I conceive the epithet of "reasoned savagery" to be strictly applicable. [1894.]

4. "L'homme n'est qu'un roseau, le plus faible de la nature, mais c'est un roseau pensant. Il ne faut pas que l'univers entier s'arme pour l'écraser. Une vapeur, une goutte d'eau, suffit pour le tuer. Mais quand l'univers l'écraserait, l'homme serait encore plus noble que ce qui le tue, parce qu'il sait qu'il meurt; et l'avantage que l'univers a sur lui, l'univers n'en sait rien.'—*Pensées de Pascal.*

5. The use of the word "Nature" here may be criticized. Yet the manifestation of the natural tendencies of men is so profoundly modified by training that it is hardly too strong. Consider the suppression of the sexual instinct between near relations.

6. A great proportion of poetry is addressed by the young to the young; only the great masters of the art are capable of divining, or think it worth while to enter into, the feelings of retrospective age. The two great poets whom we have so lately lost, Tennyson and Browning, have done this, each in his own inimitable way; the one in the *Ulysses*, from which I have borrowed; the other in that wonderful fragment "Childe Roland to the dark Tower came."

30

JOHN L. MACKIE*

The Law of the Jungle: Moral Alternatives and Principles of Evolution

WHEN PEOPLE SPEAK of "the law of the jungle," they usually mean unrestrained and ruthless competition, with everyone out solely for his own advantage. But the phrase was coined by Rudyard Kipling, in *The Second Jungle Book*, and he meant something very different. His law of the jungle is a law that wolves in a pack are supposed to obey. His poem says that "the strength of the Pack is the Wolf, and the strength of the Wolf is the Pack," and it states the basic principles of social co-operation. Its provisions are a judicious mixture of individualism and collectivism, prescribing graduated and qualified rights for fathers of families, mothers with cubs, and young wolves, which constitute an elementary system of welfare services. Of course, Kipling meant his poem to give moral instruction to human children, but he probably thought it was at least roughly correct as a description of the social behaviour of wolves and other wild animals. Was he right, or is the natural world the scene of unrestrained competition, of an individualistic struggle for existence?

Views not unlike those of Kipling have been presented by some recent writers on ethology, notably Robert Ardrey and Konrad Lorenz. These writers connect their accounts with a view about the process of evolution that has brought this behaviour, as well as the animals themselves, into existence. They hold that the important thing in evolution is

*From J. L. Mackie, "The Law of the Jungle: Moral Alternatives and Principles of Evolution," *Philosophy*, Vol. 53, no. 206 (October 1978), pp. 455–464. Reprinted with the permission of Cambridge University Press.

the good of the species, or the group, rather than the good of the individual. Natural selection favours those groups and species whose members tend, no doubt through some instinctive programming, to co-operate for a common good; this would, of course, explain why wolves, for example, behave co-operatively and generously towards members of their own pack, if indeed they do.

However, this recently popular view has been keenly attacked by Richard Dawkins in his admirable and fascinating book, *The Selfish Gene*.[1] He defends an up-to-date version of the orthodox Darwinian theory of evolution, with special reference to "the biology of selfishness and altruism." One of his main theses is that there is no such thing as group selection, and that Lorenz and others who have used this as an explanation are simply wrong. This is a question of some interest to moral philosophers, particularly those who have been inclined to see human morality itself as the product of some kind of natural evolution.[2]

It is well, however, to be clear about the issue. It is not whether animals ever behave for the good of the group in the sense that this is their conscious subjective goal, that they *aim* at the well-being or survival of the whole tribe or pack: the question of motives in this conscious sense does not arise. Nor is the issue whether animals ever behave in ways which do in fact promote the well-being of the group to which they belong, or which help the species of which they are members to survive: of course they do. The controversial issue is different from both of these: it is whether the good of the group or the species would ever figure in a correct evolutionary account. That is, would any correct evolutionary account take either of the following forms?

(i) The members of this species tend to do these things which assist the survival of this species because their ancestors were members of a sub-species whose members had an inheritable tendency to do these things, and as a result that sub-species survived, whereas other sub-species of the ancestral species at that time had members who tended not to do these things and as a result their sub-species did not survive.

(ii) The members of this species tend to do these things which help the group of which they are members to flourish because some ancestral groups happened to have members who tended to do these things and these groups, as a result, survived better than related groups of the ancestral species whose members tended not to do these things.

In other words, the issue is this: is there natural selection by and for group survival or species survival as opposed to selection by and for individual survival (or, as we shall see, gene survival)? Is behaviour that helps the group or the species, rather than the individual animal, rewarded by the natural selection which determines the course of evolution?

[1]R. Dawkins, *The Selfish Gene* (Oxford, 1976).
[2]I am among these: see p. 113 of my *Ethics, Inventing Right and Wrong* (Harmondsworth, 1977).

However, when Dawkins denies that there is selection by and for group or species survival, it is not selection by and for individual survival that he puts in its place. Rather it is selection by and for the survival of each single gene — the genes being the unit factors of inheritance, the portions of chromosomes which replicate themselves, copy themselves as cells divide and multiply. Genes, he argues, came into existence right back at the beginning of life on earth, and all more complex organisms are to be seen as their products. We are, as he picturesquely puts it, gene-machines: our biological function is just to protect our genes, carry them around, and enable them to reproduce themselves. Hence the title of his book, *The Selfish Gene*. Of course what survives is not a token gene: each of these perishes with the cell of which it is a part. What survives is a gene-type, or rather what we might call a gene-clone, the members of a family of token genes related to one another by simple direct descent, by replication. The popularity of the notions of species selection and group selection may be due partly to confusion on this point. Since clearly it is only types united by descent, not individual organisms, that survive long enough to be of biological interest, it is easy to think that selection must be by and for species survival. But this is a mistake: genes, not species, are the types which primarily replicate themselves and are selected. Since Dawkins roughly defines the gene as "a genetic unit which is small enough to last for a number of generations and to be distributed around in the form of many copies," it is (as he admits) practically a tautology that the gene is the basic unit of natural selection and therefore, as he puts it, "the fundamental unit of self-interest," or, as we might put it less picturesquely, the primary beneficiary of natural selection. But behind this near-tautology is a synthetic truth, that this basic unit, this primary beneficiary, is a small bit of a chromosome. The reason why this is so, why what is differentially effective and therefore subject to selection is a small bit of a chromosome, lies in the mechanism of sexual reproduction by way of meiosis, with crossing over between chromosomes. When male and female cells each divide before uniting at fertilization, it is not chromosomes as a whole that are randomly distributed between the parts, but sections of chromosomes. So sections of chromosomes can be separately inherited, and therefore can be differentially selected by natural selection.

The issue between gene selection, individual selection, group selection, and species selection might seem to raise some stock questions in the philosophy of science. Many thinkers have favoured reductionism of several sorts, including methodological individualism. Wholes are made up of parts, and therefore in principle whatever happens in any larger thing depends upon and is explainable in terms of what happens in and between its smaller components. But though this metaphysical individualism is correct, methodological individualism does not follow from it. It does not follow that we must always conduct our investigations and construct our explanations in terms of component parts, such as the individual members of a group or society. Scientific accounts need not be indefinitely reductive. Some wholes are obviously more accessible to us than their components. We can understand what a human being does without analysing this in terms of how each single cell in his body or his brain behaves. Equally we can often understand what a human society does without analysing this in terms of the behaviour of each of its individual members. And the same holds quite generally: we can often understand complex wholes as units, without analysing them into their parts. So if, in the account of evolution, Dawkins's concentration upon genes were just a piece of methodological individualism or reductionism, it would be inadequately motivated. But it is not: there is a special reason for it. Dawkins's key argument is that species, populations, and groups, and individual organisms too, are as genetic units too temporary to qualify for natural selection. "They are not stable through evolutionary time. Populations . . . are constantly blending with other populations and so losing their identity," and, what is vitally important, "are also subject to evolutionary change from within" (p. 36).

This abstract general proposition may seem obscure. But it is illustrated by a simple example which Dawkins gives (pp. 197–201).

A species of birds is parasitized by dangerous ticks. A bird can remove the ticks from most parts of its own body, but, having only a beak and no hands, it cannot get them out of the top of its own head. But one bird can remove ticks from another bird's head: there can be mutual grooming. Clearly if there was an inherited tendency for each bird to take the ticks out of any other bird's head, this would help the survival of any group in which that tendency happened to arise — for the ticks are dangerous: they can cause death. Someone who believed in group selection would, therefore, expect this tendency to be favoured and to evolve and spread for this reason. But Dawkins shows that it would not. He gives appropriate names to the different "strategies," that is, the different inheritable behavioural tendencies. The strategy of grooming anyone who needs it he labels "Sucker." The strategy of accepting grooming from anyone, but never grooming anyone else, even someone who has previously groomed you, is called "Cheat." Now if in some population both these tendencies or strategies, and only these two, happen to arise, it is easy to see that the cheats will always do better than the suckers. They will be groomed when they need it, and since they will not waste their time pecking out other birds' ticks, they will have more time and energy to spare for finding food, attracting mates, building nests, and so on. Consequently the gene for the Sucker strategy will gradually die out. So the population will come to consist wholly of cheats, despite the fact that this is likely to lead to the population itself becoming extinct, if the parasites are common enough and dangerous enough, whereas a population consisting wholly of suckers would have survived. The fact that the group is open to evolutionary change from within, because of the way the internal competition between Cheat and Sucker genes works out, prevents the group from developing or even retaining a feature which would have helped the group as a whole.

This is just one illustration among many, and Dawkins's arguments on this point seem pretty conclusive. We need, as he shows, the concept of an *evolutionarily stable strategy* or ESS (p. 74 *et passim*). A strategy is evolutionarily stable, in relation to some alternative strategy or strategies, if it will survive indefinitely in a group in competition with those alternatives. We have just seen that where Cheat and Sucker alone are in competition, Cheat is an ESS but Sucker is not. We have also seen, from this example, that an ESS may not help a group, or the whole species, to survive and multiply. Of course we must not leap to the conclusion that an ESS never helps a group or a species: if that were so we could not explain much of the behaviour that actually occurs. Parents sacrifice themselves for their children, occasionally siblings for their siblings, and with the social insects, bees and ants and termites, their whole life is a system of communal service. But the point is that these results are not to be explained in terms of group selection. They can and must be explained as consequences of the selfishness of genes, that is, of the fact that gene-clones are selected for whatever helps each gene-clone itself to survive and multiply.

But now we come to another remarkable fact. Although the gene is the hero of Dawkins's book, it is not unique either in principle or in fact. It is not the only possible subject of evolutionary natural selection, nor is it the only actual one. What is important about the gene is just that it has a certain combination of logical features. It is a replicator: in the right environment it is capable of producing multiple copies of itself; but in this process of copying some mistakes occur; and these mistaken copies — mutations — will also produce copies of themselves; and, finally, the copies produced may either survive or fail to survive. Anything that has these formal, logical, features is a possible subject of evolution by natural selection. As we have seen, individual organisms, groups, and species do not have the required formal features, though many thinkers have supposed that they do. They cannot reproduce themselves with sufficient constancy of characteristics. But Dawkins, in his last chapter, introduces another sort of replicators. These are what are often called cultural items or traits; Dawkins christens them *memes* —to make a term a bit like "genes" — because they replicate by memory and imitation (mimesis). Memes include tunes, ideas, fashions, and techniques. They require, as the environment in which they can replicate, a collection of minds, that is, brains that have the powers of imitation and memory. These brains (particularly though not exclusively human ones) are themselves the products of evolution by gene selection. But once the brains are there gene selection has done its work: given that environment, memes can themselves evolve and multiply in much the same way as genes do, in accordance with logically similar laws. But they can do so more quickly. Cultural evolution may be much faster than biological evolution. But the basic laws are the same. Memes are selfish in the same sense as genes. The explanation of the wide-spread flourishing of a certain meme, such as the idea of a god or the belief in hell fire, may be simply that it is an efficiently selfish meme. Something about it makes it well able to infect human minds, to take root and spread in and among them, in the same way that something about the smallpox virus makes it well able to take root and spread in human bodies. There is no need to explain the success of a meme in terms of any benefit it confers on individuals or groups; it is a replicator in its own right. Contrary to the optimistic view often taken of cultural evolution, this analogy shows that a cultural trait can evolve, not because it is advantageous to society, but simply because it is advantageous to itself. It is ironical that Kipling's phrase "the law of the jungle" has proved itself a more efficient meme than the doctrine he tried to use it to propagate.

So far I have been merely summarizing Dawkins' argument. We can now use it to answer the question from which I started. Who is right about the law of the jungle? Kipling, or those who have twisted his phrase to mean almost the opposite of what he intended? The answer is that neither party is right. The law by which nature works is not unrestrained and ruthless competition between individual organisms. But neither does it turn upon the advantages to a group, and its members, of group solidarity, mutual care and respect, and co-operation. It turns upon the self-preservation of gene-clones. This has a strong tendency to express itself in individually selfish behaviour, simply because each agent's genes are more certainly located in him than in anyone else. But it can and does express itself also in certain forms of what Broad called self-referential altruism, including special care for one's own children and perhaps one's siblings, and, as we shall see, reciprocal altruism, helping those (and only those) who help you.

But now I come to what seems to be an exception to Dawkins's main thesis, though it is generated by his own argument and illustrated by one of his own examples. We saw how, in the example of mutual grooming, if there are only suckers and cheats around, the strategy Cheat is evolutionarily stable, while the strategy Sucker is not. But Dawkins introduces a third strategy, Grudger. A grudger is rather like you and me. A grudge grooms anyone who has previously groomed him, and any stranger, but he remembers and bears a grudge against anyone who cheats him—who refuses to groom him in return for having been groomed—and the grudger refuses to groom the cheat ever again. Now when all three strategies are in play, both Cheat and Grudger are evolutionarily stable. In a population consisting largely of cheats, the cheats will do better than the others, and both suckers and grudgers will die out. But in a population that starts off with more than a certain critical proportion of grudgers, the cheats will first wipe out the suckers, but will then themselves become rare and eventually extinct: cheats can flourish only while they have suckers to take advantage of, and yet by doing so they tend to eliminate those suckers.

It is obvious, by the way, that a population containing only suckers and grudgers, in any proportions, but no cheats, would simply continue as it was. Suckers and grudgers behave exactly like one another as long as there are not cheats around, so there would be no tendency for either the Sucker or the Grudger gene to do better than the other. But if there is any risk of an invasion of Cheat genes, either through mutation or through immigration, such a pattern is not evolutionarily stable, and the higher the proportion of suckers, the more rapidly the cheats would multiply.

So we have two ESSs, Cheat and Grudger. But there is a difference between these two stable strategies. If the parasites are common enough and dangerous enough, the population of cheats will itself die out, having no defence against ticks in their heads, whereas a separate population of grudgers will flourish indefinitely. Dawkins says, "If a population arrives at an ESS which drives it extinct, then it goes extinct, and that is just too bad" (p. 200). True: *but is this not group selection after all?* Of course, this will operate only if the populations are somehow isolated. But if the birds in question were distributed in geographically isolated regions, and Sucker, Cheat and Grudger tendencies appeared (after the parasites became plentiful) in randomly different proportions in these different regions, then some populations would become pure grudger populations, and others would become pure cheat populations, but then the pure cheat populations would die out, so that eventually all surviving birds would be grudgers. And they would be able to re-colonize the areas where cheat populations had perished.

Another name for grudgers is "reciprocal altruists." They act as if on the maxim "Be done by as you did." One implication of this story is that this strategy is not only evolutionarily stable within a population, it is also viable for a population as a whole. The explanation of the final situation, where all birds of this species are grudgers, lies partly in the non-viability of a population of pure cheats. So this is, as I said, a bit of group selection after all.

It is worth noting how and why this case escapes Dawkins's key argument that a population is "not a discrete enough entity to be a unit of natural selection, not stable and unitary enough to be "selected" in preference to another population" (p. 36). Populations can be made discrete by geographical (or other) isolation, and can be made stable and unitary precisely by the emergence of an ESS in each, but perhaps different ESSs in the different regional populations of the same species. This case of group selection is necessarily a second order phenomenon: it arises where gene selection has produced the ESSs which are then persisting selectable features of groups. In other words, an ESS may be a third variety of replicator, along with genes and memes; it is a self-re-producing feature *of groups*.

Someone might reply that this is not really group selection because it all rests ultimately on gene selection, and a full explanation can be given in terms of the long-run self-extinction of the Cheat gene, despite the fact that within a population it is evolutionarily stable in competition with the two rival genes. But this would be a weak reply. The monopoly of cheating *over a population* is an essential part of the causal story that explains the extinction. Also, an account at the group level, though admittedly incomplete, is here correct as far as it goes. The reason why all ultimately surviving birds of this species are grudgers is partly that *populations* of grudgers can survive whereas *populations* of cheats cannot, though it is also partly that although a population of suckers could survive — it would be favoured by group selection, if this possibility arose, just as much as a population of grudgers — internal changes due to gene selection after an invasion of Cheat genes would prevent there being a population of suckers. In special circumstances group selection (or population selection) can occur and could be observed and explained as such, without going down to the gene selection level. It would be unwarranted methodological individualism or reductionism to insist that we not merely can but must go down to the gene selection level here. We must not fall back on this weak general argument when Dawkins's key argument against group selection fails.

I conclude, then, that there can be genuine cases of group selection. But I admit that they are exceptional. They require rather special conditions, in particular geographical isolation, or some other kind of isolation, to keep the populations that are being differentially selected apart. For if genes from one could infiltrate another, the selection of populations might be interfered with. (Though in fact in our example *complete* isolation is not required: since what matters is whether there is more or less than a certain critical proportion of grudgers, small-scale infiltrations would only delay, not prevent, the establishing of pure populations.) And since special conditions are required, there is no valid general principle that features which would enable a group to flourish will be selected. And even these exceptional cases conform thoroughly to the general logic of Dawkins's doctrine. Sometimes, but only sometimes, group characteristics have the formal features of replicators that are open to natural selection.

Commenting on an earlier version of this paper, Dawkins agreed that there could be group selection in the sort of case I suggested, but stressed the importance of the condition of geographical (or other) isolation. He also mentioned a possible example, that the prevalence of sexual reproduction itself may be a result of group selection. For if there were a mutation by which asexual females, producing offspring by parthenogenesis, occurred in a species, this clone of asexual females would be at once genetically isolated from the rest of the species, though still geographically mixed with them. Also, in most species males contribute little to the nourishment or care of their offspring, so from a genetic point of view males are wasters: resources would be more economically used if devoted only to females. So the genetically isolated population of asexual females would out-compete the normal sexually reproducing population with roughly equal numbers of males and females. So that species would in time consist only of asexual females. But then, precisely because all its members were genetically identical, it would not have the capacity for rapid adaptation by selection to changing conditions that an ordinary sexual population has. So when conditions changed, it would be unable to adapt, and would die out. Thus there would in time be species selection against any species that produced an asexual female mutation. Which would explain why nearly all existing species go in for what, in the short run, is the economically wasteful business of sexual reproduction.[3]

What implications for human morality have such biological facts about selfishness and altruism? One is that the possibility that morality is itself a product of natural selection is not ruled out, but care would be needed in formulating a plausible speculative account of how it might have been favoured. Another is that the notion of an ESS may be a useful one for discussing questions of practical morality. Moral philosophers have already found illumination in such simple items of game theory as the Prisoners' Dilemma; perhaps these rather more complicated evolutionary "games" will prove equally instructive. Of course there is no simple transition from "is" to "ought," no direct argument from what goes on in the natural world and among non-human animals to what human beings ought to do. Dawkins himself explicitly warns against any simple transfer of conclusions. At the very end of the book he suggests that conscious foresight may enable us to develop radically new kinds of behaviour. "We are built as gene machines and cultured as meme machines, but we have the power to turn against our creators. We, alone on earth, can rebel against the tyranny of the selfish replicators" (p. 215). This optimistic suggestion needs fuller investigation. It must be remembered that the human race as a whole cannot act as a unit with conscious foresight. Arrow's Theorem shows that even quite small groups of rational individuals may be unable to form coherently rational preferences, let alone to act rationally. Internal competition, which in general prevents a group from being a possible subject of natural selection, is even more of an obstacle to its being a rational agent. And while we can turn against some memes, it will be only with the help and under the guidance of other memes.

This is an enormous problematic area. For the moment I turn to a smaller point. In the mutual grooming model, we saw that the Grudger strategy was, of the three strategies considered, the only one that was

---

[3]This suggestion is made in a section entitled "The paradox of sex and the cost of paternal neglect" of the following article: R. Dawkins, "The Value Judgments of Evolution," in *Animal Economics*, edited by M.A.H. Dempster and D.J. McFarland (London and New York, forthcoming).

healthy in the long run. Now something closely resembling this strategy, reciprocal altruism, is a well known and long established tendency in human life. It is expressed in such formulae as that justice consists in giving everyone his due, interpreted, as Polemarchus interprets it in the first book of Plato's *Republic*, as doing good to one's friends and harm to one's enemies, or repaying good with good and evil with evil. Morality itself has been seen, for example by Edward Westermarck, as an outgrowth from the retributive emotions. But some moralists, including Socrates and Jesus, have recommended something very different from this, turning the other cheek and repaying evil with good. They have tried to substitute "Do as you would be done by" for "Be done by as you did." Now this, which in human life we characterize as a Christian spirit or perhaps as saintliness, is roughly equivalent to the strategy Dawkins has unkindly labelled "Sucker." Suckers are saints, just as grudgers are reciprocal altruists, while cheats are a hundred per cent selfish. And as Dawkins points out, the presence of suckers endangers the healthy Grudger strategy. It allows cheats to prosper, and could make them multiply to the point where they would wipe out the grudgers, and ultimately bring about the extinction of the whole population. This seems to provide fresh support for Nietzsche's view of the deplorable influence of moralities of the Christian type. But in practice there may be little danger. After two thousand years of contrary moral teaching, reciprocal altruism is still dominant in all human societies; thoroughgoing cheats and thoroughgoing saints (or suckers) are distinctly rare. The sucker slogan is an efficient meme, but the sucker behaviour pattern far less so. Saintliness is an attractive topic for preaching, but with little practical persuasive force. Whether in the long run this is to be deplored or welcomed, and whether it is alterable or not, is a larger question. To answer it we should have carefully to examine our specifically human capacities and the structure of human societies, and also many further alternative strategies. We cannot simply apply to the human situation conclusions drawn from biological models. Nevertheless they are significant and challenging as models; it will need to be shown how and where human life diverges from them.

31

MICHAEL RUSE AND EDWARD O. WILSON*

The Evolution of Ethics

ATTEMPTS TO LINK evolution and ethics first sprang up in the middle of the last century, as people turned to alternative foundations in response to what they perceived as the collapse of Christianity. If God does not stand behind the Sermon on the Mount, then what does? Such attempts at evolutionary ethicising became known collectively as "social Darwinism," although they owed less to Charles Darwin and more to that quintessentially Victorian man of ideas, Herbert Spencer. Finding worth in what he perceived to be the upward progress of evolution from amoeba to human, from savage to *Homo britannicus*, Spencer argued that right conduct lies in the cherishing of the evolutionary process, in order that the best or fittest be able to survive and the inadequate be rigorously eliminated.

While Spencer's ideas attracted strong support in some quarters, for example the North American barons of industry, evolutionary ethics in this mode never really caught fire. On the one hand, social Darwinism seems so immoral! Right conduct surely cannot entail stamping on widows and babies. And no amount of tinkering by revisionists, such as Prince Peter Kroptkin in the last century and Sir Julian Huxley and C. H. Waddington in this, changes the fact. On the other hand, the very basis of a Spencerian-type approach is shaky. There is no progress to evolution. In a purely Darwinian sense, an amoeba is as good as a person.

*From Michael Ruse and Edward O. Wilson, "The Evolution of Ethics," *New Scientist,* Vol. 17 (October 1985), pp. 50–52. By permission of the publisher.

Most people, therefore, have happily agreed with the 18th-century philosopher David Hume that there is an impassible gulf between matters of fact (for example, evolution) and matters of morality (disinterested help of others). To use phrasing made popular in this century by the Cambridge philosopher G. E. Moore, evolutionary ethics commits "the naturalistic fallacy" by trying to translate *is* into *ought*.

It is true that past efforts to create an evolutionary ethics have come to very little. Yet to revert to the opposite conclusion, that evolution and ethics have nothing to say to each other, is altogether too quick. Recent advances in evolutionary theory have cast a new light on the matter, giving substance to the dreams of the old theorisers, although not in the way or for the reasons they thought.

Our starting point is with the science. Two propositions appear to have been established beyond any reasonable doubt. First, the social behaviour of animals is firmly under the control of the genes, and has been shaped into forms that give reproductive advantages. Secondly, humans are animals. Darwin knew that the first claim was true, and a multitude of recent studies, from fruit flies to frogs, have affirmed it repeatedly. Darwin knew also that the second claim is true, and positive evidence continues to pour in from virtually every biological discipline. Genetically, we are a sibling species to the chimpanzee, having evolved with them for more than $3\frac{1}{2}$ billion years, parting a mere 6 million or so years ago.

What do these facts have to do with morality? A chain of reasoning leads us to a distinctly human but still biologically based ethical sense. First, note that we are not just talking about behaviour, but about *social* behaviour. Today's students of this subject, sociobiologists, know that it is often in an individual's biological self-interest to cooperate with its fellows, rather than (as traditional evolutionary ethicists thought) to fight flat out. After all, a loaf shared is better than a whole loaf, if the latter carries the risk of being killed or seriously hurt.

Secondly, and less obviously, there are ways in which nature can bring about "altruism," in the sense of self-sacrifice for the benefit of others. If those benefited are relatives, the altruist is still favouring genes identical to his own, even if he dies without leaving any direct offspring. Thus we say that the individual is altruistic but his genes are "selfish." Note that such behaviour implies nothing about good intentions or other ways of being "nice." To get altruism you can go the way of the ants. They are genetically hardwired, performing their duties in perfect cooperative harmony. They have no thoughts, at least of a human kind, only actions. Alternatively, you could go to the other extreme, and evolve super-brains, where every possible action is first weighed and assessed, and a policy of rationally assessed self-interest is always followed.

Neither of these options has proved attractive to animals like humans, and we have avoided both. If we had become hardwired in the course of evolution, we could never deviate from our course. Were something untoward to happen, we would be stuck with maladaptive behaviour. Worker ants are relatively cheap to produce, so this rigidity matters relatively little to their colonies. Humans require a great deal of parental investment, and it would be stupid in the literal sense of the word if we were to go wrong at the slightest environmental quiver. Alternatively, if we possessed super-brains, we would require even more resources than we do now; such as parental care stretched over many more years. Additionally, like those chess machines that survey every move, we would be forever making up our minds. Crises would be upon us, and we would still be thinking.

Nature's Moral Imperative

How then has nature made humans "altruistic"? The clue lies in the chess machines we just mentioned. The new breed, those that can beat grandmasters, forgo omnipotence for utility. They follow certain strategies that have proved successful. So with humans. Our minds are not tabulae rasae, but moulded according to certain innate dispositions. These dispositions, known technically as "epigenetic rules," incline us to particular courses of action, such as learning rapidly to fear heights and snakes, although they certainly do not lock us, ant-like, into undeviating behaviour.

The best studied epigenetic rules, such as those affecting fears or the avoidance of incest, appear to have been put into place because of their biological virtues. Although altruism is less well documented (there is some evidence, for example, that varying degrees of its expression have a genetic component), such behaviour is also adaptive — at least when directed in appropriate measure toward kin and allies. We need to be altruistic. Thus, we have rules inclining us to such courses of behaviour. The key question is then: how are these rules expressed in our conscious awareness? We need something to spur us against our usual selfish dispositions. Nature, therefore, has made us (via the rules) believe in a disinterested moral code, according to which we *ought* to help our fellows. Thus, we are inclined to go out and work with our fellows. In short, to make us altruistic in the adaptive, biological sense, our biology makes us altruistic in the more conventionally understood sense of acting on deeply held beliefs about right and wrong.

Such is the modern scientific account of morality; at least the one most consistent with biology. But, what has any of this to do with the concerns of the traditional evolutionary ethicist? Even if the explanation were proved to be entirely true, it does not reveal whether in some ultimate, absolute sense, evolution stands behind morality. Does the sociobiological scenario just sketched justify the same moral code that religionists believe to be decreed by God? Or that some philosophers believe to exist apart from humanity, like a mathematical theorem?

It used to be thought, in the bad old days of social Darwinism when evolution was poorly understood, that life is an uninterrupted struggle —"nature red in tooth and claw." But this is only one side of natural selection. What we have just seen is that the same process also leads to altruism and reciprocity in highly social groups. Thus the human species has evolved genuine sentiments of obligation, of the duty to be loving and kind. In no way does this materialist explanation imply that we are hypocrites consciously trying to further our biological ends and paying lip-service to ethics. We function better because we believe. In this sense, evolution is consistent with conventional views of morality.

On the other hand, the question of ultimate foundations requires a different and more subtle answer. As evolutionists, we see that no justification of the traditional kind is possible. Morality, or more strictly our belief in morality, is merely an adaptation put in place to further our reproductive ends. Hence the basis of ethics does not lie in God's will—or in the metaphorical roots of evolution or any other part of the framework of the Universe. In an important sense, ethics as we understand it is an illusion fobbed off on us by our genes to get us to cooperate. It is without external grounding. Ethics is produced by evolution but not justified by it, because, like Macbeth's dagger, it serves a powerful purpose without existing in substance.

In speaking thus of illusion, we are not saying that ethics is nothing, and should now be thought of as purely dreamlike. Unlike Macbeth's dagger, ethics is a *shared* illusion of the human race. If it were not so, it would not work. The moral ones among us would be outbred by the immoral. For this reason, since all human beings are dependent on the "ethics game," evolutionary reasoning emphatically does not lead to moral relativism. Human minds develop according to epigenetic rules that distinguish between proper moral claims like "Be kind to children" and crazy imperatives like "Treat cabbages with the respect you show your mother."

Ethical codes work because they drive us to go against our selfish day-to-day impulses in favour of long-term group survival and harmony and thus, over our lifetimes, the multiplication of our genes many times. Furthermore, the way our biology enforces its ends is by making us think that there is an objective higher code, to which we are all subject. If we thought ethics to be no more than a question of personal desires, we would tend to ignore it. Why should we base our life's plan on your love of French cuisine? Because we think that ethics is objectively based, we are inclined to obey moral rules. We help small children because it is right even though it is personally inconvenient to us.

If this perception of human evolution is correct, it provides a new basis for moral reasoning. Ethics is seen to have a solid foundation, not in divine guidance or pure moral imperatives, but in the shared qualities of human nature and the desperate need for reciprocity. The key is the deeper, more objective study of human nature, and for this reason we need to turn ethical philosophy into an applied science.

Some philosophers have argued that even if ethics could be explained wholly in such a materialist fashion, this alone would not eliminate the possibility that moral imperatives exist, sitting apart like mathematical truths. Perhaps human evolution is moving toward such celestial perfection, and the apprehension of such truths. There are biological reasons for seeing and hearing the moving train, but it still exists!

Unfortunately, the cases of mathematical principles, material objects and ethics are not parallel. Natural selection is above all opportunistic. Suppose that, instead of evolving from savannah-dwelling primates, we had evolved in a very different way. If, like the termites, we needed to dwell in darkness, eat each other's faeces and cannibalise the dead, our epigenetic rules would be very different from what they are now. Our minds would be strongly prone to extol such acts as beautiful and moral. And we would find it morally disgusting to live in the open air, dispose of body waste and bury the dead. Termite ayatollahs would surely declare such things to be against the will of God. Termite social theorists would surely argue for a stricter caste system.

Ethics does not have the objective foundation our biology leads us to think it has. But this is no negative conclusion. Human beings face incredible social problems, primarily because their biology cannot cope with the effects of their technology. A deeper understanding of this biology is surely a first step towards solving some of these pressing worries. Seeing morality for what it is, a legacy of evolution rather than a reflection of eternal, divinely inspired verities, is part of this understanding.

GOD AND BIOLOGY

32

The First Book of Moses, called Genesis

Chapter 1

IN THE BEGINNING God created the heaven and the earth.

2 And the earth was without form, and void; and darkness was upon the face of the deep. And the Spirit of God moved upon the face of the waters.

3 And God said, Let there be light: and there was light.

4 And God saw the light, that it was good: and God divided the light from the darkness.

5 And God called the light Day, and the darkness he called Night. And the evening and the morning were the first day.

6 And God said, Let there be a firmament in the midst of the waters, and let it divide the waters from the waters.

7 And God made the firmament, and divided the waters which were under the firmament from the waters which were above the firmament: and it was so.

8 And God called the firmament Heaven. And the evening and the morning were the second day.

9 And God said, Let the waters under the heaven be gathered together unto one place, and let the dry land appear: and it was so.

10 And God called the dry land Earth; and the gathering together of the waters called he Seas: and God saw that it was good.

11 And God said, Let the earth bring forth grass, the herb yielding seed, and the fruit tree yielding fruit after his kind, whose seed is in itself, upon the earth: and it was so.

12 And the earth brought forth grass, and herb yielding seed after his kind, and the tree yielding fruit, whose seed was in itself, after his kind: and God saw that it was good.

13 And the evening and the morning were the third day.

14 And God said, Let there be lights in the firmament of the heaven to divide the day from the night; and let them be for signs, and for seasons, and for days, and years:

15 And let them be for lights in the firmament of the heaven to give light upon the earth: and it was so.

16 And God made two great lights; the greater light to rule the day, and the lesser light to rule the night: he made the stars also.

17 And God set them in the firmament of the heaven to give light upon the earth,

18 And to rule over the day and over the night, and to divide the light from the darkness: and God saw that it was good.

19 And the evening and the morning were the fourth day.

20 And God said, Let the waters bring forth abundantly the moving creature that hath life, and fowl that may fly above the earth in the open firmament of heaven.

21 And God created great whales, and every living creature that moveth, which the waters brought forth abundantly, after their kind, and every winged fowl after his kind: and God saw that it was good.

22 And God blessed them, saying, Be fruitful, and multiply, and fill the waters in the seas, and let fowl multiply in the earth.

23 And the evening and the morning were the fifth day.

24 And God said, Let the earth bring forth the living creature after his kind, cattle, and creeping thing, and beast of the earth after his kind: and it was so.

25 And God made the beast of the earth after his kind, and cattle after their kind, and every thing that creepeth upon the earth after his kind: and God saw that it was good.

26 And God said, Let us make man in our image, after our likeness: and let them have dominion over the fish of the sea, and over the fowl of the air, and over the cattle, and over all the earth, and over every creeping thing that creepeth upon the earth.

27 So God created man in his own image, in the image of God created he him; male and female created he them.

28 And God blessed them, and God said unto them, Be fruitful, and multiply, and replenish the earth, and subdue it: and have dominion over the fish of the sea, and over the fowl of the air, and over every living thing that moveth upon the earth.

29 And God said, Behold, I have given you every herb bearing seed, which is upon the face of all the earth, and every tree, in the which is the fruit of a tree yielding seed; to you it shall be for meat.

30 And to every beast of the earth, and to every fowl of the air, and to every thing that creepeth upon the earth, wherein there is life, I have given every green herb for meat: and it was so.

31 And God saw every thing that he had made, and, behold, it was very good. And the evening and the morning were the sixth day.

Chapter 2

Thus the heavens and the earth were finished, and all the host of them.

2 And on the seventh day God ended his work which he had made; and he rested on the seventh day from all his work which he had made.

3 And God blessed the seventh day, and sanctified it: because that in it he had rested from all his work which God created and made.

4 These are the generations of the heavens and of the earth when they were created, in the day that the LORD God made the earth and the heavens,

5 And every plant of the field before it was in the earth, and every herb of the field before it grew: for the LORD God had not caused it to rain upon the earth, and there was not a man to till the ground.

6 But there went up a mist from the earth, and watered the whole face of the ground.

7 And the LORD God formed man of the dust of the ground, and breathed into his nostrils the breath of life; and man became a living soul.

8 And the LORD God planted a garden eastward in Eden; and there he put the man whom he had formed.

9 And out of the ground made the LORD God to grow every tree that is pleasant to the sight, and good for food; the tree of life also in the midst of the garden, and the tree of knowledge of good and evil.

10 And a river went out of Eden to water the garden; and from thence it was parted, and became into four heads.

11 The name of the first is Pison: that is it which compasseth the whole land of Hăv'-ĭ-läh, where there is gold;

12 And the gold of that land is good: there is bdellium and the onyx stone.

13 And the name of the second river is Gī'-hŏn: the same is it that compasseth the whole land of Ethiopia.

14 And the name of the third river is Hĭd'-dĕ-kĕl: that is it which goeth toward the east of Assyria. And the fourth river is Eû-phrā'-tĕś.

15 And the LORD God took the man, and put him into the garden of Eden to dress it and to keep it.

16 And the LORD God commanded the man, saying, Of every tree of the garden thou mayest freely eat:

17 But the tree of the knowledge of good and evil, thou shalt not eat of it: for in the day that thou eatest thereof thou shalt surely die.

18 And the LORD God said, It is not good that the man should be alone; I will make him an help meet for him.

19 And out of the ground the LORD God formed every beast of the field, and every fowl of the air; and brought them unto Adam to see

what he would call them: and whatsoever Adam called every living creature, that was the name thereof.

20 And Adam gave names to all cattle, and to the fowl of the air, and to every beast of the field; but for Adam there was not found an help meet for him.

21 And the Lord God caused a deep sleep to fall upon Adam, and he slept: and he took one of his ribs, and closed up the flesh instead thereof;

22 And the rib, which the Lord God had taken from man, made he a woman, and brought her unto the man.

23 And Adam said, This is now bone of my bones, and flesh of my flesh: she shall be called Woman, because she was taken out of Man.

24 Therefore shall a man leave his father and his mother, and shall cleave unto his wife: and they shall be one flesh.

25 And they were both naked, the man and his wife, and were not ashamed.

33

⚬

Act 590 of 1981, General Acts, 73rd General Assembly, State of Arkansas

AN ACT TO REQUIRE BALANCED TREATMENT OF CREATION – SCIENCE AND EVOLUTION – SCIENCE IN PUBLIC SCHOOLS; TO PROTECT ACADEMIC FREEDOM BY PROVIDING STUDENT CHOICE; TO ENSURE FREEDOM OF RELIGIOUS EXERCISE; TO GUARANTEE FREEDOM OF BELIEF AND SPEECH; TO PREVENT ESTABLISHMENT OF RELIGION; TO PROHIBIT RELIGIOUS IN-STRUCTION CONCERNING ORIGINS; TO BAR DISCRIMINATION ON THE BASIS OF CREATIONIST OR EVOLUTIONIST BELIEF; TO PROVIDE DEFINITIONS AND CLARIFICATIONS; TO DECLARE THE LEGISLATIVE PURPOSE AND LEGISLATIVE FINDINGS OF FACT; TO PROVIDE FOR SEVERABILITY OF PROVISIONS; TO PROVIDE FOR REPEAL OF CONTRARY LAWS; AND TO SET FORTH AN EFFECTIVE DATE.

Be It Enacted by the General Assembly of the State of Arkansas:
SECTION 1. *Requirement for Balanced Treatment*. Public Schools within this State shall give balanced treatment to creation – science and to evolution – science. Balanced treatment to these two models shall be given in classroom lectures taken as a whole for each course, in text-book materials taken as a whole for the sciences and taken as a whole for the humanities, and in other educational programs in public schools, to the extent that such lectures, textbooks, library materials, or educational programs deal in any way with the subject of the origin of man, life, the earth, or the universe.

SECTION 2. *Prohibition against Religious Instruction.* Treatment of either evolution – science or creation – science shall be limited to scientific evidence for each model and inferences from those scientific evidences, and must not include any religious instruction or references to religious writings.

SECTION 3. *Requirement for Nondiscrimination.* Public schools within this State, or their personnel, shall not discriminate, by reducing a grade of a student or by singling out and making public criticism, against any student who demonstrates a satisfactory understanding of both evolution – science and creation – science and who accepts or rejects either model in whole or part.

SECTION 4. *Definitions.* As used in this Act:

(a) "Creation – science" means the scientific evidences for creation and inferences from those scientific evidences. Creation – science includes the scientific evidences and related inferences that indicate: (1) Sudden creation of the universe, energy, and life from nothing; (2) The insufficiency of mutation and natural selection in bringing about development of all living kinds from a single organism; (3) Changes only within fixed limits of originally created kinds of plants and animals; (4) Separate ancestry for man and apes; (5) Explanation of the earth's geology by catastrophism, including the occurrence of a worldwide flood; and (6) A relatively recent inception of the earth and living kinds.

(b) "Evolution – science" means the scientific evidences for evolution and inferences from those scientific evidences. Evolution – science includes the scientific evidences and related inferences that indicate (1) Emergence by naturalistic processes of the universe from disordered matter and emergence of life from nonlife; (2) The sufficiency of mutation and natural selection in bringing about development of present living kinds from simple earlier kinds; (3) Emergence by mutation and natural selection of present living kinds from simple earlier kinds; (4) Emergence of man from a common ancestor with apes; (5) Explanation of the earth's geology and the evolutionary sequence by uniformitarianism; and (6) An inception several billion years ago of the earth and somewhat later of life.

(c) "Public schools" mean secondary and elementary schools.

SECTION 5. *Classification.* This Act does not require or permit instruction in any religious doctrine or materials. This Act does not require any instruction in the subject of origins, but simply requires instruction in both scientific models (of evolution – science and creation–science) if public schools choose to teach either. This **Act** does not require each individual textbook or library book to give balanced treatment to the models of evolution–science and creation–science; it does not require any school books to be discarded. This Act does not require each individual classroom lecture in a course to give such balanced treatment, but simply requires the lectures as a whole to give balanced treatment; it permits some lectures to present evolution–science and other lectures to present creation–science.

SECTION 6. *Legislative Declaration of Purpose.* This Legislature enacts this Act for public schools with the purpose of protecting academic freedom for students' differing values and beliefs; ensuring neutrality toward students' diverse religious convictions; ensuring freedom of religious exercise for students and their parents; guaranteeing freedom of belief and speech for students; preventing establishment of Theologically Liberal, Humanist, Nontheist, or Atheist religions; preventing discrimination against students on the basis of their personal beliefs concerning creation and evolution; and assisting students in their search for truth. This Legislature does not have the purpose of causing instruction in religious concepts or making an establishment of religion.

SECTION 7. *Legislative Findings of Fact.* This Legislature finds that:

(a) The subject of the origin of the universe, earth, life, and man is treated within many public school courses, such as biology, life science, anthropology, sociology, and often also in physics, chemistry, world history, philosophy, and social studies.

(b) Only evolution–science is presented to students in virtually all of those courses that discuss the subject of origins. Public schools generally censor creation–science and evidence contrary to evolution.

(c) Evolution–science is not an unquestionable fact of science, because evolution cannot be experimentally observed, fully verified, or logically falsified, and because evolution–science is not accepted by some scientists.

(d) Evolution–science is contrary to the religious convictions or moral values or philosophical beliefs of many students and parents, including individuals of many different religious faiths and with diverse moral values and philosophical beliefs.

(e) Public school presentation of only evolution–science without any alternative model of origins abridges the United States Constitution's protections of freedom of religious exercise and of freedom of belief and speech for students and parents, because it undermines their religious convictions and moral or philosophical values, compels their unconscionable professions of belief and hinders religious training and moral training by parents.

(f) Public school presentation of only evolution – science furthermore abridges the Constitution's prohibition against establishment of religion, because it produces hostility toward many Theistic religions and brings preference to Theological Liberalism, Humanism, Nontheistic religions, and Atheism, in that these religious faiths generally include a religious belief in evolution.

(g) Public school instruction in only evolution – science also violates the principle of academic freedom, because it denies students a choice between scientific models and instead indoctrinates them in evolution – science alone.

(h) Presentation of only one model rather than alternative scientific models of origins is not required by any compelling interest of the State, and exemption of such students from a course or class presenting only evolution – science does not provide an adequate remedy because of teacher influence and student pressure to remain in that course or class.

(i) Attendance of those students who are at public schools is compelled by law, and school taxes from their parents and other citizens are mandated by law.

(j) Creation – science is an alternative scientific model of origins and can be presented from a strictly scientific standpoint without any religious doctrine just as evolution – science can, because there are scientists who conclude that scientific data best support creation – science and because scientific evidences and inferences have been presented for creation – science.

(k) Public school presentation of both evolution – science and creation – science would not violate the Constitution's prohibition against establishment of religion, because it would involve presentation of the scientific evidences and related inferences for each model rather than any religious instruction.

(l) Most citizens, whatever their religious beliefs about origins, favor balanced treatment in public schools of alternative scientific models of origins for better guiding students in their search for knowledge, and they favor a neutral approach toward subjects affecting the religious and moral and philosophical convictions of students.

SECTION 8. *Short Title.* This Act shall be known as the "Balanced Treatment for Creation–Science and Evolution–Science Act."

SECTION 9. *Severability of Provisions.* If any provision of this Act is held invalid, that invalidity shall not affect other provisions that can be applied in the absence of the invalidated provisions, and the provisions of this Act are declared to be severable.

SECTION 10. *Repeal of Contrary Laws.* All State laws or parts of State laws in conflict with this Act are hereby repealed.

SECTION 11. *Effective Date.* The requirements of the Act shall be met by and may be met before the beginning of the next school year if that is more than six months from the date of enactment, or otherwise one year after the beginning of the next school year, and in all subsequent school years.

Signed on 19 March 1981 by Governor Frank White.

34

MICHAEL RUSE*

Creation Science: The Ultimate Fraud

CHARLES DARWIN WAS an undergraduate at Cambridge from 1828 to 1831. While there, he made friends with some of Britain's most influential scientists: John Henslow, the botanist, William Whewell, physicist and general man of science, and most particularly Adam Sedgwick, the best field geologist of his day. All three of these men were more than mere scientists. They were Anglican clergymen, and deeply committed Christians. Expectedly, therefore, there were strong connections between their science and their faith. For instance, they believed that all living creatures were first put here on Earth through the direct miraculous intervention of an all-powerful God.

But although these scientists mixed their religion and their science, at the time that Darwin went to university—a full 30 years before he was to publish *The Origin of Species*—they (and other active scientists) had gone beyond a crude belief in the literal truth of the *Bible*. Sedgwick openly confessed that he had given up his earlier support of a universal flood. The geological record speaks against it. And he and the others were starting to think in terms of a very old Earth. How old, no one dared say; but it certainly went back way before the traditional starting date of 4004BC—a figure calculated by the 17th-century Archbishop Ussher from the genealogies given in the *Bible*.

Given this pre-Victorian renunciation of strict Biblical literalism,

*From Michael Ruse, "Creation Science: The Ultimate Fraud," *Darwin Up To Date*, edited by Jeremy Cherfas (London: New Science Publications, 1982), pp. 7–11. By permission of the publisher.

Genesis was obviously in no position, after the appearance of *The Origin* in 1859, to mount a major threat to the advance of science in Britain. Philip Gosse, clinging desperately to his Plymouth Brethren literalist beliefs in suggesting that fossils were put in the rocks by God to test our faith, was laughed at by all. Most importantly, no concerted attempts were made to teach *Genesis* as science in the state schools. Indeed, some of those most fervently involved in the development of state education—T. H. Huxley for instance—were also those most ardently pushing evolutionism.

In the Deep South of the United States, matters were otherwise. People look to simplistic, comforting doctrines in times of stress, and the period following the Civil War certainly qualified as such a time for Southerners. The plain words of the *Old Testament*, both the factual claims of *Genesis* and the moral claims of *Leviticus*, appealed tremendously. Thus there developed a distinctively American literalist movement: a movement better known today as "fundamentalism."

This movement had its greatest success in the 1920s when several states, including Tennessee, enacted laws prohibiting the teaching of evolutionism in schools. However, as is well known, the fundamentalist victories backfired somewhat. A young teacher, John Thomas Scopes, let himself be charged for teaching evolution. With Scopes being prosecuted by William Jennings Bryan (three times a presidential candidate) and defended by the noted free-thinker and brilliant advocate Clarence Darrow, matters took on a carnival air when Darrow cross-examined Bryan on the literal truth of *Genesis*. Were the six days of Creation really exactly 24 hours each? Bryan could give no convincing reply. Although Scopes was found guilty, evolutionists won a major moral victory. Even more pertinently, the nation laughed at Tennessee, thanks to the savage reporting of H. L. Mencken, who referred to its good citizens as "anthropoid rabble." Such ridicule made other states far more wary of "monkey laws," and so the fundamentalist anti-evolution campaign ground to a halt.

Paradoxically, it was the Soviets who were responsible for its revival. Twenty-five years ago, America was shocked and frightened by Sputnik. Its very existence was taken to be clear evidence of the failure of capitalism and democracy. American science and technology were second-rate. Accordingly, huge sums of money were thrown at the problem, and this resulted in a revamping of American school science education. One cannot teach without guides and aids, so new up-to-date textbooks were commissioned and written.

Naturally enough, the biology textbooks took evolution for granted. But, given the government support of such books and their consequent success, fundamentalist flames were rekindled. And unfortunately, there proved to be plenty of dry wood on which such flames could feed. In the past decade, when we are again living in a time of perceived uncertainty—a function of such diverse phenomena as the breakdown of traditional family units and the rise in oil prices — the new literalist movement has succeeded beyond anyone's hopes or fears.

In some respects, we have moved on today from the 1920s. In 1967, the US Supreme Court finally overturned the Tennessee law, stating that it is indeed constitutional to teach evolution. Additionally, over the years, the Court has ruled that the First Amendment separation of Church and State means precisely that. One may not teach religion *as religion* in state schools. Specifically, one may not teach *Genesis* in school biology classes.

Fundamentalists have therefore been forced into a new tack. No longer is the cry to keep evolution out. Rather, the cry is to get *Genesis* in. But because *Genesis* cannot come in as religion, it must come in some other way. How can this be done? The fundamentalists think that they have the answer. They argue that all of the claims of *Genesis*, taken literally, are complemented by identical conclusions derived according to the best principles and methodology of science. In short, "creation science" (or "scientific creationism"), a body of claims that mirrors *Genesis* entirely, may legitimately be taught in the science classrooms of the nation.

Thus, the demand today is for "equal time" or "balanced treatment" for evolution and creation science in biology classes. This is a demand endorsed by many, including the incumbent president of the United States, Ronald Reagan. Indeed, so powerful is the appeal of the cry for equal time that some states of the Union have already passed laws giving the request legal backing. Arkansas was the first state in line, and thus was the first to have the constitutionality of its laws challenged in court. In fact, the Federal Judge trying the case ruled (in January 1982) that the law is unconstitutional. Nevertheless, similar laws in other states are still on the books, and there are promises of yet more. Law offers one tactic; creationists also intend to press their case less formally, by influencing schoolboards and by approaching teachers.

What is it about creation science that makes an evolutionist like myself think it a dangerous sham? It is not its religious origins *per se*. Rather, it is the dishonest way in which religion is being fobbed off on us as science, and the wretched arguments used to achieve this end.

Take the most striking feature of the whole body of creation science literature, namely the distinctively peculiar fashion in which virtually all of the arguments proceed. In orthodox science, one is expected at some point to go out and to do some research of one's own. Creation scientists rarely or never do this. Rather, they take the arguments, side comments and off-hand remarks of evolutionists, showing that in fact they support creationism rather than evolutionism. Everything is parasitic on the work of real scientists. Indeed, reading the major works of creation scientists (such as *Scientific Creationism* edited by Henry M. Morris, or, *Evolution: The Fossils Say No!* by Duane T. Gish) one would think that the main spokesmen for the fundamentalist cause are Ernst

Mayr, G. G. Simpson, Theodosius Dobzhansky and Stephen Jay Gould! This, in itself, smacks of impoverished thought, verging on the dishonest. If Mayr and Gould are so reliable in their side thoughts, why are their main evolutionary conclusions to be ignored? Moreover, at times, creation scientists move right out into blatant fraud. For instance, one recent work, *Creation: The Facts of Life* by Gary E. Parker, refers constantly to "noted Harvard geneticist [Richard] Lewontin's" view that the hand and the eye are the best evidence of God's design.

Can this really be so? Has the distinguished author of *The Genetic Basis of Evolutionary Change* really forsaken Darwin for Moses? Not quite. Looking at the actual context from which the quote was lifted (a brilliant article on adaptation in the September 1978 issue of *Scientific American*), we find that Lewontin says that *before Darwin*, people believed the hand and the eye to be the best evidence of God's design. *Now* we know them to be the products of evolution through natural selection. But you would never learn this from Parker's book, which makes Lewontin say the very opposite.

As one might expect, given their sources, most of the creation science arguments are less arguments for creationism and more arguments against evolutionism: the assumption is that if you are against the one, you are for the other. This, incidentally, was an assumption that got the Attorney General of Arkansas into hot water in the recent trial of the "balanced treatment" law's constitutionality. Astrophysicist Chandra Wickramasinghe is against orthodox evolutionism, and the Attorney General called him as a witness for the state. However, although he is certainly outside the mainstream of modern thought, Wickramasinghe is no Biblical creationist: he believes rather that Earth was seeded by life from outer space. On the witness stand, Wickramasinghe confessed that he could not see how anyone could hold to the views of the creationists, especially with regard to the supposedly short history of the Earth. The judge, like everyone else, was amazed at the damaging effect this had on the state's case.

Many of the creationists' anti-evolutionary arguments have a familiar ring. For instance, in virtually all of their works that old chestnut about natural selection being a tautology is dusted off and trotted out, with all sorts of citations to Sir Karl Popper and Sir Peter Medawar to back up its authority. This has been refuted many times, and Sir Karl Popper himself has recanted in print. Then again, creationists constantly argue that a natural beginning to life is an impossibility, for such a process would violate the second law of thermodynamics. The coming of new life requires the natural production of order from disorder. The second law states that the process always goes the other way: you can scramble an egg but you cannot unscramble it. Therefore, life could never have appeared here on Earth, without the aid of the Great Designer in the sky.

No amount of pointing out the fallacious nature of this argument will budge the creationists. But, as any sixth-former ought to be able to tell you, the appeal to the second law in this context is quite irrelevant. The second law says nothing on the matter of new life, for the law applies only to closed systems. If one has the right kind of energy to do so, one can always make order out of disorder. The law states that in a closed (isolated) system, this usable energy will eventually be exhausted. However, this Earth of ours is an open system. Usable energy is coming in from the Sun all the time, and has been from the beginning of Earth's history. Hence, the possibility of order coming from non-order at some point here on Earth is in no way barred by thermodynamics.

The creationists spend a great deal of time on the fossil record. This in itself is a misleading tactic, for the fossil record has always been but one part of the evolutionary spectrum. Picking out the record for such extended discussion gives a false picture of evolutionary studies. In fact, evolutionists turn for insight and support to all areas of biology: biogeography, morphology, embryology, systematics and so forth. And yet, creationists hurry us through these fields, barely mentioning them, if at all. Consider, for instance, the finches of the Galapagos, which from Darwin on have always been a major brick in the evolutionary edifice. Creationists usually ignore them entirely — undoubtedly a deliberate omission, given the prominence of Darwin's finches in evolutionary writings. When they are mentioned, it is claimed that they are all the same species. A flat lie.

Most of the arguments based on the fossil record were devised even before the *Origin* was published, and not surprisingly, are somewhat dated. For instance, much is made of the supposed absence of Precambrian organisms. Why do trilobites appear in the fossil record full-blown, with no predecessors? Of course, the answer is that they do not. There is now massive Precambrian evidence, taking life back at least three-quarters of Earth's 4.6 thousand million year history. Creationists gloss over this.

Similarly, the creationist claim that gaps in the fossil record prove the impossibility of evolution no longer has any scientific validity. Bridging fossils have been found for all of the famous gaps. Take, for instance, the hiatus that existed between reptiles and birds. Even before the *Origin* appeared, a fossil feather was discovered that was shortly to be identified as belonging to *Archaeopteryx*, the organism that sits right on the fence between reptiles and birds. (Full specimens of *Archaeopteryx* were found early in the 1860s.) One could not hope for a more perfect example of a "missing link."

So, what do creationists do in the face of such evidence? They argue that *Archaeopteryx* had feathers (true), that evolutionists classify it as a bird (true), and that hence it is no link (false). Obviously, this line of argument is just a verbal sleight of hand. *Archaeopteryx* has feathers and a fused wishbone, like a bird. Like a reptile, it has a tail, teeth, small brain, separate digits, but lacks a developed sternum. It is right on the border, and has frequently been classified as a reptile by those ignorant of its feathers.

Perhaps the most incredible creationist argument concerns the roughly "progressive" nature of the fossil record. That one finds such a progression in the record, from relatively simple to relatively complex, was first noted at the beginning of the 19th century, by the French father of comparative anatomy, Baron Georges Cuvier. In the *Origin*, naturally Darwin explained progression in terms of evolution up from humble forms, although by then all scientists recognised that the record is more than a simple progression: there is much branching through the course of ages.

Creationists cannot deny the record. They explain it all away as an artefact of Noah's Flood. Trilobites, being slow and sluggish, got trapped at the bottom of the rising waters. Dinosaurs managed to scramble half-way up, before the waters overtook and drowned them. Humans, most agile of all, got to the tops of hills before they met their watery fate (excluding any lucky enough to float away). Thus, we get the illusion of progression. But, say the creationists, it is no more than that: an illusion.

Comment is hardly needed. The creationists' discussions read like extracts from *Alice in Wonderland*. All we need to find are fossil remains of the Mock Turtle. (We already have the Dodo.) Just think! No trilobite was light enough, quick enough or lucky enough to swim to the top. No human was heavy enough, slow enough or sufficiently unlucky as to get trapped at the bottom. How improbable. Apart from anything else, knowing the average Englishman's fondness for his pets, do you really think that none would have taken his tame trilobites to the top? Or, that there would have been none so stubborn and bloody-minded as to stay in his cottage at the bottom of the valley, come hell or high water (or, as the creationists no doubt presume, come both)?

Even the creationists recognise the implausibility of their position, for they add that some human marks are found down the fossil record, along with the dinosaurs. But, no evolutionist need fear. The supposed "human" traces turn out to have a very fishy (or dinosaurish) look about them. Apart from anything else, the traces generally consist of "footprints" much larger than any human could make today. The lame excuse offered (as in *Genesis Flood* by John Whitcomb Jr and Henry Morris) is that in the olden days there were giants. But what authority is there for this? None other than our old friend, *Genesis* (6:4). The non-scientific circular notion of creation science is there for all to see.

Moving on to their final arguments, Creationists deal quickly with the age of the Earth. It is boiled down to around a millionth of what we learn from various radiometric dating techniques. (Think of it. The age of the Earth is a *million* times shorter than most believe.) How can this reduction be achieved? Quite simply. If one wishes to deny conclusions, then either one denies premises or one denies inferences. Creationists do both. For instance, they argue that one cannot assume that rates of radioactive decay are constant. The reasons why tend not to be specified too clearly.

In parallel with negative arguments against conventional dating techniques, the creationists for once offer a positive argument of their own. They suggest that if we extrapolate back from current human population numbers and growth rates, we arrive at a sole Adam and Eve around the time of God's original creative act (about 6000 years ago). If we accept evolution, assuming that *Homo sapiens* arrived here on Earth about a million years ago, then given population growth we would today have more humans than there are electrons in the world. An obvious impossibility.

I hardly need say what a silly argument this is. The whole point of evolutionary theory is that population growth does not occur unchecked. Disease, famine, war, all lead to the Malthusian struggle for existence, which leads in turn to selection and evolution. Put matters this way. If there were no checks, we should be knee-deep in houseflies and herrings long before humans began to bother us. And in any case, the figures brandished by the creationists are phoney. They cite a human population growth rate of 2 per cent; but we have only started to edge up to this rate in recent years. Written records tell us that, in earlier times, growth rates were far lower than they are today.

Finally, there is the question that hovers over all discussions about origins. Not without reason was Darwin's theory known as the "monkey theory." What about us? What about *Homo sapiens*? Creationists flatly deny that the increasingly well-known hominid record has any relevance at all. There are no proto-human intermediaries. What about the much-feted *Australopithecus*? With respect, they were no human forefathers, for they had the brain of an ape and walked on their knuckles like an ape. Even *Homo erectus* is really no more than a rather run-down specimen of *Homo sapiens*.

If you believe this, you will believe anything. Suffice it to say that the recent discoveries of *Australopithecus afarensis* in Ethiopia put the total lie to the creationists. This small creature has an ape-size brain, and yet it walked around on its two hind legs, just like you and me. Even if one sat down and deliberately drew up a hypothetical ideal human ancestor, one could not do a better job than nature herself. Only by ignoring a massive amount of empirical evidence can one deny our simian ancestry.

Creation science is not science. It is crude dogmatic religion. For this reason, it is as offensive to the true believer as it is to the scientist. It can be maintained only through a perversion of our God-given reason. Interestingly, in their more candid moments, even the creationists admit the non-scientific nature of their case.

For instance, in *The Genesis Flood*, John C. Whitcomb Jr and Henry Morris write:

> But during the period of Creation, God was introducing order and organisation and energisation into the universe in a very high degree, even to life itself! *It is thus quite plain that the processes used by God in creation were utterly different from the processes which now operate in*

the universe! The Creation was a unique period, entirely incommensurate with this present world. This is plainly emphasised and re-emphasised in the divine revelation which God has given us concerning Creation, which concludes with these words:

And the heavens and the earth were *finished*, and *all* the host of them. And on the seventh day God *finished* His work which He had made, and He *rested* on the seventh day from *all His work* which He had made. And God blessed the seventh day, and hallowed it; because that in it He *rested* from *all* his work which God had created and made.

In view of these strong and repeated assertions, is it not the height or presumption for man to attempt to study Creation in terms of present processes? [pp. 223–24. Their italics.]

I could not put the non-scientific nature of creation science better myself.

Incidentally, I am often asked why it was that the Attorney General of Arkansas never used the bigger guns in the creationist arsenal, such as Morris and Gish. It is suggested that he threw the case by not putting them on the stand. My hunch is that he realised he simply could not use them, for they have left such a trail of incriminating statements in their writings over the years.

Can you imagine a bright young lawyer cross-examining Morris on the above paragraphs? The Attorney General was forced to use people who had written virtually nothing. The one exception was Wickramasinghe, and look what a mess he made of the case for the defence.

One final point. The creation scientists like to portray themselves as martyrs. They claim that they do not want to exclude evolution. All they want is the right to have their own ideas brought before the world—or, more accurately, brought into the classroom. Hence, creationists argue that it is not they who are the bigots. It is the evolutionists. It is the established scientists who exclude ideas, insisting that no views other than their own may be taught in the classroom. But, conclude the creationists, this dogmatic approach obviously is contrary to the principles of sound education. Children should be exposed to all ideas, and then allowed to choose for themselves the one they prefer.

Nevertheless, smooth as it looks, this argument is perniciously fallacious. Sound education has never implied an indifferent purveying of wares before children, letting them pick and choose for themselves. The good teacher has always given to his or her children the truest and the best, as he or she knows it. A good physics teacher teaches astronomy not astrology, even though far more people look to the stars for their destiny than for confirmation of the Copernican revolution. Similarly, the good biology teacher teaches biology, not dogmatic religion. This is not dogmatic bigotry but professionalism.

Sound education is one thing. Teaching lunatic ideas to children is quite another. Creation science is plainly and unmistakably a fraud, and should not be taught alongside evolution.

35

ARTHUR PEACOCKE*

Welcoming the 'Disguised Friend'— Darwinism and Divinity

"Darwinism appeared, and, under the disguise of a foe, did the work of a friend. It has conferred upon philosophy and religion an inestimable benefit, by showing us that we must choose between two alternatives. Either God is every-where present in nature, or He is nowhere." (Aubrey Moore, in the 12th edition of *Lux Mundi,* 1891, p. 73)

IT WOULD, NO doubt, come as a surprise to many of the biologically cultured 'despisers of the Christian religion' to learn that, as increasingly thorough historical investigations are showing,[1] the nineteenth-century reaction to Darwin in theological and ecclesiastical circles was much more positive and welcoming than the legends propagated by both popular and academic bio-logical publications are prepared to admit. Furthermore, the scientific reaction was also much more negative than usually depicted, those skeptical of Darwin's ideas including initially *inter alios* the leading comparative anatom-ist of his day, Richard Owen (a Cuverian), and the leading geologist, Charles Lyell. Many theologians deferred judgment, but the proponents of at least one strand in theology in nineteenth-century England chose to inter-twine their insights closely with the Darwinian—I refer to that 'catholic' revival in the Church of England of a stress on the doctrine of the Incarna-tion and its extension into the sacraments and so of a renewed sense of the

*This is a revised and shortened form of a paper presented at the Vatican Observa-tory-CTNS Conference on "Scientific Perspectives on Divine Action" (Evolution and Molecular Biology) held in July 1996; it was also delivered as an Idreos Lecture at Harris-Manchester College, Oxford, in May 1997. By permission of the author.

sacramentality of nature and God's immanence in the world. More of the nineteenth-century theological reaction to Darwin was constructive and reconciling in temper than practically any biological authors today will admit.

That is perhaps not surprising in view of the background to at least T. H. Huxley's aggressive propagation of Darwin's ideas and his attacks on Christianity, namely, that of clerical restriction on, and opportunities for, biological scientists in England in the nineteenth century. His principal agenda was the establishment of science as a profession independent of ecclesiastical control–and in this we can sympathize. So it is entirely understandable that the present, twentieth-century *zeitgeist* of the world of biological science is that of viewing 'religion' as the opposition, if no longer in any way a threat. This tone saturates the writings of the biologists Richard Dawkins and Stephen Gould and many others–and even philosophers such as Daniel Dennett. So-called Creationists, of course, continue to reinforce this prejudice.

This has polarized the scene, but what I do find even more surprising, and less understandable, is the way in which the 'disguised friend' of Darwinism, more generally of evolutionary ideas, has been admitted (if at all) only grudgingly, with many askance and sidelong looks, into the parlors of Christian theology. (Only in 1996 did the Pope finally admit that evolution is "more than a hypothesis"). I believe it is vital for this churlishness to be rectified in this last decade of the twentieth century if the Christian religion (indeed any religion) is to be believable and have intellectual integrity enough to command even the attention, let alone the assent, of thoughtful people in the beginning of the next millennium.

Biological Evolution and God's Relation to the Living world

Let us consider various features and characteristics of biological evolution and any reflections about God's relation to the living world to which they may give rise.

Continuity and emergence

A notable aspect of the scientific account of the natural world in general is the seamless character of the web that has been spun on the loom of time: the process appears as continuous from its cosmic 'beginning' , in the 'hot big bang', to the present and at no point do modern natural scientists have to invoke any nonnatural causes to explain their observations and inferences about the past–including the origin of life.

The processes that have occurred can be characterized as those of *emergence,* for new forms of matter, and a hierarchy of organization of these forms themselves, appear in the course of time. To these new organizations of matter it is, very often, possible to ascribe new levels of what can only be called 'reality': in other words new kinds of reality may be said to 'emerge' in time. Notably, on the surface of the earth, new forms of *living* matter (that is, living organisms) have come into existence by this continuous process—that is what we mean by evolution.

What the scientific perspective of the world, especially the living world, inexorably impresses upon us is a *dynamic* picture of the world of entities and structures involved in continuous and incessant change and in process without ceasing. The scientific perspective of a cosmos, and in particular that of the biological world, as in development all the time must reintroduce into our understanding of God's creative relation to the world a dynamic element which was always implicit in the Hebrew conception of a 'living God', dynamic in action—even if obscured by the tendency to think of 'creation' as an event in the past. Any notion of God as Creator must now take into account that God is continuously creating, continuously giving existence to, what is new; that God is *semper Creator*; that the world is a *creatio continua*. The traditional notion of God *sustaining* the world in its general order and structure now has to be enriched by a dynamic and creative dimension—the model of God sustaining and giving continuous existence to a process which has an inbuilt creativity, built into it by God. God is creating at every moment of the world's existence in and through the perpetually endowed creativity of the very stuff of the world. God indeed makes 'things make themselves,' as Charles Kingsley put it in *The Water Babies*.

Thus it is that the scientific perspective, and especially that of biological evolution, impels us to take more seriously and more concretely than hitherto the notion of the immanence of God-as-Creator—that God is the Immanent Creator *creating in and through the processes of the natural order*. I would urge that all this has to be taken in a very strong sense. If one asks where do we see God-as-Creator during, say, the processes of biological evolution, one has to reply: "The processes themselves, as unveiled by the biological sciences, *are* God-acting-as-Creator, God *qua* Creator." (This is not pantheism for it is the *action* of God that is identified with the creative processes of nature, not God's own self.) The processes are not themselves God, but the *action* of God-as-Creator. God gives existence in divinely created time to a process that itself brings forth the new: thereby God is crea*ting*. This means we do not have to look for any extra supposed gaps in which, or mechanisms whereby, God might be supposed to be acting as Creator in the living world.

The model of musical composition for God's activity in creation is here, I would suggest, particularly helpful. There is no doubt of the 'transcendence' of the composer in relation to the music he or she creates—the composer gives it existence and without the composer it would not be at all. So the model properly reflects, as do all those of artistic creativity, that transcendence of God as Creator of all-that-is which, as the 'listeners' to the music of creation, we wish to affirm. Yet, when we are actually listening to a musical work, say, a Beethoven piano sonata, then there are times when we are so deeply absorbed in it that, for a moment we are thinking Beethoven's musical thoughts with him. In such moments the

> music is heard so deeply
> That it is not heard at all, but you are the music
> While the music lasts.[2]

Yet if anyone were to ask at that moment, "*Where* is Beethoven now?" we could only reply that Beethoven-*qua*-composer was to be found only in the music itself. The music would in some sense be Beethoven's inner musical thought kindled in us and we would genuinely be encountering Beethoven-*qua*-composer. This very closely models, I am suggesting, God's immanence in creation and God's self-communication in and through the processes by means of which God is creating. The processes revealed by the sciences, especially evolutionary biology, are in themselves God-acting-as-Creator. There is no need to look for God as some kind of *additional* factor supplementing the processes of the world. God, to use language usually applied in sacramental theology, is 'in, with, and under' all-that-is and all-that-goes-on.

The mechanism of biological evolution

There appear to be no serious biologists who doubt that natural selection is a factor operative in biological evolution–and most would say it is by far the most significant one. At one end of the spectrum authors like Richard Dawkins argue cogently for the all-sufficiency of natural selection in explaining the course of biological evolution. However, other biologists are convinced that, even when all the subtleties of natural selection are taken into account, it is not the whole story and some even go so far as to say that natural selection alone cannot account for speciation, the formation of distinctly new species. Some other considerations which, it is claimed, are needed to be taken into account are said to be: the 'evolution of evolvability', the constraints and selectivity effected by self-organizational principles, how an organism might evolve is a consequences of itself, 'genetic assimilation'; the innovative behavior of an individual living creature in a particular environment; 'top-down causation' (or 'whole-part constraint') in evolution through a flow of information; much molecular evolutionary change is immune to natural selection; *group* selection* (after all!); long-term changes resulting from 'molecular drive' (gene-hopping); effects of the context of adaptive change, and *stasis*.

What is significant about all these proposals is that they are all operating entirely within a naturalistic framework and assume a basically Darwinian process to be operating, even when they disagree about its speed and smoothness. That being so, it has to be recognized that the history of life on Earth involves chance in a way unthinkable before Darwin. There is a creative interplay of 'chance' and law apparent in the evolution of living matter by natural selection.

*The hypothesis that natural selection operates at the level of a group sharing certain features

The original mutational events are random with respect to the future of the biological organism, even its future survival; but these changes have their consequences in a milieu that has regular and lawlike features. For the biological niche in which the organism exists then filters out, by the processes of natural selection, those changes in the DNA that enable the organisms possessing them to produce more progeny.

The interplay between 'chance', at the molecular level of the DNA, and 'law' or 'necessity' at the statistical level of the population of organisms tempted Jacques Monod, in *Chance and Necessity*[3] to elevate 'chance' to the level almost of a metaphysical principle whereby the universe might be interpreted. As is well known, he concluded that the 'stupendous edifice of evolution' is, in this sense, rooted in 'pure chance' and that *therefore* all inferences of direction or purpose in the development of the biological world, in particular, and of the universe, in general, must be false.

The responses to this thesis and attack on theism came mainly, it is interesting to note, from theologically informed scientists, and some philosophers, rather than from theologians. They have been well surveyed by D. J. Bartholomew[4] and their relative strengths and weaknesses analyzed. I shall here pursue what I consider to be the most fruitful line of theological reflection on the processes that Monod so effectively brought to the attention of the twentieth century–a direction that I took[5] in response in actual debate with Monod and which has been further developed by the statistically informed treatment of Bartholomew.

There is no reason why the randomness of molecular event in relation to biological consequence has to be given the significant metaphysical status that Monod attributed to it. The involvement of what we call 'chance' at the level of mutation in the DNA does not, of itself, preclude these events from displaying regular trends and manifesting inbuilt propensies at the higher levels of organisms, populations, and ecosystems. To call the mutation of the DNA a 'chance' event serves simply to stress its randomness with respect to biological consequence. As I have earlier put it:

> Instead of being daunted by the role of chance in genetic mutations as being the manifestation of irrationality in the universe, it would be more consistent with the observations to assert that the full gamut of the potentialities of living matter could be explored only through the agency of the rapid and frequent randomization which is possible at the molecular level of the DNA.[6]

This role of 'chance', or rather randomness (or 'free experiment') at the micro-level is what one would expect if the universe were so constituted that all the potential forms of organizations of matter (both living and nonliving) which it contains might be thoroughly explored. This interplay of chance and law is in fact creative within time, for it is the combination of the two which allows new forms to emerge and evolve–so that natural selection appears to be opportunistic. As in many games, the consequences of the fall of the dice depend very much on the rules of the game.[7] It has become increasingly apparent that it is chance operating within a lawlike framework that is the basis of the inherent creativity of the natural order, its ability to generate new forms, patterns, and organizations of matter and energy. If all were governed by rigid law, a repetitive and uncreative order would prevail; if chance alone ruled, no forms, patterns, or organizations would persist long enough for them to have any identity or real existence and the universe could never have been a cosmos and susceptible to rational inquiry. It is the combination of the two which makes possible an ordered universe capable of developing within itself new modes of existence.

This combination for a theist can only be regarded as an aspect of the God-endowed features of the world. The way in which what we call 'chance' operates within this 'given' framework to produce new structures, entities, and processes can then properly be seen as an eliciting of the potentialities that the physical cosmos possessed *ab initio*. One might say that the potential of the 'being' of the world is made manifest in the 'becoming' that the operation of chance makes actual. God is the ultimate ground and source of both law ('necessity') and 'chance.'

To return to our musical model—for a theist, God must now be seen as acting rather like a composer extemporizing a fugue to create in the world *through* what we call 'chance' operating within the created order, each stage of which constitutes the launching pad for the next. The Creator, it now seems, is unfolding the divinely endowed potentialities of the universe, in and through a process in which these creative possibilities and propensities (see next section), inherent by God's own intention within the fundamental entities of that universe and their interrelations, become actualized within a created temporal development shaped and determined by those selfsame God-given potentialities.[8]

Trends in evolution?

Given an immanentist understanding of God's presence 'in, with, and under' the processes of biological evolution adopted up to this point, can God be said to be implementing any purpose in biological evolution? Or is the whole process so haphazard, such a matter of happenstance, such a matter of what Monod and Jacob called *bricolage* (tinkering), that no meaning, least of all a divinely intended one, can be discerned in the process? That is, of course, the position adopted by many nontheist biologists. However, Popper[9] has pointed out that the realization of possibilities, which may be random, depends on the total situation within which the possibilities are being actualized so that "there exist weighted possibilities which are *more than mere possibilities*, but tendencies or propensities to become real"[10] and that these "propensities in physics are properties of *the whole situation* and sometimes even of the particular way in which a situation changes. And the same holds of the propensities in chemistry, in biochemistry, and in biology."[11]

I suggest[12] that the evolutionary process is characterized (because they enhance the chance of an organism being 'naturally selected') by *propensities* toward increase in complexity, information-processing and -storage, consciousness, sensitivity to pain, and even self-consciousness (a necessary prerequisite for social development and the cultural transmission of knowledge down the generations). *Some* successive forms, along *some* branch or 'twig' (à la Gould), have a distinct probability of manifesting more and more of these characteristics. However, the actual physical form of the organisms in which these propensities are actualized and instantiated is contingent on the history of the confluence of disparate chains of events, including the survival of the mass extinctions that have occurred (96 percent of all species in the Permo-Triassic one,[13] 225 million years ago). So it is not surprising that recent reinterpretation of the fossils of very early (ca. 530 million years ago) soft-bodied fauna found in the Burgess shale of Canada show that, had any a larger proportion of these survived and prevailed, the actual forms of contemporary, evolved creatures would have been very much more disparate in anatomical *plans* than those now observed to exist—albeit with a very great diversity in the few surviving designs.[14] But even had these particular organisms, unique to the Burgess shale, been the progenitors of subsequent living organisms, the same propensities toward complexity, etc., would also have been manifest in their subsequent evolution, for these 'propensities' simply reflect the advantages conferred in natural selection by these features. The same considerations apply to the arbitrariness and contingency of the mass extinctions, which Gould also strongly emphasizes. So that, providing there had been enough time, a complex organism with consciousness, self-consciousness, social and cultural organization (that is, the basis for the existence of 'persons') would have been likely eventually to have evolved and appeared on the earth (or on some other planet amenable to the emergence of living organisms), though no doubt with a physical form very different from *homo sapiens*. There can, it seems to me (*pace* Stephen Gould[15]), be overall direction and implementation of divine purpose through the interplay of chance and law without a deterministic plan fixing all the details of the structure(s) of what emerges possessing personal qualities. Hence the emergence of self-conscious persons capable of relating personally to God can still be regarded as an intention of God continuously creating through the processes of that to which God has given an existence of this contingent kind and not some other. It certainly must have been possible since it actually happened—with us!

I see no need to postulate any *special* action of God—along the lines, say, of some divine manipulation of mutations at the quantum level—to ensure that persons emerge in the universe, and in particular on Earth. Not to coin a phrase, 'I have no need of that hypothesis'![16]

The diversity of life

The natural world is immensely variegated in its hierarchies of levels of entities, structures and processes, in its 'being'; and it abundantly diversifies with a cornucopian fecundity in its 'becoming' in time. From the unity in this diversity and the richness of the diversity itself, one may adduce,[17] respectively, both the essential oneness of its source of being, namely, the one God the Creator, and the unfathomable richness of the unitive Being of that Creator God. But now we must reckon more directly with the diversity itself. The forms even of nonliving matter throughout the cosmos as it appears to us is even more diverse than what we can now observe immediately on the earth. Furthermore the multiply branching bush (as Gould appropriately calls it) of terrestrial biological evolution appears to be primarily opportunist in the direction it follows and, in so doing, produces the enormous variety of biological life on this planet

We can only conclude that, if there is a Creator, least misleadingly described in terms of 'personal' attributes, then that Creator intended this rich multiformity of entities, structures, and processes in the natural world and, if so, that such a Creator God takes what, in the personal world of human experience, could only be called 'delight' in this multiformity of what has been created. The existence of the *whole* tapestry of the created order, in its warp and woof, and in the very heterogeneity and multiplicity of its forms must be taken to be the Creator's intention. We can only make sense of that, utilizing our resources of personal language, if we say that God may be said to have something akin to 'joy' and 'delight' in creation. We have a hint of this in the satisfaction attributed to God as Creator in the first chapter of Genesis: "And God saw everything he had made, and behold, it was very good."[18] This naturally leads to the idea of the 'play' of God in creation, on which I have expanded elsewhere,[19] in relation to Hindu thought as well as to that of Judaism and Christianity. But now the darker side of evolution.

The ubiquity of pain, suffering, and death

The ability for information-processing and -storage is indeed the necessary, if not sufficient, condition for the emergence of consciousness. This sensitivity to, this sentience of, its surroundings inevitably involves an increase in its ability to experience pain, which constitutes the necessary biological warning signals of danger and disease. So that it is impossible readily to envisage an increase of information-processing ability without an increase in the sensitivity of the signaling system of the organism to its environment. In other words, an increase in 'information-processing' capacity, with the advantages it confers in natural selection, cannot (*of necessity* cannot) but have as its corollary an increase, not only in the level of consciousness, but also in the experience of pain. Insulation from the surrounding world in the biological equivalent of three-inch nickel steel would be a sure recipe for preventing the development of consciousness.

New patterns can only come into existence in a finite universe ('finite' in the sense of the conservation of matter-energy) if old patterns dissolve to make place for them. This is a condition of the creativity of the process–that is, of its ability to produce the new–which at the biological level we observe as new forms of life only through death of the old. For the death of individuals is essential for release of food resources for new arrivals, and species simply die out by being ousted from biological 'niches' by new ones better adapted to survive and reproduce in them. So there is a kind of *structural* logic about the inevitability of living organisms dying and of preying on each other–for we cannot conceive, in a lawful, non-magical universe, of any way whereby the immense variety of developing, biological, structural complexity might appear, except by utilizing structures already existing, either by way of modification (as in biological evolution) or of incorporation (as in feeding).[20] The structural logic is inescapable: new forms of matter arise only through the dissolution of the old; new life only through death of the old. So that biological death of the individual is the prerequisite of the creativity of the biological order, that creativity which eventually led to the emergence of human beings.

But death not only of individuals but of whole species has also occurred on Earth during the periods of mass extinctions which are now widely attributed to chance extraterrestrial collisions of the planet with comet showers, asteroids, or other bodies. These could be cataclysmic and global in their effects and have been far more frequent than previously imagined. This adds a further element of sheer contingency to the history of life on the earth.

Hence pain, suffering, and death, which have been called 'natural evil'– the features of existence inimical to biological life, in general, and human flourishing, in particular–appear to be inevitable concomitants of a universe that is going to be creative of new forms, some of which are going to be conscious and self-conscious.

Even so, the theist cannot lightly set aside these features of the created order. The spontaneity and fecundity of the biological world is gained at the enormous price of universal death and of pain and suffering during life,[21] even though individual living creatures, other than humans, scarcely ever commit suicide in any way that might be called intentional.

For any concept of God to be morally acceptable and coherent, the ubiquity of pain, suffering, and death as the means of creation through biological evolution entails that if God is also immanently present in and to natural processes, in particular those that generate conscious and self-conscious life, then we cannot but infer that–in some sense hard to define–*God*, like any human creator, suffers in, with, and under the creative processes of the world with their costly unfolding in time.

Rejection of the notion of the impassibility of God has, in fact, been a feature of the Christian theology of recent decades. There has been an increasing assent to the idea that it is possible "to speak consistently of *a God who suffers eminently and yet is still God, and a God who suffers universally. . . .*"[22] As Paul Fiddes points out in his survey and analysis of this change in theological perspective, among the factors that have promoted the view that God suffers are new assessments of "the meaning of love [especially, the love of God], the implications of the cross of Jesus, the problem of [human] suffering, and the structure of the world."[23] It is this last-mentioned–the "structure of the world"–on which the new perspectives of the biological sciences bear by revealing the world processes, as already described, to be such that involvement in them by the immanent Creator has to be regarded as involving suffering on the Creator's part. God, we find ourselves having tentatively to conjecture, suffers in the 'natural' evils of the world along with ourselves because—again we can only hint at this stage—God purposes to bring about a greater good thereby, namely, the kingdom of free-willing, loving persons in communion with God and with each other.[24]

Other aspects of human evolution

That *homo sapiens* represents only an extremely small, and very recently arrived, fraction of living organisms that have populated the earth raises the question of God's purposes in creating such a labyrinth of life. To assume it was all there simply to lead to us clearly will not do. Hence my attribution to God of a sheer exuberance in creativity for its own sake. Moreover, since biological death was present on the earth long before human beings arrived on the scene and was the prerequisite of our coming into existence through the processes of biological evolution, when St. Paul says that "sin pays a wage, and the wage is death,"[25] that cannot possibly mean for us now *biological* death. It can only mean 'death' in some other sense, such as the death of our relation to God consequent upon sin. I can see no sense in regarding biological death as the consequence of that very real alienation from God that is sin, because God had already used biological death as the means for creating new forms of life, including ourselves, long before human beings appeared on Earth. This means those classical Christian formulations of the theology of the redemptive work of Christ that assume a causal connection between biological death and sin urgently need reconsidering.

Moreover, the biological-historical evidence is that human nature has emerged only gradually by a continuous process from other forms of primates and that there are no sudden breaks of any substantial kind in the sequences noted by paleontologists and anthropologists. Moreover, there is *no* past period for which there is reason to affirm that human beings possessed moral perfection existing in a paradisal situation from which there has been only a subsequent decline. We appear to be rising beasts[26] rather than fallen angels! Although there was no golden age, no perfect past, no original perfect, individual "Adam" from whom all human beings have now declined, what *is* true is that humanity manifests aspirations to a perfection not yet attained, a potentiality not yet actualized, but no "original righteousness" so beloved by some theologians. Sin as alienation from God, humanity, and nature is real and appears as the consequence of our very possession of that *self*-consciousness which always places ourselves at the egotistical center of the 'universe' created by our consciousness. Sin is about a falling short from, about not having realized, what God intends us to be and is part and parcel of our having evolved into self-consciousness, freedom, and intellectual curiosity. The domination of Christian theologies of redemption by classical conceptions of the 'Fall' urgently needs, it seems to me, to be rescinded and what we mean by redemption to be rethought if it is to make any sense to our contemporaries.

However, we all have an awareness of the tragedy of our failure to fulfill our highest aspirations; of our failure to come to terms with finitude, death, and suffering; of our failure to realize our potentialities and to steer our path through life. Freedom allows us to make the wrong choices, so that sin and alienation from God, from our fellow human beings and from nature are real features of our existence. So the questions of not only "Who are we?" but, even more acutely, "What should we be becoming–where should we be going?" remain acute for us. Christians find the clue to the answers to these questions in the person of Jesus of Nazareth and what he manifested of God's perennial expression in creation (as the *Logos* of the cosmos, as 'God incarnate'). This leads me now to consider specifically Christian affirmations in the light of the foregoing reconsiderations of God's creating through an evolutionary process (and, it must be said, other religions and religious figures need also to be similarly examined).

The significance of Jesus the Christ in a possible Christian evolutionary perspective

... [I]n scientific language, the Incarnation may be said to have introduced a new species into the world–the Divine man transcending past humanity, as humanity transcended the rest of the animal creation, and communicating His vital energy by a spiritual process to subsequent generations....
(J. R. Illingworth, in the 12th edition of *Lux Mundi*, 1891, p. 132)

Jesus' resurrection convinced the disciples, notably Paul, that it is the union of *his* kind of life with God which is not broken by death and capable of being taken up into God. For he manifested the kind of human life which, it was believed, can become fully life with God, not only here and now, but eternally beyond the threshold of death. Hence his imperative "Follow me" constitutes a call for the transformation of humanity into a new kind of human being and becoming. What happened to Jesus, it was thought, *could* happen to all.

In this perspective, Jesus the Christ, the whole Christ event, has, I would suggest, shown us what is possible for humanity. The actualization of this potentiality can properly be regarded as the consummation of the purposes of God already incompletely manifested in evolving humanity. In Jesus there was a *divine* act of new creation because the initiative was *from God* within human history, within the responsive human will of Jesus inspired by that outreach of God into humanity designated as 'God the Holy Spirit'. Jesus the Christ is thereby seen, in the context of the whole complex of events in which he participated as the paradigm of what God intends for all human beings, now revealed as having the potentiality of responding to, of being open to, of becoming united with, God. In this perspective, he represents the consummation of the evolutionary creative process which God has been effecting in and through the world.

Darwinism and Divinity may, after all, join hands!

NOTES

1. E.g., J. R. Moore, *The Post-Darwinian Controversies: A Study of the Protestant Struggle to Come to Terms with Darwin in Great Britain and America* (Cambridge: Cambridge University Press, 1979); J. R. Lucas, "Wilberforce and Huxley: A Legendary Encounter," *The Historical Journal* 22, no. 2 (1979): 313–30; J. V. Jensen, "Return to the Huxley-Wilberforce Debate," *Brit. J. Hist. Sci.* 21 (1988): 161–79.

2. T. S. Eliot, "The Dry Salvages," *The Four Quartets*, ll. 210–12.

3. Jacques Monod, *Chance and Necessity* (London: Collins, 1972).

4. David J. Bartholomew, *God of Chance* (London: SCM Press, 1984).

5. A. R. Peacocke, "Chance, Potentiality and God," *The Modern Churchman* 17 (New Series 1973): 13–23; and in *Beyond Chance and Necessity*, ed. J. Lewis (London: Garnstone Press, 1974,), pp. 13–25; "Chaos or Cosmos," *New Scientist* 63 (1974): 386–89; A. R. Peacocke, *Creation and the World of Science* [henceforth *CWS*] (Oxford: Clarendon Press, 1979), chap. 3.

6. *CWS*, p. 94; see also by this author, *Theology for a Scientific Age: Being and Becoming Natural, Divine and Human* [henceforth *TSA*] (London: SCM Press, 1993, 2d enlarged ed.; reprint 1995), pp. 115–21.

7. R. Winkler and M. Eigen, *Laws of the Game* (New York: Knopf, 1981; London: Allen Lane, 1982).

8. Cf. *CWS*, pp. 105–106, and *TSA*, pp. 115–21.

9. Karl R. Popper, *A World of Propensities* (Thoemmes, Bristol, 1990).

10. Ibid., p. 12.

11. Ibid., p. 17.

12. It is interesting to note that Richard Dawkins, too, in his *River Out of Eden* (London: Wiedenfeld and Nicolson, 1995), includes (pp. 151ff.) among the thresholds that will be crossed *naturally* in "a general chronology of a life explosion on any planet, anywhere in the universe . . . thresholds that any planetary replication bomb can be expected to pass," those for: high-speed information-processing (no. 5), achieved by possession of a nervous system; consciousness (no. 6, concurrent with brains); and language (no. 7).

13. Stephen Jay Gould, *Wonderful Life: the Burgess Shale and the Nature of History* (London: Penguin Books, 1989), p. 306, citing David M. Raup.

14. Ibid., p. 49.

15. Ibid., p. 51 and passim.

16. If there are any such influences by God shaping the direction of evolutionary processes at specific points—for which I see no evidence (how could we know?) and no theological need—I myself could only envisage them as being through God's whole-part constraint on all-that-is affecting the confluence of what, to us, would be independent causal chains. Such specifically directed constraints I would envisage as possible by being exerted upon the whole interconnected and interdependent system of the whole earth in the whole cosmos which is in and present to God, who is therefore its ultimate Boundary Condition and therefore capable of shaping the occurrence of particular patterns of events, if God chooses to do so. For a fuller exposition of this approach, see *TSA*, pp. 157–65; and, more particularly and recently, "God's Interaction with the World–The Implications of Deterministic 'Chaos' and of Interconnected and Interdependent Complexity," in *Chaos and Complexity*, eds. R. J. Russell, N. Murphy, and A. R. Peacocke (Vatican City State: Vatican Observatory Publications; Berkeley, Calif.: Center for Theology and the Natural Sciences, 1995–distributed by the University of Notre Dame Press), pp. 263–87.

17. *TSA*, chap. 8, section 1.

18. Genesis 1, v. 31.

19. *CWS*, pp. 108–11.

20. The depiction of this process as "nature, red in tooth and claw" (a phrase of Tennyson's that actually predates Darwin's proposal of evolution through natural selection) is a caricature for, as many biologists have pointed out (e.g., G. G. Simpson in *The Meaning of Evolution* [New Haven: Bantam Books, Yale University Press, 1971], p. 201), natural selection is not even in a figurative sense the outcome of struggle, as such. Natural selection involves many factors that include better integration with the ecological environment, more efficient utilization of available food, better care of the young, more cooperative social organization—and better capacity of surviving such 'struggles' as do occur (remembering that it is in the interest of any predator that their prey survive as a species!).

21. Cf., Dawkins's epithet, "DNA neither cares nor knows. DNA just is. And we dance to its music" (*River Out of Eden*, p. 133).

22. Paul S. Fiddes, *The Creative Suffering of God* (Oxford: Clarendon Press, 1988) p. 3 (emphasis in the original).

23. Ibid., p. 45 (see also all of chap. 2).

24. I hint here at my broad acceptance of John Hick's 'Irenaean' theodicy in relation to 'natural' evil (q.v., "An Irenaean Theodicy," in *Encountering Evil*, ed. Stephen T. Davis (Edinburgh: T. & T. Clark, 1981), pp. 39–52; and his earlier *Evil and the God of Love* (London: Macmillan, 1966), especially chapters 15 and 16) and the position outlined by Brian Hebblethwaite in chapter 5 ("Physical Suffering and the Nature of the Physical World") in his *Evil, Suffering and Religion* (London: Sheldon Press, 1976).

25. Rom. 6 v. 23, REB.

26. Rising from an amoral (and in that sense) innocent state to the capability of moral, and immoral, action.

CLONING

36

PHILIP HEFNER*

Cloning as Quintessential Human Act

Cloning Reveals the Human Situation Today

THE RECENT APPEARANCE of Dolly, the cloned sheep, as well as both the reports and the debate about cloning humans, provide us glimpses into the quintessential character of the human situation today. We are created co-creators (some will say created by God; others, by nature): creatures of nature who themselves intentionally enter into the process of creating nature in startling ways. We face even the prospect of creating ourselves, in ways that are startling and troubling.

The Significance of This Revelation:

(1) The character of the cloners

What is the significance of cloning as revelation of the human situation? In the first place, the scientific knowledge that underlies cloning and the technological ability to clone are no more and no less morally charged than is our basic human nature itself. Cloning is neither unnatural nor bizarre. Rather, it is in principle an unsurprising exemplification of what we have known for a long time about ourselves. A creature that can, through genetic

*From Philip Hefner, "Cloning as Quintessential Human Act," *Insights,* Vol. 9, no. 1 (August 1997), pp. 18–21. Reprinted by permission of the author.

engineering, totally rearrange the life-forms that constitute our agricultural enterprises could also, almost predictably, be expected to learn eventually how to make itself. The most significant revelation deriving from Dolly will prove to be what it tells about the cloners, ourselves. In cloning, we are in fact addressing ourselves, and it is about ourselves that we have the greatest questions.

We often talk as if to clone is "playing God," when, in fact, it is to playing the role of human co-creators, and we have no clear ideas about what that entails. A salutary place to begin would be to ask: What sorts of persons ought to be allowed to control the cloning process? What character and what set of virtues would we want clones to possess?

(2) Cloned humans are real persons

The life we engineer in the laboratory is really life. Inasmuch as the cloner of humans is itself a natural creature, cloning must be considered to be a process of nature. As such, cloning humans would finally be more like nature's process of creating identical twins than some Frankenstein horror story. A cloned person would grow and develop through the fundamental processes that govern the development of every other person: a genotype giving rise to a phenotype, a person emerging through the interaction with physical environment and culture. Parenting and schooling would continue to be critical in the development of a personal identity, since no human being can survive by genes alone. Those of us who believe that God started this whole process in creation and sustains it from day to day would conclude that this cloned person is, like the rest of us, in the image of God and possesses a soul (given that there are several different theological notions of what it means to have a soul).

Man persons will have difficulty accepting these ideas, primarily because they are so alienated from nature that they cannot understand their humanity within the parameters of nature as we know it today. Cloning is another occasion for us to learn that whatever else we may be, we are thoroughly natural creatures.

(3) Where do we fit in?

In my book, *The Human Factor,* I not only suggest the concept of created co-creator, but I also argue that since humans have emerged within the evolutionary processes of nature, the purpose and meaning of human being must be considered with reference to our place in nature and our potential contribution to the rest of nature. In theological terms, whatever the meaning of human existence is, it has something to do with the rest of creation, of which we are part. In the public discussion, it is clear that for the most part cloning of humans and animals is considered from the angle of benefiting humans and virtually no other perspective is articulated. Ian Wilmut's testimony to Congress is marked by this view.

In my opinion, this becomes a major issue that is revealed in the cloning discussion. Is this little speck of a species on a speck of a planet, living for a minuscule period of time, really supposed to learn how to manipulate the entire biosphere for the purposes of healing every human disease or producing milk more efficiently for the human population? Is this what cows and pigs and sheep are for, to be resources for improving milk production, or for modeling human diseases? Apparently Ian Wilmut and virtually every other person in our society thinks so, too. Is it possible for us to entertain the notion that our ability to do genetic engineering (as well as our other technological abilities) might be intended for the benefit of the entire natural order, as well as for ourselves? What would this look like? Wilmut says that the potential benefits of his research must outweigh any disadvantage of harm to the animals involved. Could there be a criterion of benefits to the animals?

(4) What is natural? What is artificial?

The [recent] cloning developments . . . are shown, thereby, to be also revelations of nature, as well as of human beings. Nature is like this, at least on planet earth–it clones sheep and perhaps even humans. Familiar dualisms are empty and unhelpful. The dualism between pristine nature and the nature that bears irretrievably the marks of human intervention is almost fully obsolete. Our concept of "nature" today is closely related to our image of the cyborg–the pristine and the artificial in one reality. Cloned humans are natural persons.

Since natural children inherit genes from both a father and a mother, cloned humans are not natural children of those whose cells are cloned, but rather of that person's parents.

(5) Multiple contexts of morality

Since being human is intrinsically a moral enterprise, the discussion of cloning must include from the outset questions of motivations, purposes, and consequences. The moral and spiritual issues have to do with why we clone, what interests it serves, and the moral status of those interests. These also include the question of what the purpose of cloning, or of any human life, is, and what the consequences of cloning might be. In the nonhuman realm, under the assumption that plants and animals exist to serve us, including our need for food, clothing, and medicines and cosmetics, we have concluded, without much public discussion, that all genetic engineering is desirable, if it serves our wellbeing. In other words, plant and animal nature has been defined as commodity or resource for human purposes. What confidence can we have that human cloning will not proceed under the same commodity rubric?

The practice of human cloning will be marked by a dauntingly complex set of motives and values, in a wide range of contexts: *personal* contexts (individual motives and values will determine whether people choose to be cloned), *professional* contexts (medical professionals will administer cloning and legal professionals will argue individual cases and draw up contracts), *free market* contexts (the biotechnology companies will make cloning profitable), and the *government* context (regulating or deregulating cloning for the common good). In all of these contexts, decisions are morally charged, marked by conflict of opinion and clash of competing interests. The complexity arises when any specific cases of cloning come to mind. Think of the difference in how we might regard a wealthy person cloning an offspring as a potential source of transplantable body parts and a childless couple who see good reason to practice cloning as an alternative to *in vitro* fertilization, or as a form of adoption.

We are about to enter once again into the whole question of what human life is all about and how it should be conducted—cloning has simply raised the stakes in this discussion. When we consider that the cloning discussion is simultaneous with the discussions of abortion and doctor-assisted death, we sense the enormous burden that our individual and societal psyches must bear in these days.

Theological Axioms

My Christian tradition offers some very general, but yet pertinent, guidelines for this discussion: (1) all life is a gift of God, including the life that is able to discover how to clone; (2) we are expected to be good stewards of God's gifts, which means (3) that we are both free and accountable to God and our fellow humans for what we do; we have a vocation from God to be free and accountable; (4) all of this takes place under the conditions of sin—we are finite, fallible, motivated by self-interest that can include greed and desire for power; sin is not an excuse for withdrawing from life, but rather a reminder of what sort of realistic vigilance must accompany our decisions of how to live. My Lutheran tradition asserts that we are saints and sinners at the same time. This describes precisely the condition of humans as cloners of other animals and of themselves.

Practical Suggestions

We should recognize that there is no quick fix to the "cloning question," nor will we ever arrive at a pure or perfect resolution to all the issues that cloning raises. Our public policies governing all nonhuman cloning should be carefully scrutinized for their motives and consequences. Our policies concerning the cloning of humans should be designed to do the following: (1) allow considerable time for public discussion and reflection before authorizing or financing such cloning; (2) give adequate attention to the complex sets of interests and values that will impinge upon this issue; (3) bring policy to bear in a subtle and multifaceted manner, appropriate to the situations in which cloning might be carried out; (4) factor into our policy the likelihood that future developments will reveal how little we really know at this time and how fallible even our best judgments are. . . .

37

RONALD A. LINDSAY*

Taboos without a Clue:
Sizing Up Religious Objections to Cloning

THE FUROR FOLLOWING the announcement of recent experiments in cloning, including the cloning of the sheep Dolly, has prompted representatives of various religious groups to inform us of God's views on cloning. Thus, the Reverend Albert Moraczewski of the National Conference of Catholic Bishops has announced that cloning is "intrinsically morally wrong" as it is an attempt to "play God" and "exceed the limits of the delegated dominion given to the human race." Moreover, according to Reverend Moraczewski, cloning improperly robs people of their uniqueness. Dr. Abdulaziz Sachedina, an Islamic scholar at the University of Virginia, has declared that cloning would violate Islam's teachings about family heritage and eliminate the traditional role of fathers in creating children. Gilbert Meilander, a Protestant scholar at Valparaiso University in Indiana, has stated that cloning is wrong because the point of the clone's existence "would be grounded in our will and desires" and cloning severs "the tie that united procreation with the sexual relations of a man and woman." On the other hand, Moshe Tendler, a professor of medical ethics at Yeshiva University, has concluded that there is religious authority for cloning, pointing out that respect for "sanctity of life would encourage us to use cloning if only for one individual . . . to prevent the loss of genetic line."

This is what we have come to expect from religious authorities: dogmatic pronouncements without any support external to a particular religious

*Reprinted from Ronald A. Lindsay, "Taboos without a Clue: Sizing Up Religious Objections to Cloning," *Free Inquiry*, Vol. 17, no. 3 (Summer 1997), pp. 15–17.

tradition, self-justifying appeals to a sect's teachings, and metaphor masquerading as reasoned argument. And, of course, the interpreters of God's will invariably fail to agree among themselves as to precisely what actions God would approve.

Given that these authorities have so little to offer by way of impartial, rational counsel, it would seem remarkable if anyone paid any attention to them. However, not only do these authorities have an audience, but their advice is sought out by the media and government representatives. Indeed, President Clinton's National Bioethics Advisory Commission devoted an entire day to hearing testimony from various theologians.

Questionable Ethics

The theologians' honored position reflects our culture's continuing conviction that there is a necessary connection between religion and morality. Most Americans receive instruction in morality, if at all, in the context of religious belief. As a result, they cannot imagine morality apart from religion, and when confronted by doubts about the morality of new developments in the sciences—such as cloning—they invariably turn to their sacred writings or to their religious leaders for guidance. Dr. Ebbie Smith, a professor at Southwestern Baptist Theological Seminary, spoke for many Americans when he insisted that the Bible was relevant to the cloning debate because "the Bible contains God's revelation about what we ought to be and do, if we can understand it."

But the attempt to extrapolate a coherent, rationally justifiable morality from religious dogma is a deeply misguided project. To begin, as a matter of logic, we must first determine what is moral before we decide what "God" is telling us. As Plato pointed out, we cannot deduce ethics from "divine" revelation until we first determine which of the many competing revelations are authentic. To do that, we must establish which revelations make moral sense. Morality is logically prior to religion.

Moreover, most religious traditions were developed millennia ago, in far different social and cultural circumstances. While some religious precepts retain their validity because they reflect perennial problems of the human condition (for example, no human community can maintain itself unless basic rules against murder and stealing are followed), others lack contemporary relevance. The world of the biblical patriarchs is not our world. Rules prohibiting the consumption of certain foods or prescribing limited, subordinate roles for women might have some justification in societies lacking proper hygiene or requiring physical strength for survival. But they no longer have any utility and persist only as irrational taboos. In addition, given the limits of the world of the Bible and the Koran, their authors simply had no occasion to address some of the problems that confront us, such as the ethics of *in vitro* fertilization, genetic engineering, or cloning. To pretend otherwise, and to try to apply religious precepts by extension and analogy to these novel problems is an act of pernicious self-delusion.

To underscore these points, let us consider some of the more common objections to cloning that have been voiced by various religious leaders:

Cloning is playing god

This is the most common religious objection, and its appearance in the cloning debate was preceded by its appearance in the debate over birth control, the debate over organ transplants, the debate over assisted dying, etc. Any attempt by human beings to control and shape their lives in ways not countenanced by some religious tradition will encounter the objection that we are "playing God." To say that the objection is uninformative is to be charitable. The objection tells us nothing and obscures much. It cannot distinguish between interferences with biological process that are commonly regarded as permissible (for example, use of analgesics or antibiotics) and those that remain controversial. Why is cloning an impermissible usurpation of God's authority, but not the use of tetracycline?

Cloning is unnatural because it separates reproduction from human sexual activity

This is the flip side of the familiar religious objection to birth control. Birth control is immoral because it severs sex from reproduction. Cloning is immoral because it severs reproduction from sex. One would think that allowing reproduction to occur without all that nasty, sweaty carnal activity might appeal to some religious authorities, but apparently not. In any event, the "natural" argument is no less question-begging in the context of reproduction without sex than it is in the context of sex without reproduction. "Natural" most often functions as an approbative and indefinable adjective; it is a superficially impressive way of saying, "This is good, I approve." Without some argument as to why something is "natural" and "good" or "unnatural" or "bad," all we have is noise.

Cloning robs persons of their God-given uniqueness and dignity

Why? Persons are more than the product of their genes. Persons also reflect their experiences and relationships. Furthermore, this argument actually demeans human beings. It implies that we are like paintings or prints: the more copies that are produced, the less each is worth. To the contrary, each clone will presumably be valued as much by their friends, lovers, and spouses as individuals who are produced and born in the traditional manner and not genetically duplicated.

Beyond Biology

All the foregoing objections assume that cloning could successfully be applied to human beings. It is worth noting that this issue is not entirely free from doubt since Dolly was produced only after hundreds of attempts. And although in principle the same techniques should work in humans, biological experiments cannot always be repeated across different species.

Of course, if some of the religious have their way, the general public may never know whether cloning would work in humans, as research into applications of cloning to human beings could be outlawed or driven underground. This would be an unfortunate development. Quite apart from the obvious, arguably beneficial, uses of cloning, such as asexual reproduction for those incapable of having children through sex, there are potential spin-offs from cloning research that could prove extremely valuable. Doctors, for example, could develop techniques to take skin cells from someone with liver disease, reconfigure them to function as liver cells, clone them, and then transplant them back into the patient. Such a procedure would avoid the sometimes fatal complications that accompany genetically nonidentical transplants as well as problems caused by the chronic shortage of available organs for transplant.

This is not to discount the potential for harm and abuse that would result from the development of cloning technology, especially if we also master techniques for manipulating DNA. If we are able to modify a human being's genetic composition to achieve a predetermined end and can then create clones from the modified genetic structure, we could, theoretically, create a humanlike order of animals that would be more intelligent than other animals but less intelligent and more docile than (other?) human beings. Sort of ready-made slaves.

But religious precepts are neither necessary nor sufficient for avoiding such dangers. What we require is a secular morality based on our needs and interests and the needs and interests of other sentient beings. In considering the example just given, it is apparent that harmful consequences to normal human beings could result from the creation of these humanoid slaves, as many could be deprived of a means of earning their livelihood. It would also lead to an enormous and dangerous concentration of power in the hands of those who controlled these humanoids. And, although in the abstract we cannot decide what rights these humanoids would have, it is probable that, as sentient beings with at least rudimentary intelligence, they would have a right to be protected from ruthless exploitation and, therefore, we could not morally permit them to be treated as slaves. Even domesticated animals have a right to be protected from cruel and capricious treatment.

Obviously, I have not listed all the factors that would have to be considered in evaluating the moral implications of my thought experiment. I have not even tried to list all the factors that would have to be considered in assessing the many other ways—some of them now unimaginable—in which cloning technology might be applied. My point here is that we have a capacity to address these moral problems as they arise in a rational and deliberate manner if we rely on secular ethical principles. The call by many of the religious for an absolute ban on cloning experiments is a tacit admission that their theological principles are not sufficiently powerful and adaptable to guide us through this challenging future.

I want to make clear that I am not saying we should turn a deaf ear to those who offer us moral advice on cloning merely because they are religious. Many bioethicists who happen to have deep religious convictions have made significant, valuable contributions to this field of moral inquiry. They have done so, however, by offering secular and objective grounds for their arguments. Just as an ethicist's religious background does not entitle her to a special deference, so, too, her religious background does not warrant her exclusion from the debate, provided she appeals to reason and not supernatural revelation.

BIBLIOGRAPHIC ESSAY

IN THE PAST two decades, the philosophy of biology has grown from the meager output of but a few workers tangentially interested in the subject to a fully thriving discipline. At the overview level, still well worth reading is a work that was an inspiration to many in the field, Morton Beckner's *The Biological Way of Thought* (New York: Columbia University Press, 1959). More recent and more elementary is David Hull's *The Philosophy of Biological Science* (Englewood Cliffs: Prentice-Hall, 1974). Even more recent, although rather less elementary, is Alexander Rosenberg, *The Structure of Biological Science* (Cambridge: University of Cambridge Press, 1985).

Going now systematically through the sections of this collection, as far as the nature of life is concerned, most informative is a volume edited by the evolutionist Theodosius Dobzhansky and his student Francisco Ayala, *Studies in the Philosophy of Biology* (Berkeley: University of California Press, 1974). The contributions are a little uneven, but the best (including those of the editors) are very good. The classic work on vitalism is Henri Bergson, *Creative Evolution* (London: Macmillan, 1913), a book with many fine insights even if one rejects the overall thesis. In our own times, there is a vitalistic flavour to several of the contributions to a volume edited by Steven Rose, *Towards a Liberatory Biology* (London: Allison and Busby, 1982). Anything written by J.B.S. Haldane is worth reading. Start with a recently edited selection by his student, John Maynard Smith, *On Being the Right Size and Other Essays* (Oxford: Oxford University Press, 1985).

Moving next to Darwin and natural selection, the place to begin is with the *Origin* itself. It is remarkably readable for a "Great Book." Most recommended is the first edition, which is uncluttered by all the qualifications Darwin added later. Ernst Mayr has edited a facsimile edition (Cambridge, Mass.: Harvard University Press, 1964). Alternatively, with an extremely good introduction by the historian John Burrow, an excellent buy is the (inexpensive) Penguin reprint. For those who want to read more broadly in Darwin's work, a recent selection by Mark Ridley, *The Essential Darwin* (London: Unwin Hyman, 1987) should satisfy. In any case, primary reading should be backed up by looking at the collection edited by Philip Appleman, *Darwin* (New York: Norton, 1979).

A recent and very thorough history of the whole topic of evolution is Peter Bowler, *Evolution: The History of An Idea* (Berkeley: University of California Press, 1984). Ernst Mayr's, *The Growth of Biological Thought* (Cambridge, Mass.: Harvard University Press, 1982) is a brilliant exposition of his subject's history as seen by one biologist, although it is somewhat Wagnerian in dimension. More specifically directed to Darwin, trying to bring out some of the philosophical undercurrents, is my *The Darwinian Revolution: Science Red in Tooth and Claw* (Chicago: University of Chicago Press, 1979). Less historical, but a most stimulating view of the Darwinian picture, is a book from which I have included an extract: Richard Dawkins, *The Blind Watchmaker* (New York: Norton, 1986). The American paperback now comes with software so one can run one's own selection programs. For balance, look also at Michael Denton, *Evolution: A Theory in Crisis* (London: Burnett, 1985).

Thinking about the nature of evolutionary theory as a whole, if one has no biological training, one should read first an introduction to the subject. The best is by John Maynard Smith, *The Theory of Evolution* (Harmondsworth, Middlesex: Penguin, third edition, 1975). Back this up with a book which is extracted in a later section, George C. Williams, *Adaptation and Natural Selection* (Princeton: Princeton University Press, 1966), and, of course, any of Stephen Jay Gould's collections, *Ever Since Darwin* (New York: Norton, 1977), *The Panda's Thumb* (New York: Norton, 1980), *Hen's Teeth and Horse's Toes* (New York: Norton, 1983), and *The Flamingo's Smile* (New York: Norton, 1985). Reprinted from *Natural History*, they touch provocatively on all aspects of evolutionary thought.

The literature on the tautology question continues to grow. The best known philosophical critic was Sir Karl Popper, especially in his *Objective Knowledge* (Oxford: Oxford University Press, 1972) and his autobiographical introduction to P. Schilpp, ed., *The Philosophy of Karl Popper* (La Salle, Ill.: Open Court, 1974). Popper has, however, recanted somewhat, for instance in his "Natural selection and the emergence of mind", *Dialectica*, 32 (1978), 339–355. There is good material pertinent to the question in a recent collection edited by J. Pollard, *Evolutionary Theory: Paths into the Future* (Great Britain: John Wiley and Sons, 1984), and the matter is raised in a collection edited by myself, *But Is It Science? The Philosophical Question in the Evolution/ Creation Controversy* (Buffalo: Prometheus, 1988). This includes contributions by Popper, Dawkins, Gould, and others.

In his article on Darwinism and its putative expansion, Gould suggests that it is inappropriate to regard evolutionary theorizing as fitting a formal mathematical structure. Rather, it seems to him to be more a set of related models. The formal picture, which supposes that scientific theories are axiomatic or "Hypothetico-Deductive", is the common philosophical view—it is presupposed by Flew—and as applied to neo-Darwinism is defended in my *Philosophy of Biology* (London: Hutchinson, 1973). But this is challenged by E. A. Lloyd, *The Structure and Confirmation of Evolutionary Theory* (Westport, Conn.: Greenwood Press, 1988) and Paul Thompson, *The Structure of Biological Theories* (Albany: SUNY Press, 1988). A brilliant example of the model approach in scientific play is given by R. C. Lewontin, *The Genetic Basis of Evolutionary Change* (New York: Columbia University Press, 1974). Confirming its seminal influence, Morton Beckner's, *The Biological Way of Thought* had already raised some of these issues.

Punctuated equilibria theory is well discussed in two books by its co-formulator, Niles Eldredge, *Time Frames: The Rethinking Darwinian Evolution and the Theory of Punctuated Equilibria* (New York: Simon and Schuster, 1985) and *Unfinished Synthesis* (New York: Oxford University Press, 1985). Pertinent to this and to other logical issues in evolutionary thought is a philosophically oriented collection edited by Elliott Sober: *Conceptual Issues in Evolutionary Biology* (Cambridge, Mass.: MIT Press, 1984). This reprints Gould's most extended attack on adaptationism, co-authored with Richard Lewontin: "The Spandrels of San Marco and the Panglossian paradigm: a critique of the adaptationist programme," *Proceedings of the Royal Society*, 13, 205 (1979), 581– 598. Those who want to take their philosophical understanding of selection further—much further—simply cannot miss Sober's own, *The Nature of Selection* (Cambridge, Mass.: MIT Press, 1985). This is a seminal study. The pre-evolutionary roots of today's differences between evolutionists are explored in my, *The Darwinian Paradigm: Selected Essays on Its History, Philosophy and Religious Implications* (London: Routledge, 1989).

Almost any issue of the journal *Systematic Zoology* carries something on the species problem, and there is a thorough symposium with contributions and interchanges between Ernst Mayr and Michael Ghiselin in a recent issue of *Biology and Philosophy*, 2 (2), 1987, 127–226. Additionally, Sober in the above-mentioned collection reprints much useful material. With respect to the debate between different schools of thought concerned with overall problems of taxonomy, a still useful introduction is David Hull's, "Contemporary Systematic Philosophies," *Annual Review of Ecology and Systematics*, 1 (1970), 19–54. Supplement this with Niles Eldredge and Joel Cracraft, *Phylogenetic Patterns and the Evolutionary Process* (New York: Columbia University Press, 1980) and Mark Ridley, *Evolution and Classification: The Reformation of Cladism* (London: Longman, 1986).

Many people have written on teleology. A good recent collection is N. Rescher, *Current Issues in Teleology* (Lanham, Md.: University Press of America, 1986). But, I do not think you should miss the writings on the subject by the late Ernest Nagel, especially the pertinent sections in his magisterial, *The Structure of Science* (New York: Harcourt, Brace, Jovanovich, 1961, chapter 12), and his later review of the subject, "Teleology revisited," *Journal of Philosophy*, 74 (1977), 261–308. This review was reprinted in his *Teleology Revisited and Other Essays* (New York: Columbia University Press, 1979). Two works exclusively on teleology are Andrew Woodfield, *Teleology* (Cambridge: Cambridge University Press, 1976) and Larry Wright, *Teleological Explanations* (Berkeley: University of California Press, 1976). The former is a good review of the literature and the latter a sprightly defense of the teleological way of thinking.

Molecular biology and its implications have also had extended treatment. Historically, the coming of molecules has been well-treated, both in a popular account by Horace Freeland Judson, *The Eighth Day of Creation: The Makers of the Revolution in Biology* (New York: Touchstone, 1979), and in the sparkling, near libellous memoirs of James Watson, *The Double Helix* (New York: Atheneum, 1968). On the question of the relationship between old and new genetics, there is a vigorous interchange between David Hull, Kenneth Schaffner, Michael Ruse, and William Wimsatt, reprinted in the Sober collection. A recent, detailed, but very powerful analysis is given by Alex Rosenberg in the already-mentioned, *The Structure of Biological Science*. To pick up on some of the recent scientific developments towards which Schaffner's essay directs us, look at Manfred Eigen and others, "The origin of genetic information," *Scientific American*, 244 (4) (1981), 88–118. On the social issues, particularly useful (and very clear on the scientific questions) is Clifford Grobstein, *A Double Image on the Double Helix* (San Francisco: Freeman, 1979). For those with a taste for metaphysical speculation, particularly about the ultimate meaning of the molecular revolution, Jacques Monod, *Chance and Necessity* (London: Fontana, 1974) should not be missed but this should be balanced with the crisp common sense of Mary Midgley, *Evolution as a Religion: Strange Hopes and Stranger Fears* (London: Methuen, 1985).

The sociobiology controversy, especially as it applies to humans, has been written about at length. E. O. Wilson's key books are *Sociobiology: The New Synthesis* (Cambridge, Mass.: Harvard University Press, 1975) and *On Human Nature* (Cambridge, Mass.: Harvard University

Press, 1978). Then, for discussion, start with Arthur Caplan's collection, *The Sociobiology Debate* (New York: Harper and Row, 1978), and go on to a special double issue of the *Philosophical Forum*, 13(1981), 2/3. Also, not to be missed is a remarkably balanced collection, reporting on an American Association for the Advancement of Science symposium: G. W. Barlow and J. Silverberg (eds.), *Sociobiology: Beyond Nature/Nurture?* (Boulder, Col.: Westview, 1980). Philosophical works that take very contrasting positions are (anti) Philip Kitcher, *Vaulting Ambition* (Cambridge, Mass.: MIT Press, 1985) and (pro) Michael Ruse, *Sociobiology: Sense or Nonsense?* (Dordrecht: Reidel, 2nd ed., 1985).

The place to start the philosophical investigation of extraterrestrials is with Edward Regis, *Extraterrestrials: Science and Alien Intelligence* (Cambridge: Cambridge University Press, 1985). A very informative historical overview by Lewis White Beck introduces this collection. At the more scientific level, *Communication with Extraterrestrial Intelligence*, edited by Carl Sagan (Cambridge, Mass: MIT Press, 1973) contains much useful information. Major historical discussions are S.J. Dick, *Plurality of Worlds: The Origins of the Extraterrestrial Life Debate from Democritus to Kant* (Cambridge: Cambridge University Press, 1983) and Michael Crowe, *The Extraterrestrial Life Debate, 1750–1900: The Idea of a Plurality of Worlds from Kant to Lowell* (Cambridge: Cambridge University Press 1986).

At some point, anyone interested in extraterrestrials will obviously have to face questions to do with life's origins. The *locus classicus* on this topic is by the Russian scientist, A. I. Oparin, *The Origin of Life* (London: Macmillan, 1938, reprinted New York: Dover, 1953). Set this in context with John Farley, *The Spontaneous Generation Controversy from Descartes to Oparin* (Baltimore: Johns Hopkins University Press, 1977), and Loren Graham, *Science, Philosophy and Human Behaviour in the Soviet Union* (New York: Columbia University Press, 1987). Dawkins, *The Blind Watchmaker* has a good update on modern thought, a topic also covered in my *Darwinism Defended: A Guide to the Evolution Controversies* (Reading, Mass.: Addison-Wesley, 1982).

Of the many books about Social Darwinism, the definitive starting point is Richard Hofstadter, *Social Darwinism in American Thought* (New York: Braziller, 1959). Useful additions and correctives can be found in Cynthia Eagle Russett, *Darwin in America* (San Francisco: Freeman, 1976). The English scene is well-covered by Greta Jones in *Social Darwinism and English Thought* (Brighton, Sussex: Harvester, 1980). Scholars divide over whether Darwinism contributed to the rise of National Socialism. Daniel Gasman, *The Scientific Origins of National Socialism: Social Darwinism in Ernst Haeckel and the German Monist League* (New York: Elsevier, 1971) thinks it did. Alfred Kelley, *The Descent of Darwin: The Popularization of Darwinism in Germany, 1860–1914* (Chapel Hill: University of North Carolina Press, 1981) is not so sure.

For T. H. Huxley, look at the volume published by his grandson Julian Huxley: *Evolution and Ethics* (London: Pilot Press, 1947). It contains contributions by both generations, and makes a fascinating contrast. The evolution and ethics relationship, with special respect to human sociobiology, is discussed, in detail, fairly, but ultimately critically, in two exceptionally lucid books: Peter Singer, *The Expanding Circle: Ethics and Sociobiology* (New York: Farrar, Straus and Giroux, 1981) and Roger Trigg, *The Shaping of Man* (Oxford: Blackwell, 1982). A vitriolic attack on Mackie was launched by Mary Midgley, "Gene-juggling," *Philosophy*, 54 (1979), 439–458. The case for the defense is made in a spirited discussion by Jeffrey Murphy, *Evolution, Morality and the Meaning of Life* (Totowa, N.J.: Rowman and Littlefield, 1982).

Filling out the scientific understanding of some of the issues, particularly today's emphasis on the individual as the key unit in selection, one must look at Richard Dawkins, *The Selfish Gene* (Oxford: Oxford University Press, 1976). This should be supplemented with an excellent collection by Robert Brandon and Richard Burian, *Genes, Organisms, Populations* (Cambridge, Mass.: MIT Press, 1984), where one gets both science and philosophy brought to bear on the issues. E. O. Wilson's views on ethics can be found in his *On Human Nature*, and in a recent part-autobiographical volume, *Biophilia* (Cambridge, Mass.: Harvard University Press, 1984). My own views on morality and its foundations (real or apparent) can be found in *Taking Darwin Seriously: A Naturalistic Approach to Philosophy* (Oxford: Blackwell, 1986).

And so to religion. Two first-class collections dealing with historical aspects of the science/religion relationship are Ernan McMullin *Evolution and Creation* (Notre Dame: University of Notre Dame Press, 1985) and David Lindberg and Ronald Numbers, *God and Nature* (Berkeley: University of California Press, 1986). In the former, the essay by McMullin on the religious side to the relationship and, in the latter, the essay by Numbers on the history of Creationism, are quite exceptional. Creationism itself is shown in its true light in works by (the biologist) Douglas Futuyma, *Science on Trial* (New York: Pantheon, 1983) and by (the philosopher) Philip Kitcher, *Science Abused* (Cambridge, Mass.: MIT Press, 1982). A good collection, with details of the Arkansas trial is A. Montague, *Science and Creationism* (Oxford: Oxford University Press, 1984). Then, looking at more profitable ways of exploring the biology/religion connection today, most valuable is Arthur Peacocke's, *God and the New Biology* (London: Longman, 1986). Less sophisticated but also useful is R. J. Berry, *God and Evolution* (London: Hodder and Stoughton, 1988). For those interested in these issues, an invaluable source is the journal *Zygon*, which focuses exclusively on the ground shared by the scientist and the believer.

In addition to the various books and articles just detailed, interested students will want to consult the *Encyclopedia of Philosophy* (New York: Macmillan, 1967), edited by Paul Edwards. It is a remarkably rich source of information. See especially the lively essay on Vitalism by Morton Beckner and the informative discussions of biology and Darwinism by the same author. The major philosophy of science journals, especially *Philosophy of Science* and the *British Journal for the Philosophy of Science,* are now carrying much more material on biology. Also there is a recently started journal focused exclusively on philosophical issues in biology: *Biology and Philosophy*. Well worth consulting also are more popular publications like *New Scientist, BioScience,* and the *American Scientist*. They often include discussions on the border between biology and philosophy.

Finally, if more detailed information or references are needed, they can probably be found in my handbook, *Philosophy of Biology Today* (Albany: SUNY Press, 1988).

The ten years since the first edition of this book was published have been incredibly fertile and exciting for anyone interested in the philosophical aspects of the biological sciences. What follows is just a small sampling of some of the more useful writings and should not in any sense be considered definitive.

First, if you are going to consider seriously the nature of modern evolutionary biology, you really must look at some of the truly excellent historical studies that are appearing. I recommend strongly the one-volume biography of Charles Darwin, A. Desmond, and J. Moore, *Darwin: The Life of a Tormented Evolutionist* (London: Michael Joseph, 1991), as well as the first (and so far only) volume of the major study by J. Browne, *Charles Darwin: Voyaging. Volume 1 of a Biography* (New York: Knopf, 1995). My own bibliographic survey should also prove useful, The Darwin Industry: A Guide," *Victorian Studies* 39, no. 2 (1996): 217–35, as well as a very large history by me which covers just about the whole of evolutionary theory's past and present: *Monad to Man: The Concept of Progress in Evolutionary Biology* (Cambridge, Mass.: Harvard University Press, 1996). A very lively analysis of the very idea of evolution is R. J. Richards, *The Meaning of Evolution: The Morphological Construction and Ideological Reconstruction of Darwin's Theory* (Chicago: University of Chicago Press, 1992).

More philosophical treatments of the evolutionary question include the pro-Darwinian, H. Cronin, *The Ant and the Peacock* (Cambridge: Cambridge University Press, 1991); R Dawkins, *A River Out of Eden* (New York: Basic Books, 1995); *Climbing Mount Improbable* (New York: Norton, 1996); and D. C. Dennett, *Darwin's Dangerous Idea* (New York: Simon and Schuster, 1995). Somewhat more critical are the writings of Stephen Jay Gould, *Wonderful Life: The Burgess Shale and the Nature of History* (New York: W. W. Norton Co., 1989), and *Full House: The Spread of Excellence from Plato to Darwin* (New York: Paragon, 1996). Evolutionary theory is also a major topic in the introductory text by E. Sober, *Philosophy of Biology* (Boulder, Colo.: Westview Press, 1993). You might also take a look at S. A. Kauffman, *The Origins of Order: Self-Organization and Selection in Evolution* (Oxford: Oxford University Press, 1993). Good collections of papers include E. Sober, *From a Biological Point of View* (Cambridge: Cambridge University Press, 1994); D. Hull, *The Metaphysics of Evolution* (Albany, N.Y.: SUNY Press, 1989); and R. N. Brandon, *Concepts and Methods in Evolutionary Biology* (Cambridge: Cambridge University Press, 1996). Look also at Brandon's excellent book *Adaptation and Environment* (Princeton: Princeton University Press, 1990).

Classification is treated well in the collection by M. Ereshefsky, ed., *The Units of Evolution: Essays on the Nature of Speciation* (Cambridge, Mass.: M.I.T. Press, 1992). Molecular and related issues are treated in K. Schaffner, *Discovery and Explanation in Biology and Medicine* (Chicago: University of Chicago Press, 1993). Recombinant DNA research is less of a frightening topic these days, although the basic issues remain. The Human Genome Project is a hot topic and you should start with P. Kitcher, *The Lives to Come: The Genetic Revolution and Human Possibilities* (New York: Simon and Schuster, 1996). Human sociobiology continues to get attention. Look at the recent excellent S. Pinker, *How the Mind Works* (New York: Norton, 1997). To get some background on why Wilson argued as he did, look at his autobiography, *Naturalist* (Washington, D.C.: Island Press, 1994).

Evolutionary ethics is a very hot topic these days: among other works I recommend M. Bradie, *The Secret Chain: Evolution and Ethics* (Albany, N.Y.: SUNY Press, 1994), and P. L. Farber, *The Temptations of Evolutionary Ethics* (Berkeley and Los Angeles: University of California Press, 1994). Pertinent writings by myself include *The Darwinian Paradigm: Essays on Its History, Philosophy and Religious Implications* (London: Routledge, 1985), and *Evolutionary Naturalism: Selected Essays* (London: Routledge, 1994). For an exciting view of some of the social background to evolutionary theory look at P. Crook, *Darwinism: War and History* (Cambridge: Cambridge University Press, 1994).

Finally, the god-and-evolution question has come right back up to boiling. An excellent general discussion is I. Barbour, *Religion in an Age of Science* (New York: Harper and Row, 1990). Articulate defenses of the creationist position can be found in P. E. Johnson, *Darwin on Trial* (Washington, D.C.: Regnery Gateway, 1991), and in A Plantinga, "When Faith and Reason Clash: Evolution and the Bible," *Christian Scholar's Review* 21, no. 1 (1991): 8–32. Good responses are found in E. McMullin, "Plantinga's Defense of Special Creation," *Christian Scholar's Review* 21, no. 1 (1992): 55–79, and in P. L. Pennock, *Tower of Babel: Scientific Evidence and the New Creationism* (Cambridge, Mass.: M.I.T. Press, 1998). Historical background can be found in R. Numbers, *The Creationists* (New York: A. A. Knopf, 1992).

A very enjoyable introduction to some of the main players in the field can be found in W. Callebaut, *Taking the Naturalist Turn* (Chicago: University of Chicago Press, 1993). Finally, *Biology and Philosophy* continues to thrive. There is much discussion of just about every imaginable topic and plenty of suggestions for further research. Go to it!